PEDOLOGY, WEATHERING, AND GEOMORPHOLOGICAL RESEARCH

Pedology, Weathering, and Geomorphological Research

PETER W. BIRKELAND

Department of Geological Sciences
University of Colorado, Boulder, Colorado

New York Oxford University Press London 1974 Toronto

Library of Congress Catalogue Card Number: 73-83939
Printed in the United States of America

To My Family

Preface

Although there are many textbooks on soils, there are few that serve the needs of geomorphologists and sedimentary petrologists working in Quaternary research. This book is an attempt to fill that gap. The term "pedology," in the title, refers to the study of soils in their natural setting—the field—since field studies are most significant to Quaternary studies. Much of the research that geomorphologists undertake involves the use of soils to date deposits on the basis of soil development and to reconstruct the environment during soil formation. This book focuses on these problems, but other related problems are also discussed. I feel that one cannot adequately use soils for any purpose without understanding the processes and factors that control their formation. Hence, the over-all organization of the book is, first, a discussion of soil morphology, weathering, and soil-forming processes and, then, variation in soils with variation in the soil-forming factors (climate, organisms, topography, parent material, and time). My discussion of soil classification is brief and generalized because I feel that it is more important to understand the genesis of a soil than it is to classify a soil. Only the new U.S. soil classification system is used. The books ends with a short discussion of the use of soils in Quaternary stratigraphic studies. For those readers beginning work in soils, I want to stress the importance of sound field work, because pedology is a field science. Laboratory studies are necessary, but they are only as good as the field work and sampling upon which they are based.

Much of this book is based on material presented at the University of Colorado in a beginning soils course for graduate students and for fourth-year undergraduates. Much of my approach to pedology was gained from my first course in pedology, taught by Hans Jenny. He was able to bring together those aspects of geology, chemistry, and biology that are important to an understanding of the genesis of soils, and I hope I have succeeded in doing that here. The background I expect for the course and for a clear understanding of this text is one year of college chemistry and one year of physical geology.

I have relied heavily on my own experiences for examples of different aspects of pedology in writing this book. I did this because I feel that one has to work with something or see it in its field setting before one really understands it and can explain it to others. Other aspects of pedology could have been covered, but I have had no personal experience with them. Hence, I often used as my examples in the text either work I have done or work I have seen in the field on numerous formal and informal field trips over the years with various colleagues. The literature cited is not too extensive, but I have tried to include both the relevant geological and pedological literature in attacking problems common to both disciplines. A more encyclopedic treatment of the literature would have to have been at the cost of deleting materials I consider more important in conveying important principles to the reader.

The metric system is used throughout. In all the cited literature using English units, I have converted the data to the metric system. This results, in places, in uneven numbers plotted on the x and y axes in the figures, because I converted directly from the English to the metric, keeping the original placement of the English-system tickmarks on the figure axes.

Two appendices appear at the end of the book. Appendix 1 is designed to give the reader information on the soil data that have to be collected in order to describe a soil profile adequately. Appendix 2 presents climatic maps for the United States. These are included because the examples used in the text generally are taken from many different climatic regions in the United States. For each of these examples one can get a general idea of the local or regional climate by referring to these maps.

I would like to acknowledge the assistance of many people who, over the years, have helped me crystallize many of the ideas put forth here. J. Hoover Mackin first kindled my interest in Quaternary studies, and this was soon followed by a summer of fieldwork for the U.S. Geological Survey, assisting H. H. Waldron and D. R. Mullineaux in the Puget Sound Lowland. My graduate work was undertaken at Stanford University, and while there I especially benefited from conversations with A. D. Howard, S. N. Davis, R. R. Compton, and D. M. Hopkins. I then went to the Department of Soils and Plant Nutrition at the University of California, Berkeley, where I received my initial soils training. I am grateful to H. Jenny, R. J. Arkley, I. Barshad, and E. P. Perry of that department and to E. L. Begg and K. D. Gowans of the same department on the Davis Campus for their help and patience while I was learning about soils. Several people then in the Department of Geology and Geophysics on the Berkeley Campus were also extremely helpful to me in the field and in the office discussions. These include C. Wahrhaftig, R. L. Hay, and R. J. Janda. At the University of Colorado, field trips and discussions with J. T. Andrews, W. C. Bradley, D. D. Runnells, and T. R. Walker have helped clear my thinking on various aspects of soils and weathering. I also would like to acknowledge helpful discussions and field trips with R. B. Morrison and D. R. Crandell of the U.S. Geological Survey. As my work progressed, I traveled to the midcontinent to view the critical deposits and soils of that area. Many of the trips I have taken there were with the Friends of the Pleistocene, but a few were with colleagues who took time off from their work to show me around. For these courtesies I am grateful to J. C. Frye, H. B. Willman, J. P. Kempton, and D. L. Gross of the Illinois State Geological Survey and to R. L. Hooke and M. Singer of the University of Minnesota. Finally, I would like to thank the students in my various courses here for being the sounding board for many of these ideas, for challenging the ones they thought were weak, and for providing me with their ideas, field companionship, and citations in the literature that I had overlooked.

My family has been especially helpful in all of my scientific endeavors. We always go into the field as a group, and each of them, Suzanne, Karl, and Robin, in their own way, has pitched in and helped make each trip a success.

I am grateful to several individuals for critically reading preliminary drafts of the manuscript prior to editorial review. H. E. Wright, Jr., of the University of Minnesota, read the entire manuscript, J. L. Clayton, of the U.S. Forest Service, read all but Chapters 11 and 12, D. D. Runnells, of the University of Colorado, read Chapters 3 and 4, and R. B. Morrison, of the U.S. Geological Survey, read Chapter 2. These reviews helped me immeasureably in the preparation of the manuscript; any shortcomings in the book, however, fall directly on my shoulders.

I want to thank several people in the Department of Geological Sciences for their help with the manuscript preparation. Mrs. Paulina Franz, Mrs. Edith Ellis, and Miss Ann Cairns all worked on the first rough draft of the manuscript, and Mrs. Franz typed the final draft. R. M. Burke, B. H. Gray, R. Leech, T. C. Meierding, and R. R. Shroba helped with the proofreading. Mr. Norm Nielsen was responsible for preparing the photographs for use in the book. In addition, Mr. Frank Murillo of the University of California drew Fig. 8–5.

In several places in the book I have used unpublished soil laboratory analyses. These analyses were run through the courtesy of Mr. Rolf Kihl of the Institute for Arctic and Alpine Research Sedimentology Laboratory, University of Colorado.

P.W.B.

Boulder, Colorado
November 1973

Contents

PEDOLOGY, WEATHERING, AND GEOMORPHOLOGICAL RESEARCH

CHAPTER 1

The soil profile, horizon nomenclature, and soil characteristics

Soils differ from geologic deposits generally, but in some places the two are so similar in appearance that they are difficult to tell apart. In this section we shall focus on the soil profile, and the properties that should be recognized for an adequate soil description, and discuss the origin of some of these properties. Furthermore, because buried soils are widely used by geologists, some guidelines will be offered for recognizing them. Procedures for describing a soil profile are given in Appendix 1.

Soil profile

The term "soil" has many definitions, depending upon who is using the term. For example, to engineers "soil" is unconsolidated surficial material, whereas to many soil scientists it is mainly the medium for plant growth. A definition of soil that serves our purpose well is a slight modification of that given by Joffe[16]: a soil is described as a natural body consisting of layers or horizons of mineral and/or organic constituents of variable thicknesses, which differ from the parent material in their morphological, physical, chemical, and mineralogical properties and their biological characteristics (Fig. 1–1). Soil horizons generally are unconsolidated, but some contain sufficient amounts of silica, carbonates, or iron oxides to be cemented.

A soil profile consists of the vertical arrangement of all the soil horizons down to the parent material. In studying a soil, therefore, the investigator must be able to identify the parent material from which the soil formed. This is no easy task and requires a good deal of experience in geology and pedology. However, once the parent material is recognized, and its original properties estimated, one can begin to determine departures in the properties of the original material and identify these materials as soil horizons.

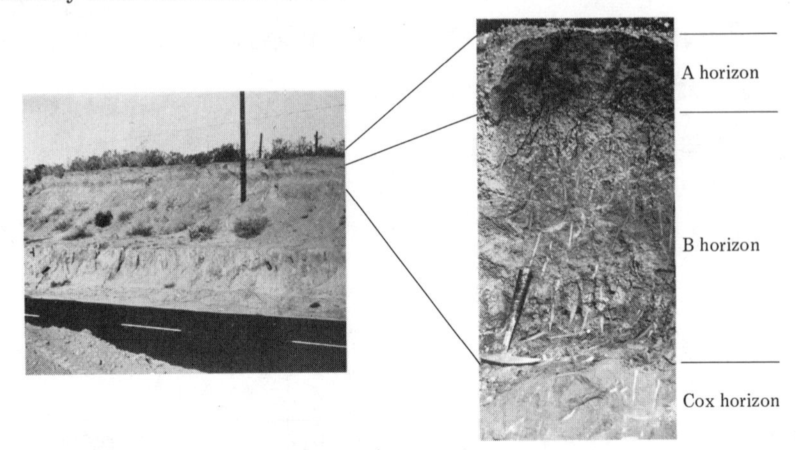

Fig. 1–1 Soil formed on Quaternary marine terrace deposits near San Diego, California.

Some geologists distinguish between a soil profile and a weathering profile.[19,28] Where this is done the soil profile is generally considered to make up the upper part of the much thicker weathering profile. However, because it is difficult to separate these two profiles on the basis of the processes involved, I will not make this distinction. What some would call the weathering profile beneath the soil profile probably would qualify as a Cox soil horizon under the horizon nomenclature used here. A problem with thick weathered zones (say thicker than 30 m) is the separation of the products of soil formation from those of diagenesis. The boundary between the two is difficult to define, and no doubt all gradations exist. Fortunately, most soils are not thick enough for this to be a major problem.

Soil horizon nomenclature

In recent years, the Soil Conservation Service of the U.S. Department of Agriculture has introduced a new set of names for soil horizons[33,34] because the more traditional terms, such as A horizon and B horizon, were neither precisely defined nor used in the same sense by all workers. The new terms, such as mollic epipedon or argillic horizon, are precisely defined, so much so that at times recognition of such a horizon may require laboratory analysis. The boundaries between these units and those of the A, B, and C horizons in a profile do not necessarily coincide. In fact, in some soils the surface horizon of the new classification can include both the A and B horizons as determined in the field. This creates problems because the ABC notations are still used in field descriptions. Because there is merit in both systems, they are combined into one system here (Table 1–1). This system is a modified

Table 1–1
Soil horizon nomenclature*

MASTER HORIZONS

O horizon Surface accumulation of organic material overlying a mineral soil. Lower limits are 30 per cent organic matter if the mineral fraction contains more than 50 per cent clay or 20 per cent organic matter if the mineral fraction has no clay. The horizon is O1 if most vegetative matter is recognizable, O2 if the original form of plant or animal matter is not recognizable.

A horizon Accumulation of humified organic matter mixed with mineral fraction. Occurs at the surface or below an O horizon. Organic matter contents are less than those required for the O horizon. If data are available, the A horizon can be subdivided into the following: *Mollic A horizons* are dark colored (chroma of 4.0 or less, value darker than 3.5, when moist), contain at least 1 per cent organic matter (0.58 per cent organic carbon), and have a base saturation of over 50 per cent. Generally associated with grassland vegetation.
Umbric A horizons are similar to mollic A horizons except that the base saturation is less than 50 per cent. Generally associated with forest vegetation.
Ochric A horizons are too light in color and low in organic matter to be mollic or umbric A horizons. Generally associated with young soils and/or semiarid vegetation.

E horizon Underlies an O or A horizon and is characterized by less organic matter, and/or less sesquioxides,** and/or less clay than the underlying horizon. Horizon is light colored due mainly to the color of primary mineral grains because secondary coatings on grains are absent. Also known as an A2 or albic horizon.

B horizon Underlies an O, A, or E horizon and shows little or no evidence of the original rock structure.
Argillic B horizons have more silicate clay than the A or E horizon and/or the assumed parent material. In other words, silicate clays have been translocated into the B from overlying horizons, or they have formed in place within the B, or both. Clay translocation is recognized in the field by oriented clay films that coat either mineral grains or small channels or ped surfaces. Clay films can be destroyed during subsequent pedogenesis.
Natric B horizons meet the requirements of the argillic B horizons, have columnar or prismatic structure, and more than 15 per cent

saturation with exchangeable sodium in some subhorizons.

Spodic B horizons generally occur beneath an E horizon and are characterized by a concentration of organic matter and sesquioxides that have been translocated downward from the E horizon.

Oxic B horizons are highly weathered subsurface horizons that are characterized by hydrated oxides of iron and aluminum, 1:1 lattice clays, and a low cation-exchange capacity. Few primary silicate minerals remain with the exception of quartz, which is quite resistant to weathering.

Cambic B horizons lie in the position of the B horizon and are characterized by at least enough pedogenic alteration to eradicate most rock structure, form some soil structure, and remove or redistribute primary carbonate. Their color has higher chromas or redder hues than does the color of the underlying horizons. [Although there are restrictions on the texture of this horizon, there is little reason for such restrictions. If this horizon designation is based primarily on color (the color B), I prefer to see it based on a hue redder than that of oxidized C horizon rather than on just a higher chroma.] Similar horizons can be found in other soil profiles, for example, below argillic B horizons; these horizons do not qualify as cambic B horizons, however, because they are not in the position of the B horizon. Although there is some difficulty in identifying the cambic B horizon, it is a useful term for those B horizons that are characterized primarily by the development of soil structure and/or fairly intense oxidation.

K horizon A subsurface horizon so impregnated with carbonate that its morphology is determined by the carbonate.[14] Authigenic carbonate coats or engulfs all primary grains in a continuous medium and makes up 50 per cent or more by volume of the horizon. The uppermost part of the horizon commonly is laminated. If cemented, the horizon corresponds with some caliches and calcretes.

C horizon A subsurface horizon, excluding bedrock, like or unlike material from which the soil formed or is presumed to have formed. Lacks properties of A and B horizons, but includes weathering as shown by mineral oxidation, accumulation of silica, carbonates, or more soluble salts, and gleying.

R horizon Consolidated bedrock underlying soil.

SELECTED SUBORDINATE DEPARTURES

The following symbols are used with the master horizon designation to denote special features. They follow the master horizon designation, as well as any numbers (e.g., B2t).

b Buried soil horizon. May be deeply buried and not affected by subsequent pedogenesis or shallow and part of a younger soil profile.

ca Accumulation of carbonates of alkaline earths, usually calcium, in amounts greater than the parent material is presumed to have had. Occurs in A, B, C, or R horizons.

cs Accumulation of gypsum in amounts greater than the parent material is presumed to have had. Occurs in A, B, C, or R horizons.

g — Horizon is characterized by strong gleying or reduction of iron, so that colors approach neutral, with or without mottles. Occurs in A, B, or C horizons.

h — Illuvial concentration of humus, appearing as coatings on grains or as silt-size pellets.

ir — Illuvial concentration of iron, appearing as coatings on grains or as silt-size pellets. A spodic B horizon characterized by both illuvial humus and iron is designated Birh.

m — Strong irreversible cementation, for example by accumulation of iron, calcium carbonate, or silica.

P — Indicates horizon of an exhumed soil.[29] Notation precedes that of the master horizon.

sa — Accumulation of salts more soluble than gypsum, in amounts greater than the parent material is presumed to have had. Occurs in A, B, C, or R horizons.

si — Cementation by silica, as nodules or as a continuous medium. If cementation is continuous, the notation sim is used. Such horizons are also known as duripans or silcrete.

t — Accumulation of translocated clay, such as in an argillic B horizon.

x — Denotes subsurface horizon characterized by a bulk density greater than that of the overlying soil, hard to very hard consistence, and seemingly cemented when dry.

ox, n — In many unconsolidated Quaternary deposits, the C horizon consists of an oxidized C overlying a seemingly unweathered C. The oxidized C does not meet the requirements of the cambic B horizon. In stratigraphic work, it is important to differentiate between these two kinds of C horizons. It is suggested that Cox be used for oxidized C horizons,[4] and Cn for unweathered C horizons.[17]

v — Denotes spherically shaped voids or vesicles in ochric A horizons in desert regions.[35]

*In the new classification there are restrictions on some horizons located beneath the A horizon that require a minimum thickness for the horizon in question or a minimum content of various salts. Because many of these restrictions seem to serve little purpose and are inappropriate for many geomorphological studies, they are not used here, nor are geomorphologists advised to follow them. (Table modified from references 13, 33, and 34.)
**Compounds of iron and aluminum.

form of the nomenclature used by the FAO/UNESCO in their international system of soil horizon nomenclature and soil classification used in constructing a soil map of the world.[13] The definitions in Table 1–1 are less precise than those given by the Soil Survey Staff,[33,34] and one should refer to the latter for more detailed descriptions and for laboratory analyses. Soil-profile depth functions of various key properties are helpful in visualizing the more common horizons (Fig. 1–2).

Most soil profiles can be divided into several master or most promin-

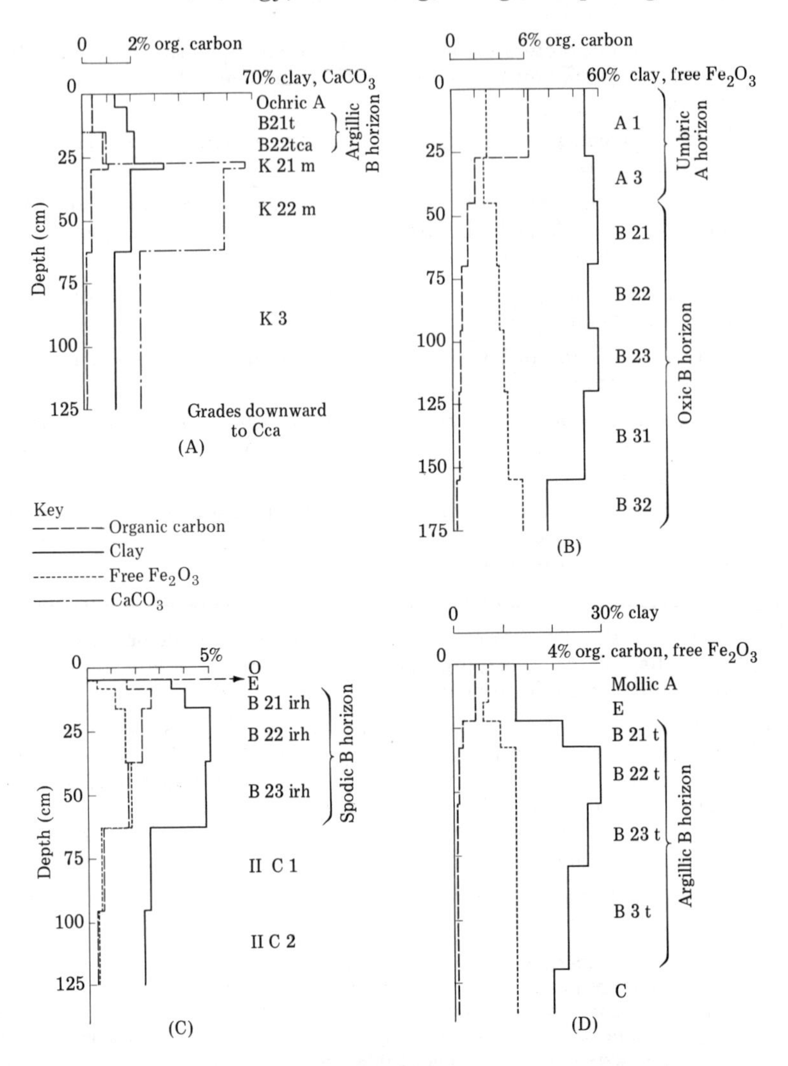

Fig. 1–2 Laboratory data on soil profiles that illustrate properties commonly associated with various soil horizons. (A) Parent material is gravelly alluvium. Per cent clay is for the noncarbonate fraction. (Terino soil from Gile and others,[14] Table 2, © 1966, The Williams & Wilkins Co., Baltimore.) (B) Parent material is serpentine. (Profile No. 27.[33]) (C) Parent material is glacial outwash. (Profile No. 20.[33]) (D) Parent material is loess (Profile No. 11.[33])

ent horizons (Table 1–1). Surface or near-surface horizons relatively high in organic matter are designated O and A horizons, the difference between the two being determined by the amount of organic matter present. Beneath the O or A horizon, in some environments, there is a light-colored horizon relatively leached of iron compounds called the E horizon; this is identical to the A2 horizon of some classifications. The B horizon commonly is beneath the surface horizon or horizons. This horizon encompasses a multitude of soil characteristics relative to those of the assumed parent material. Among the B horizon characteristics are clay accumulation, the production of red color, the accumulation of iron compounds with or without organic matter, and the residual concentration of resistant materials following the removal of more soluble constituents under conditions of intensive weathering and leaching. The slightly weathered C horizon commonly is beneath the B horizon and beneath that is unweathered bedrock, the R horizon. In desert environments, carbonate buildup plays an important role in soil morphology and genesis, and horizons high in carbonate are designated K. Any of the master horizons can be further subdivided on the basis of subordinate properties (Table 1–1).

Soil horizons can be further subdivided by adding a number to the master horizon designation. A3 and B1 are used for horizons transitional from the A to the B, with A3 having properties closer to those of the A, and B1 having properties closer to the B. Transitional horizons can also be designated AB and AC if more detailed subdivision is not possible or warranted. B2 is used for that part of the B horizon that displays the maximum expression of the properties upon which the B horizon is defined (e.g., B2t for the maximum expression of an argillic B horizon). A B3 horizon still retains many properties of the B2 horizon but is transitional to the underlying horizon. Within the C horizon, numbers are used to denote a vertical sequence of layers (C1, C2, etc.). A second number can be added for even further subdivision (B21t, B22t, and B31), based on subtle changes in such properties as color or texture.

Many unconsolidated deposits of Quaternary age consist of depositional layers of contrasting texture and/or lithology, and the soil profile extends through more than one layer. Examples are loess/till, floodplain silt/gravelly outwash, and colluvium/outwash. Such primary differences in texture and/or lithology are important in any soil-profile description. Each different geologic layer is so noted by a Roman numeral, counting from the top down. The numerals precede the master horizon designation, and the numeral for the uppermost layer (I) is omitted (Fig. 1–3).

Soils can be classified by their position in a stratigraphic section and in the landscape.[21,25] Three soils are recognized (Fig. 1–3). Relict soils are those that have remained at the land surface since the time of initial formation; they may or may not have acquired most of their properties over some time interval in the past. Buried soils are those that were formed on some ancient land surface, were subsequently buried by a younger deposit, and generally are far enough below the present land surface not to be affected by present pedogenic processes. Exhumed soils were formerly buried but subsequently exposed to current pedogenesis with erosional removal of the overlying material. The letter "P" preceding the master horizon designation denotes exhumed soil horizons.

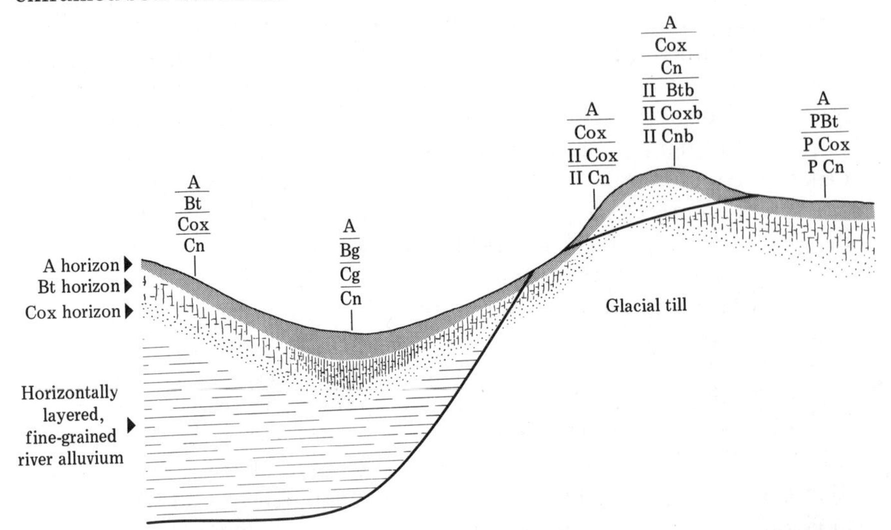

Fig. 1–3 Lateral variation in soil profiles due to lateral variation in environmental conditions and lithology and age of parent materials. Soil horizons parallel the land surface. Those horizons that are developed from river alluvium truncate depositional layering, hence soil properties are for the most part independent of properties of the depositional layers. The soil formed on glacial till was partly truncated by erosion before burial by loess; it grades laterally to the right from a buried to an exhumed soil, because the loess cover has been removed by erosion in that direction.

Soil horizons are more or less parallel to the land surface below which they formed, and indeed this is one criterion for differentiating soil layers from depositional layering (Fig. 1–3). Various properties of the horizons may change laterally, however, due to changing environmental conditions, and this results in lateral change in the overall soil profile. In most cases, the lateral change from one kind of profile to another is gradational; in contrast, lateral changes in parent material can be quite abrupt.

Soil characteristics

An undisturbed soil sample consists of a matrix of inorganic and organic solid particles in association with interconnected voids (Fig.

1–4). Depending on local conditions, varying amounts of soil water and gases occupy the voids. I will discuss the main physical and chemical characteristics of the soil, along with an outline of water movement. Only those properties most important to field studies will be dealt with here. These and other properties are treated in more detail elsewhere,[3,5,9,24] and much of the following information comes from the references just cited. Laboratory quantification of some of these properties is desirable, and one should consult Black[6] for appropriate analytical methods.

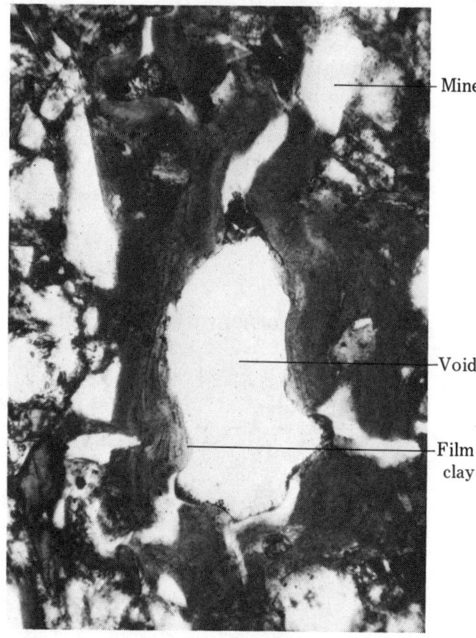

Fig. 1–4 Thin section of a soil sample to show inorganic particles and void distribution.

Color

Color is a valuable aid, if used with caution, in qualitatively recognizing processes that are or have been operating in a soil. Indeed, with buried soils color is the property that first catches one's attention. Dark brown to black colors in near-surface horizons reflect an accumulation of humified and/or nonhumified organic matter. Dark colors may also result from the accumulation of MnO_2, but these usually have a bluish cast and are not always close to the surface. Grayish colors (chromas* near 1) and/or hues* bluer than 10Y indicate reducing

*Color terms follow Munsell notation (see Appendix I).

conditions (gleying), and the color is due mostly to ferrous iron compounds. Yellow-brown to red colors result from the presence of iron oxides and hydroxides and are characteristic of B and C horizons. White or light-gray colors above the B horizon characterize an E horizon and suggest enough leaching by vertically or laterally moving water so that most of the grains are free of colloidal coatings of oxides and hydroxides of aluminium and iron. The same colors below a B horizon are usually due to concentrations of $CaCO_3$.

Intensity of color gives a measure of the amount of pigmenting material present, but not a very accurate one. The reason for this is that the texture of the material greatly influences the amount of surface area that has to be coated to impart a certain color on the material. For example, coarse-grained material has a much lower surface area per unit volume than does fine-grained material (Table 1–2), and so it would take much less pigment to impart a certain color to a coarse-grained soil than it would to a fine-grained soil.

Table 1–2

Particle size classes used in pedology and some of their properties

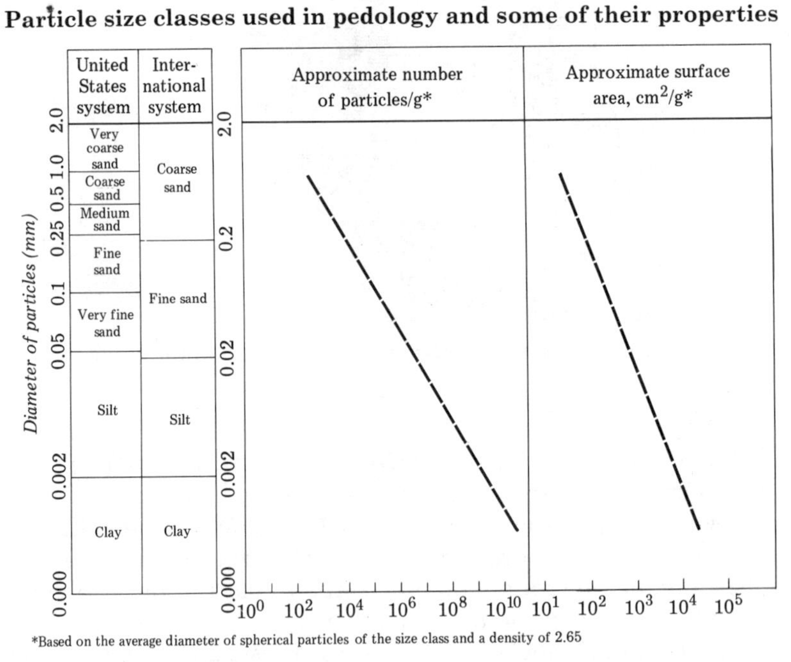

*Based on the average diameter of spherical particles of the size class and a density of 2.65

(Modified from Black,[5] © 1957, John Wiley and Sons)

---

*Color terms follow Munsell notation (see Appendix 1).

Texture

Soil texture depends upon the proportion of sand, silt, and clay sizes, as based on the inorganic soil fraction that is less than 2 mm in diameter. Particle-size classes used in pedology are given in Table 1–2; the U.S. system will be used here. Specific combinations of sand, silt, and clay define soil textural classes (Fig. 1–5). Mechanical analysis aids in textural classification, but an approximate classification can be made by the use of simple field tests (Appendix 1).

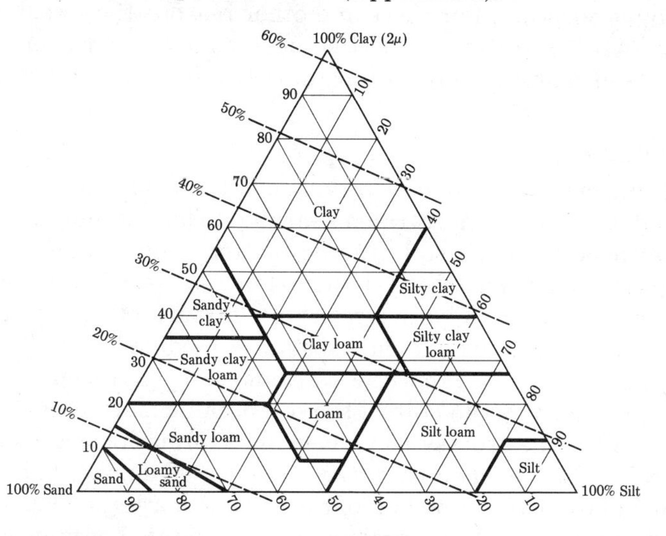

Fig. 1–5 Soil textural classes plotted on a triangular diagram.[33] The dashed lines are values for moisture equivalent calculated from the equation, moisture equivalent = 0.023 sand + 0.25 silt + 0.61 clay. (Taken from Bodman and Mahmud,[7] Fig. 3, © 1932, The Williams & Wilkins Co., Baltimore, and Jenny and Raychaudhuri,[15] Fig. 5.) The moisture equivalent approximates field capacity; thus the plot gives a general relationship between soil-texture classes and field capacity for soils low in organic matter.

The variation in particle size originates several ways. Sand and most silt are made up of minerals released by initial weathering or inherited from the parent material, although they may have been reduced in size by weathering. Some of the sand and silt in oxic B horizons, however, occurs as aggregates of clay-size material.[34] The clay-size fraction consists of both layer-lattice clay minerals and other crystalline and amorphous materials; this fraction originates in the parent material, is inherited later (e.g., eolian influx), or is formed within the soil. Because clay refers to both a size fraction and a suite of minerals, the term "clay" will be used here for all material that is less than 0.002 mm in diameter, and "clay mineral" will be used for the layer-lattice clay minerals.

Texture is one of the more important characteristics of a soil profile. The variation in texture from horizon to horizon can be used to decipher the pedogenic and geologic history of a soil. The fine-grained fraction also affects many processes operating within a soil, because the surface area per unit volume increases markedly as particle size decreases (Table 1–2). Soils with large internal surface areas are more chemically active than are soils with low surface areas because of their greater charge per unit volume and their capacity to hold greater amounts of water by adsorption. For these and other reasons they weather more rapidly. Many other soil properties, such as organic matter, nutrient content, and degree of aeration, are closely related to soil texture.

Organic matter

Organic matter is found in varying amounts in mineral soils and is almost always most concentrated near the surface. A wide spectrum of material makes up the soil organic matter, which ranges from undecomposed plant and animal tissue to humus, the latter being defined as "... a complex and rather resistant mixture of brown and dark brown amorphous and colloidal substances modified from the original tissues or synthesized by the various soil organisms."[9] Humus commonly makes up the bulk of the soil organic matter, and its chemistry as well as the processes involved in its formation are exceedingly complex.[18,24] Large amounts of CO_2 are evolved during its formation. Carbon makes up over one-half of the organic matter, and carbon content is commonly used to characterize the amount of organic matter in soils. Generally, the per cent of organic matter in a soil is 1.724 times the per cent of organic carbon. The C:N ratio is a rough measure of the amount of decomposition of the original organic material and is related to environmental conditions. The ratio is high (> 20) in plant tissue and low (< 10) in humus.

Soil organic matter is important to many soil properties, especially to the formation of surface soil structures, and to reactions that go on during pedogenesis. It increases considerably both the water-holding capacity of mineral soils and the cation exchange capacity. The organic acids that are produced promote weathering and form chelating compounds that increase the solubility of some ions in the soil environment. The CO_2 that is evolved builds up to reach concentrations higher than those in the atmosphere; this results in the formation of abundant carbonic acid, which lowers the soil pH and thus promotes weathering.

Structure

Structure involves a bonding together into aggregates of individual soil particles. Individual aggregates, termed peds, are classified into

Table 1–3
Description and probable origin of soil structure

TYPE	SKETCH* AND DESCRIPTION	PROBABLY ORIGIN[3,5,24,38]	USUAL ASSOCIATED SOIL HORIZON
Granular *Crumb*	Spheroidally shaped aggregates with faces that do not accommodate adjoining ped faces; the two differ only in that crumb is porous	Colloids, mainly organic, bind the particles together; clay and Fe and Al hydroxides may be responsible for some binding, and flocculating capacity of some ions, such as Ca^{2+}, may be helpful; periodic dehydration helps form more stable aggregates	Mollic or Umbric A
Angular blocky *Subangular blocky*	Approximately equidimensional blocks with planar faces that are accommodated to adjoining ped faces; face intersections are sharp with angular blocky, rounded with subangular blocky	Many faces may be intersecting shear planes developed during swelling and shrinkage that accompany changes in soil moisture	Argillic B
Prismatic *Columnar*	Particles are arranged about a vertical line, and ped is bounded by planar, vertical faces that accommodate adjoining faces; prismatic has a flat top, and columnar a rounded top	Faces develop as a result of tensional forces during times of dehydration; rounded column tops may be due to some combination of erosion by percolating water and greater amounts of upward swelling of column centers upon wetting	Argillic B (prismatic); Natric B (columnar)
Platy	Particles are arranged about a horizontal plane	May be related to particle-size orientation inherited from parent material or induced by freeze-thaw processes May be related to layering in cementing material, induced during its precipitation (carbonate, silica, Fe hydroxides)	E, or those with fragipan Km, Csim, Spodic B

*Taken from Soil Survey Staff[32]

several types on the basis of shape (Table 1–3). Although structure type is associated with soil horizon, details of the origins of many types are rather poorly known. Organic matter is important in the formation of spheroidally shaped structures, as is clay content in the formation of blocky, prismatic, and columnar structures. Some soils in tropical regions, however, are low in organic matter, yet they are well aggregated; this appears to result from cementation by iron hydroxides.

Structure is important to the movement of water through the soil and to surface erosion. A-horizon structures, although they vary from soil to soil, tend to produce larger-sized pores than would be the case for a structureless surface soil. These larger pores allow the soil to take up large amounts of rain water over a short period of time, and thus the possibility of runoff and surface erosion is reduced. Remove the A horizon, however, and erosion due to greater runoff may ensue if the infiltration rate in the exposed B horizon is less than that of the A horizon. The fact that many structural aggregates are water stable is important because it means that the percolating waters are fairly free of clay particles. However, aggregates unstable in water can break down and contribute clay to the percolating soil water. This tends to plug some of the pores, with a concomitant decrease in infiltration rate. B-horizon structures provide avenues for the translocation of

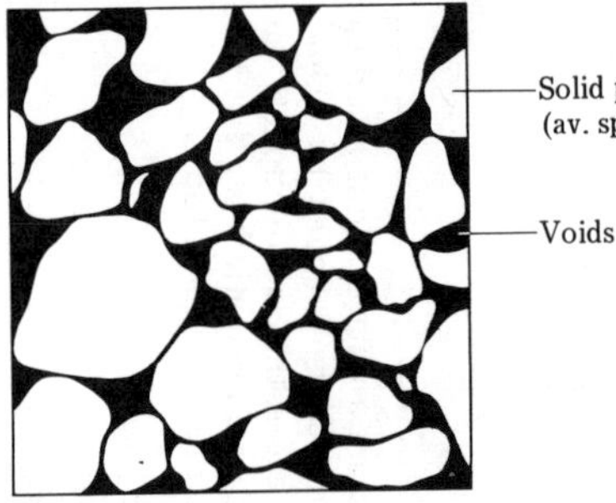

Fig. 1–6 Sketch of soil sample to show solid particle and void space distribution. The mineral grains in many soils are mainly quartz and feldspar, so 2.65 is an adequate average mineral specific gravity for the sand fraction. Bulk density and porosity are calculated as follows:

Weight of oven-dry soil: 63 g
Volume of soil in field: 35 cc

$$\text{Bulk density} = \frac{\text{weight}}{\text{volume}} = \frac{63}{35} = 1.8 \text{ g/cc}$$

$$\text{Porosity (\%)} = \left(1 - \frac{\text{bulk density}}{\text{particle density}}\right) \times 100$$

$$= \left(1 - \frac{1.8}{2.65}\right) \times 100 = 32$$

water and any contained solids along the ped interfaces. Indeed, this is where most clay films are located.

Bulk density

Bulk density is a measure of the weight of the soil per unit volume (g/cc), usually given on an oven-dry (110°C) basis (Fig. 1–6). Variation in bulk density is attributable to the relative proportion and specific gravity of solid organic and inorganic particles and to the porosity of the soil. Most mineral soils have bulk densities between 1.0 and 2.0. Although bulk densities are seldom measured, they are important in quantitative soil studies, and measurement should be encouraged. Such data are necessary, for example, in calculating soil moisture movement within a profile and rates of clay formation and carbonate accumulation. Even when two soils are compared qualitatively on the basis of their development for purposes of stratigraphic correlation,

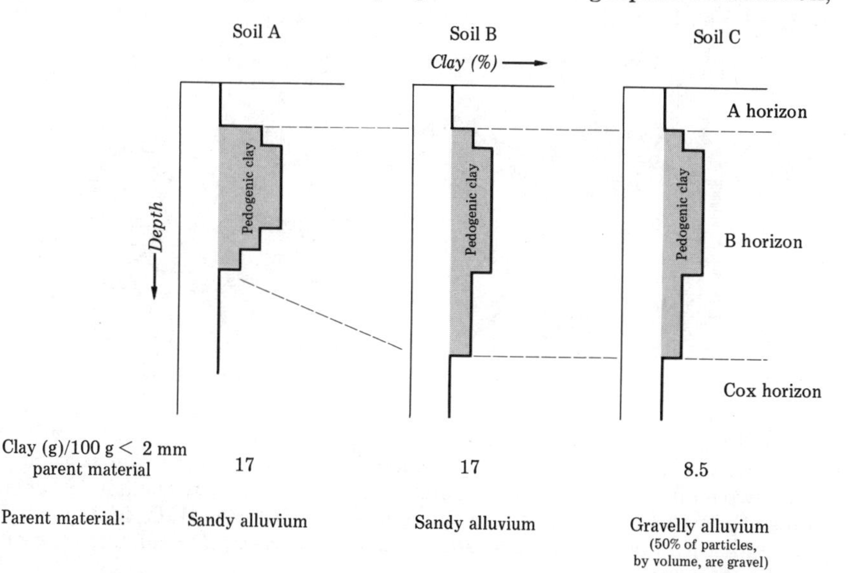

Fig. 1–7 Comparison of three soil profiles based on both per cent clay and weight of clay formed. Assume no clay present in all three profiles at the time of deposition. Soils A and B have identical parent materials, but clay is concentrated in a thinner horizon in soil A than in soil B. Soil A, therefore, probably would be classified as more strongly developed than soil B, although the same amount of clay has formed in each. Soils B and C have identical profiles of per cent clay, but soil C has formed from a 50 per cent gravel parent material (textural data usually are given for the < 2 mm fraction). Since gravel concontributes little to clay formation, and makes up one-half of the soil solids, the amount of clay formed in soil C is one-half that in soil B.

one can make more accurate comparisons on the basis of total weight of clay formed from 100 g of parent material than on per cent of clay alone (Fig. 1–7). To convert per cent to weight per unit volume, per cent is multiplied by bulk density.

Soil moisture retention and movement

Various amounts of water occupy the pore spaces in a soil. This water is held in the soil by adhesive forces between organic and inorganic particles and water molecules and cohesive forces between adjacent water molecules. Thin films of water are held tightly to the particle surfaces and are relatively immobile, whereas thick films are more mobile, and water can migrate from particle to particle both laterally and vertically.

Several soil moisture states are recognized (Fig. 1–8). One can start with a soil devoid of water and begin to add water from the top. Initially the soil may have its pores saturated with water, but because the outer

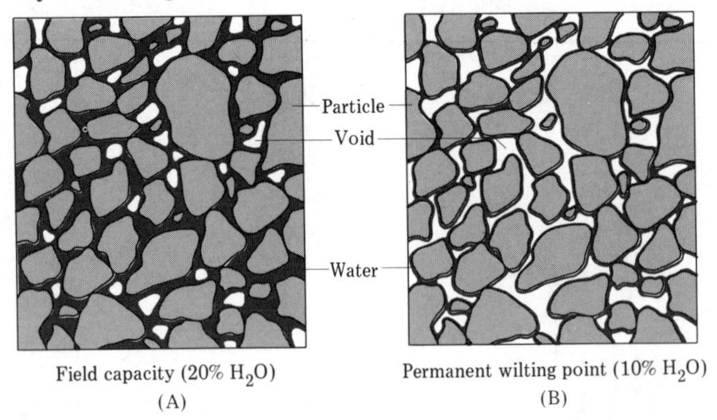

Fig. 1–8 Diagram of water in soil at field capacity (A) and permanent wilting point (B) and an example of how to calculate the amount of water present at these two moisture states. The calculation of amount of water held in 100 cm of soil at the two moisture states in the figure is as follows. (Bulk density = 1.5; P_w = moisture percentage; D = soil horizon thickness; d = amount of water held in soil (cm).

At field capacity (20 g H_2O/100 g soil or 20% H_2O)

$$d = \frac{P_w}{100} \times \text{bulk density} \times D = \frac{20}{100} \times 1.5 \times 100 = 30 \text{ cm } H_2O$$

(30 cm of water are held in 100 cm of this soil at field capacity)

At permanent wilting point (10 g H_2O/100 g soil or 10% H_2O)

$$d = \frac{10}{100} \times 1.5 \times 100 = 15 \text{ cm } H_2O$$

(15 cm of H_2O are held in 100 cm of this soil at permanent wilting point)

Available water-holding capacity is the difference between the above two water contents, or 15 cm of H_2O.

edge of the water film is under low surface tension, this more loosely held water migrates downward, under the influence of gravity, as a wetting front that more or less parallels the ground surface. After two or three days, re-distribution of water ceases for the most part, and the forces that hold water films on the particle surfaces equal the force of downward gravitational pull. At this point the soil is said to have a water content at field capacity (Fig. 1–8). At field capacity, water can be removed from the soil by evaporation from the surface and transpiration through vegetation, the latter probably being the more important mechanism for water removal from the soil profile. As roots remove water from the pores, the water film becomes thinner and is held by ever stronger forces of attraction until a point is reached at which the water is held so tightly by the particles that the roots can no longer extract it. The water content under these conditions is the permanent wilting point. Many soils under field conditions seldom obtain a water content less than that of permanent wilting point. Available water-holding capacity is the difference between field capacity and permanent wilting point; if given in water depth units, it is the amount of water required to wet a given thickness of soil from permanent wilting point to field capacity.

It is difficult to determine the moisture content under the above conditions in the field. Joint laboratory and field studies, however, suggest that water held at 15-atmos. tension approximates permanent wilting point for a variety of broad-leaved plants and that held at $\frac{1}{3}$-atmos. tension approximates field capacity, although this is not always the case.[30] These are the usual moisture contents reported in the soils literature, and the values are given as per cent (P_w)

$$P_w = \frac{H_2O\ (g)}{\text{Oven-dry soil (g)}} \times 100$$

Using moisture per cent and bulk density data, one can estimate the depth to which water will wet a soil (Fig. 1–9). Arkley[1] has developed a method for calculating annual water movement within a soil, taking into account the water-holding properties of the soil and the seasonal distribution and amount of precipitation and potential evapotranspiration (see Fig. 11–3). These are important data because percolating water redistributes clay and more soluble constituents downward in the soil or through the entire soil. Thus, many soil properties relate to the depth of wetting.

Soil-moisture retention and movement are strongly related to the surface area per unit volume of the soil mass, and this in turn is related to the clay and organic-matter contents. Buckman and Brady[9] give the following approximate available water-holding capacities for a 10-cm-thick layer of soil: 1.4 cm for clay, 1.7 cm for silt loam, and 1 cm for sandy loam. Thus, for a given rainfall, sandy soils are wetted to greater depths than are more heavily textured soils. Gravelly sands wet even more deeply than sands because gravel-sized particles have a low

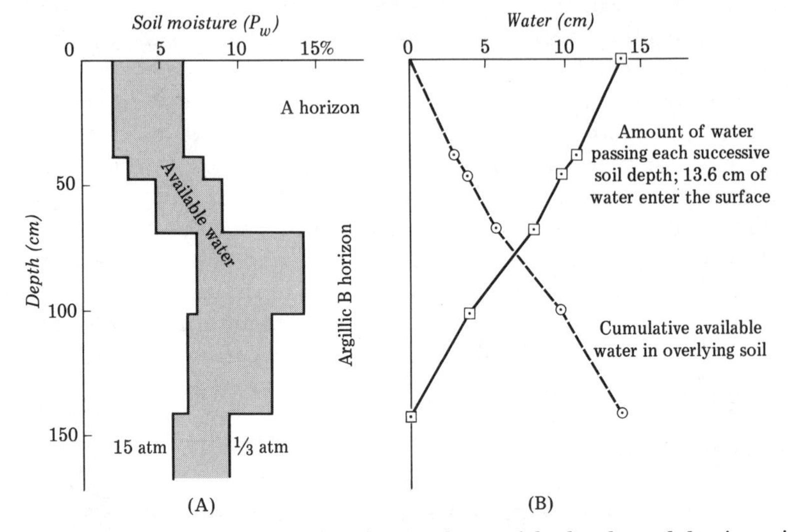

Fig. 1–9 (A) Variation in soil-moisture data with depth and horizon in a Snelling sandy clay loam, California (data from Arkley[1]). (B) Plot of cumulative available water and the amount of water passing through the soil for a 13.6 cm rainfall, calculated by the method presented in the legend for Fig. 1–8. It can be shown that if the soil is at permanent wilting point and begins to receive water from the surface, 13.6 cm of water is necessary to bring the top 142 cm of soil to field capacity, or an influx of 13.6 cm of water (assuming negligible losses) would percolate downward to 142 cm depth. Downward migration would be as a diffuse, near-horizontal wetting front, with the soil above near field capacity and the soil below near permanent wilting point.

surface area per unit volume. Because of this close association of moisture content with content of colloidal-size material, many workers have tried to correlate the two with varying success[15,22,31]; it might be possible to use some of these data to approximate soil-moisture conditions, provided textural data are available (Fig. 1–5).

Capillary rise from a shallow water table may deliver water and soluble salts to the overlying horizons. In theory, the smaller the pores the higher the water will rise, but irregularly-shaped soil pores hardly make ideal capillary tubes. Rode[24] suggests such rise may be 1 m or less in sands and 3 to 4 m in clays. Parts of the soil profile at distances above the water table greater than these distances should not be influenced by capillary rise. Soluble salts or iron or manganese compounds might accumulate at the top of either the water table or the capillary fringe.

Few criteria have been developed, however, to determine if such accumulations are produced by upward- or downward-moving water. If the accumulations are parallel with the ground surface and bear some consistent relationship with the other soil horizons, downward-moving water is suggested. If, however, the accumulations cut across soil horizons, and there is no reasonable depth relationship with the ground surface, derivation from a water table is suggested.

It is important to note that field capacity and permanent-wilting-point moisture conditions describe the movement of water and not necessarily the amount of water involved in weathering reactions within the soil. Weathering occurs at all water contents, because a thin water film is always in contact with mineral grains. If field capacity is often reached, the ions in the water film released by weathering may be constantly flushed from the soil. If, however, the film is thin and field capacity is seldom reached, ionic concentrations in the water film may approach saturation, and this would inhibit further weathering unless periodic flushing occurs to lower ionic concentrations.

Cation exchange capacity, exchangeable cations, and per cent base saturation

Most soil colloids, both inorganic and organic, have a net negative surface charge, the origin of which is discussed in Chapter 4. Cations are attracted to these charged surfaces. The strength of cation attraction varies with the colloid and the particular cation, and some cations may exchange for others. The total negative charge on the surface is called the cation exchange capacity, and it is expressed in milliequivalents per 100 g oven-dry material. The exchangeable cations are those that are attracted to the negatively charged surfaces. Base saturation is expressed as the per cent of base ions (non-hydrogen) that make up the total exchangeable cations. Thus, there generally is a close relationship between base saturation and pH. Exchangeable sodium percentage is the amount of sodium relative to the total exchangeable cations, and this is important in defining a natric B horizon.

Soil pH

Soil pH's have an extreme range of 2 to 11, but most soil pH's range from 5 to 9. Soil pH is dependent on the ionic content and concentration in both the soil solution and the exchangeable cation complex adsorbed to the surfaces of colloids. In general, the ionic concentration increases from a low in the soil solution to a high at the colloid surface (Fig. 5–5). Furthermore, there is an equilibrium between the ions in solution and the exchangeable ions, and thus the ions are present in about the same proportions in both environments. Ions commonly present are Ca^{2+}, Mg^{2+}, K^+, Na^+, H^+, Cl^-, NO_3^-, SO_4^{2-}, HCO_3^-, CO_3^{2-}, and OH^-.

The relative proportion of these ions determines the soil pH (Fig. 1–10). Hydrogen ions are derived from rainfall and from organic and inorganic acids produced within the soil. There are few data on rainfall pH, but the available data show a variation from 3.0 to 9.8.[11] Pure water, in equilibrium with atmospheric CO_2 at 25°C, should have a pH of 5.7.

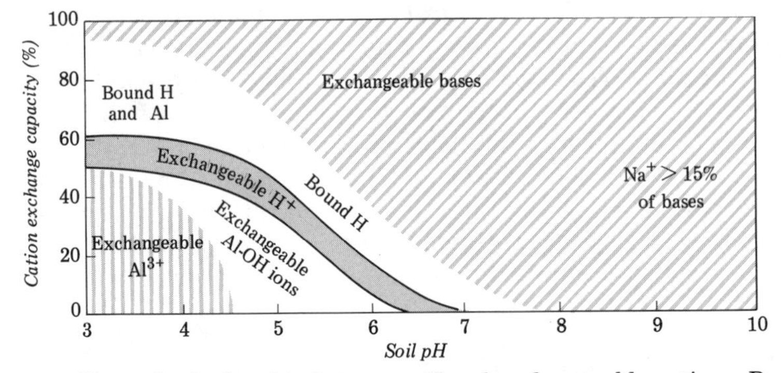

Fig. 1–10 General relationship between pH and exchangeable cations. Precise values vary from soil to soil for many reasons. Bound H^+ is that H^+ held so tightly to colloid surfaces that little of it is exchangeable. (Taken mostly from Buckman and Brady,[9] Fig. 14.1, © 1969, The Macmillan Co.)

Values lower than this are thought to be due partly to atmospheric pollution, whereas higher values are attributed to salts derived both from wind-blown sea water along coasts or from wind-blown dust. Carbonic acid is formed within the soil by the combination of CO_2 and water. A pH below 5.7 is possible because CO_2 content up to 10 or more times greater than atmospheric is possible because of respiration by plant roots and by microorganisms. An important H^+ source is the wide variety of organic acids produced within the soil.[18] Still other sources are the exchangeable Al^{3+} and Al-hydroxy ions, which release H^+ during hydrolysis

$$Al^{3+} + H_2O \rightarrow Al(OH)^{2+} + H^+ \text{ (under very acid conditions)}$$
$$Al(OH)_2^+ + H_2O \rightarrow Al(OH)_3 + H^+ \text{ (under less acid conditions)}$$

The exchange complex in acidic soils is dominated by H^+, Al^{3+}, and Al-hydroxy ions, and base content is low. As cation content increases, however, they replace H^+ and Al-hydroxy ions, and OH^- concentration and pH increase. The alkalinity will depend on the strength of the base formed. For example, $Ca(OH)_2$ is formed in the presence of $CaCO_3$ and the resulting pH can approach 8.5. In contrast, $NaCO_3$ and $NaHCO_3$ form NaOH, a stronger base, and this results in pH's over 8.5.

One final point on pH is that the pH of a given soil is not uniform throughout. There may be slight variations from place to place due to slight variations in CO_2 or organic acid concentrations, the content and composition of the exchangeable bases, or presence of nearby roots, since these commonly contain adsorbed H^+. Thus, although most work on weathering and mineral stabilities uses soil pH values, it should be remembered that these values reflect average conditions and may not reflect conditions where minerals are being weathered or synthesized.

Classification of soil-profile development

Degree of soil-profile development is used as a qualitative measure of the amount of change that has taken place in the parent material. It is commonly used in Quaternary stratigraphy for soils formed from unconsolidated deposits.[20,21,23] The ranking is generally on a relative scale, based on the diagnostic properties of a sequence of soils in an area. It is felt, however, that a more quantitative scale would be useful, because soil of the same age may vary in its development from place to place due to variations in soil-forming factors. Such variations in soil characteristics are important in both soil studies and soil-stratigraphic studies. The following scheme is modified from Birkeland.[4]

A *weakly developed* soil profile is one with an A-Cox and/or Cca or an A-cambic B-Cox and/or Cca soil horizon sequence. Soils with other B horizons are excluded. If carbonate horizons are present, development is probably stage I (see Appendix 1).

A *moderately developed* soil profile is one with an A- or A and E-B-Cox and/or Cca soil horizon sequence. The B horizon may be argillic, natric, spodic, or oxic. If carbonates are present, they may display stage II development.

A *strongly developed* soil profile is similar to a moderately developed one with these exceptions: the B horizon in the strongly developed profile is generally thicker and redder, contains more clay, and has a more strongly developed structure; if carbonates are present, they probably would form a K horizon (stages III and IV).

The distinctions between a moderately and a strongly developed profile are qualitative, but they can be quantified to a degree on color and texture. This is difficult to do, however, as both vary with the soil-forming factors. For example, it is difficult to compare soil development on a loess with 20 per cent primary clay with that on a gravelly outwash with 5 per cent primary clay. Clay content ratios might be used here, and, although soils workers seem to prefer A horizon:B horizon ratios, geologists might prefer B horizon:C horizon ratios. Or increases in the amount of clay (weight or per cent) from one horizon to the other (C to B) could be used to distinguish between moderately and strongly developed soils. As regards color, again there is no fast rule. It seems, however, that each development rank might be accompanied by a different hue. Thus, if a moderately developed soil has a 10YR hue, a strongly developed soil would have stronger color, perhaps a 7.5YR hue.

Although this classification of soil development has been used mainly for soils formed from unconsolidated deposits, it can be used for soils formed on bedrock.

Recognition of buried soils

Some buried soils are so obvious that few people would argue as to their pedogenic origin (Fig. 1–11); others, however, are quite difficult to differentiate from geologic deposits. Many people have had the experience in field conferences that one man's soil becomes another's geologic deposit. Part of the problem might be that some people have not spent enough time studying surface soils to really understand their properties and profile characteristics.

The same criteria used to recognize and describe surface soils can be used for buried soils.[39] The first test is to trace the material laterally in outcrop to be sure it is a soil and not a deposit. The relationship between the soil horizons and bedding should be deciphered because soil horizons can truncate geologic bedding. A problem comes up, however, in places where a soil forms from a surface that parallels geologic bedding. A common example is the soils formed on alluvial fan deposits in the semiarid southwestern United States. Whenever deposition stopped for a long enough time, a soil could form, but it could be buried later during renewed deposition. In this case the soil horizons would be parallel to the bedding in the fan deposit. Some depositional layers in fan deposits can be poorly sorted, however, and resemble soils in their thickness, color, and texture. Thus, one has to look for pedologic features in the zone to be sure soil formation has taken place. In general, the organic matter in the A horizon does not persist after burial,[36] but the mineral part of the A horizon may still be present and recognizable by a slightly lower clay content than that of the B horizon. Generally the buried B horizon is the most important horizon for recognizing buried soils. If it is an argillic B, it will have greater clay content, redder or browner colors, and better developed structure than the C horizon. Thin-section study of the horizon may disclose features that could only be produced by pedogenesis.[8,10,12] In drier regions, the presence of a carbonate-enriched horizon beneath the

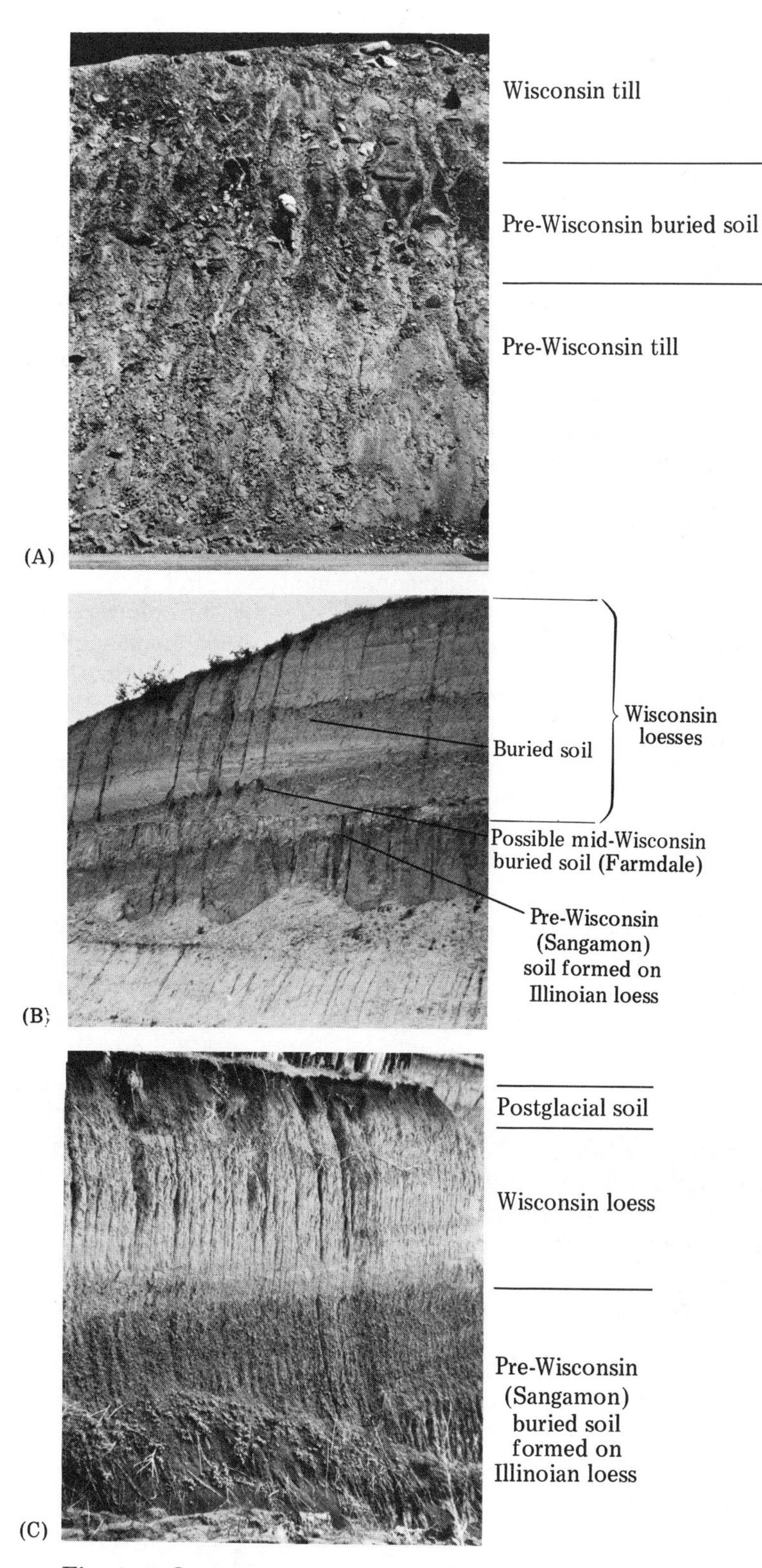

Fig. 1–11 Quaternary deposits and buried soils in (A) the Sierra Nevada, (B) Missouri, and (C) Nebraska.

B horizon is helpful in identifying a buried soil. If a K horizon is present, it may be laminated in its uppermost part. Moreover, carbonate horizons should have a distinct relationship to the buried land surface so that ground-water origin can be ruled out.[26]

One other criterion that is helpful in the recognition of a buried soil is the abruptness of the horizon boundaries (Fig. 1–2). Quite commonly the upper boundary of a soil horizon is sharper than the lower boundary, and this need not be the case in geologic layering. For example, with depth in a well-developed soil profile the transition from a low-clay-content A horizon to an argillic B horizon may take place over a few centimeters, whereas with greater depth in the B horizon there is a more gradual decrease in clay content toward the C horizon. The same criterion appears to hold for pedogenic carbonate horizons, that is, the upper horizon boundary is sharp and marked by a thin transition zone between the overlying non-carbonate material and the carbonate-enriched horizon. If a color B horizon is present, the color is redder or browner in the upper part of the horizon, and chroma gradually diminishes with depth. In contrast, post-burial diagenetic alteration within a deposit may result in a gradual decrease in color both upward and downward from the zone of maximum alteration.

Another criterion that is helpful in the recognition of buried soils is the mineralogical characteristics of the profile. Some nonclay silicate minerals become weathered and/or etched upon weathering (Fig. 8–5). If the zone studied is a B horizon, weathering and etching may be greater there than in the underlying C horizon and overlying deposit. Mineral depletion during weathering may be reflected by resistant-to-less-resistant mineral ratios that have a consistent relationship with depth (Figs. 8–3 and 8–4). Clay minerals may also give a clue on the pedogenic origin of a horizon. Quite commonly the clay minerals that form during soil formation vary in type with depth in a profile (Ch. 4), provided such variation is not due to variations in the parent material. If clay minerals were originally in the deposit and later underwent weathering within a soil they may have been selectively altered with depth to other clay minerals (Fig. 8–15).

A more difficult problem in the recognition of buried soils comes when the upper part of the soil has been removed by erosion, leaving only that part of the profile that was below the B horizon. Here, oxidation colors help in identifying the material as part of a buried soil, as long as post-burial, ground-water alteration can be ruled out. In the midcontinent, evidence for such a history is shown by the carbonate content of superimposed loesses. The older loess may have been leached of carbonate during an interval of soil formation, the soil B horizon may have been removed during a subsequent period of erosion, and the leached loess may then have been buried by carbonate-bearing loess.[27,37] Thus, the major remaining evidence for an unconformity and an interval of soil formation is the presence of carbonate-bearing loess overlying loess that had been leached of its carbonate prior to burial.

It is difficult to classify buried soils to the same degree of accuracy as surface soils. This is because during burial changes take place in properties critical to such classification. For example, A horizons are critical to classification yet are rare in buried soils. Upon burial, other changes take place in pH and base saturation, and these are important to classification. One should try, however, to classify buried soils on as many of the same criteria as are used to classify surface soils.

REFERENCES

1. Arkley, R. J., 1963, Calculations of carbonate and water movement in soil from climatic data: Soil Sci., v. 96, p. 239–248.
2. ——1964, Soil survey of the eastern Stanislaus area, California: U.S. Dept. Agric., Soil Surv. Series 1957, no. 20. 160 p.
3. Baver, L. D., 1956, Soil physics: John Wiley and Sons, New York, 489 p.
4. Birkeland, P. W., 1967, Correlation of soils of stratigraphic importance in western Nevada and California, and their relative rates of profile development, p. 71–91 *in* R. B. Morrison and H. E. Wright, Jr., eds., Quaternary soils: Internat. Assoc. Quaternary Research, VII Cong., Proc. v. 9, 338 p.
5. Black, C. A., 1957, Soil–plant relationships: John Wiley and Sons, New York, 332 p.
6. ——1965, Methods of soil analysis (parts 1 and 2): Amer. Soc. Agron., Madison, Series in Agron., no. 9. 1572 p.
7. Bodman, G. B., and Mahmud, A. J., 1932, The use of the moisture equivalent in the textural classification of soils, Soil Sci., v. 33, p. 363–374.
8. Brewer, R., 1964, Fabric and mineral analysis of soils: John Wiley and Sons, New York, 407 p.
9. Buckman, H. O., and Brady, N. C., 1969, The nature and properties of soils: The Macmillan Co., Toronto, 653 p.
10. Cady, J. G., 1965, Petrographic microscope techniques, p. 604–631 *in* C. A. Black, ed., Methods of soil analysis (part 1): American Soc. Agron., Madison, Series in Agron., no. 9, 770 p.
11. Carroll, D., 1962, Rainwater as a chemical agent of geologic processes—a review: U.S. Geol. Surv. Water-Supply Pap. 1535-G, 18 p.
12. Dalrymple, J. B., 1964, The application of soil micromorphology to the recognition and interpretation of fossil soils in volcanic ash deposits from the North Island, New Zealand, p. 339–349 *in* A. Jongerius, ed., Soil micromorphology: Elsevier Publ. Co., New York, 540 p.
13. Dudal, R., 1968, Definitions of soil units for the soil map of the world: World soil resources reports no. 33, World soil resources office, FAO, Rome, 72 p.

14. Gile, L. H., Peterson, F. F., and Grossman, R. B., 1966, Morphological and genetic sequences of carbonate accumulation in desert soils: Soil Sci., v. 101, p. 347–360.
15. Jenny, H., and Raychaudhuri, S. P., 1960, Effect of climate and cultivation on nitrogen and organic matter reserved in Indian soils: Indian Council of Agricultural Research, New Delhi, 126 p.
16. Joffe, J. S., 1949, Pedology: Pedology Publ., New Brunswick, N.J., 662 p.
17. Kohl, F., ed., 1965, Die bodenkarte: Niedersäschsisches Landesamt für Bodenforschung, Hanover, 124 p.
18. Kononova, M. M., 1966, Soil Organic Matter : Pergamon Press, New York, 554 p.
19. Leighton, M. M., and MacClintock, P., 1962, The weathered mantle of glacial tills beneath original surfaces in north-central United States: Jour. Geol., v. 70, p. 267–293.
20. Morrison, R. B., 1964, Lake Lahontan: Geology of the southern Carson Desert, Nevada: U.S. Geol. Surv. Prof. Pap. 401, 165 p.
21. ____ 1967, Principles of Quaternary soil stratigraphy, p. 1–69 *in* R. B. Morrison and H. E. Wright, Jr., eds., Quaternary soils: Internat. Assoc. Quaternary Res., VII Cong., Proc. v. 9, 338 p.
22. Nielsen, D. R., and Shaw, R. H., 1958, Estimation of the 15-atmosphere moisture percentage from hydrometer data: Soil Sci., v. 86, p. 103–105.
23. Richmond, G. M., 1962, Quaternary stratigraphy of the La Sal Mountains, Utah: U.S. Geol. Surv. Prof. Pap. 324, 135 p.
24. Rode, A. A., 1962, Soil Science: Israel Program for Scientific Translations, Jerusalem, 517 p.
25. Ruhe, R. V., 1965, Quaternary paleopedology, p. 755–764 *in* H. E. Wright, Jr., and D. G. Frey, eds., The Quaternary of the United States: Princeton Univ. Press, Princeton, 922 p.
26. ____ 1967, Geomorphic surfaces and surficial deposits in southern New Mexico: New Mex. Bur. Mines and Min. Resources, Memoir 18, 65 p.
27. ____ 1968, Identification of paleosols in loess deposits in the United States, p. 49–65 *in* C. B. Schultz and J. C. Frye, eds., Loess and related eolian deposits of the world: Internat. Assoc. Quaternary Res., VII Cong., Proc. v. 12.
28. ____ 1969, Quaternary landscapes in Iowa: Iowa State Univ. Press, Ames, 255 p.
29. ____ and Daniels, R. B., 1958, Soils, paleosols, and soil-horizon nomenclature: Soil Sci. Soc. Amer. Proc., v. 22, p. 66–69.
30. Salter, P. J., and Williams, J. B., 1965, The influence of texture on the moisture characteristics of soils: I. A critical comparison of techniques for determining the available-water capacity and moisture characteristic curve of a soil: Jour. Soil Sci., v. 16, p. 1–15.
31. ____ and ____ 1967, The influence of texture on the moisture characteristics of soils: IV. A method of estimating the available-water capacities of profiles in the field: Jour. Soil Sci., v. 18, p. 174–181.
32. Soil Survey Staff, 1951, Soil survey manual: U.S. Dept. Agri. Handbook no. 18, 503 p.

33. —— 1960, Soil Classification, a comprehensive system (7th approximation): U.S. Dept. of Agri., Soil Cons. Service, 265 p.
34. —— 1967, Supplement to soil classification system (7th approximation): U.S. Dept. of Agri., Soil Cons. Service, 207 p.
35. Springer, M. E., 1958, Desert pavement and vesicular layer of some soils of the desert of the Lahontan Basin, Nevada: Soil Sci. Soc. Amer. Prov., v. 22, p. 63–66.
36. Stevenson, F. J., 1969, Pedohumus: Accumulation and diagenesis during the Quaternary: Soil Sci., v. 107, p. 470–479.
37. Willman, H. B. and Frye, J. C., 1970, Pleistocene stratigraphy of Illinois: Illinois State Geol. Surv. Bull. 94, 204 p.
38. White, E. M., 1966, Subsoil structure genesis: Theoretical consideration: Soil Sci., v. 101, p. 135–141.
39. Yaalon, D. H. (chairman), 1971, Criteria for the recognition and classification of Paleosols, p. 153–158 *in* D. H. Yaalon, ed., Paleopedology: Israel Univ. Press, Jerusalem, 350 p.

CHAPTER 2

Soil classification

We classify soil by common properties for the purposes of systematizing knowledge about soils and determining the processes that control similarity within a group and dissimilarities among groups. The numbers of individual soils in a group are a function of the limits one allows in the defining properties. Thus, there are many soil series and types at the lowest category of classification because many restrictions are imposed by the limits of the diagnostic properties. Different series can be grouped together at higher levels of classification, and these in turn can be regrouped until one ends up at the highest category with a few orders, each with wide limits on allowable variations in differentiating properties. In spite of the large number of members, soils in each order should share many properties in common, because they have been formed under a somewhat similar set of pedogenetic processes. To be really useful, however, classification has to be based on soil properties and not geologic, climatic, or vegetational properties. Maps of soils at any category of classification are useful for geomorphic interpretation.

Soil classification at lower levels

Soil maps are prepared by the study of many soil profiles. With time, it becomes apparent that the soils can be grouped on the basis of similar profile characteristics and that these characteristics can change with changes in the soil-forming factors. In general, boundaries between mapping units, although commonly gradational, will be governed by one or more of the factors.[8,9] The mapping of soils is thus complex, because as mapping proceeds one must keep the multitude of factors in mind when predicting and drawing boundaries around the mapping units. Most mapping nowadays is done with the aid of aerial photographs; these increase mapping effectiveness because subtle tonal changes are many times the clue to lateral changes in soil properties and therefore the mapping unit.

The soil series is the basic classification unit used in soil mapping,[15,16] and in a sense it is similar in concept to geologic formations. A soil series is a group of soil profiles with somewhat similar profile characteristics such as the kind, thickness, arrangement, and properties of soil horizons. The properties used to differentiate series must be observable in the field and must have pedogenetic significance. Series, therefore, are conceptual and are defined according to permissible ranges in the properties. Series names are derived from a local place name, such as a town or county. Many times, changes in parent material lithology and texture are reflected in basic soil-profile differences, and these have been used to differentiate series. This is not always the case, and present practice bases most of the differentiating criteria on observable soil properties. The texture of the A horizon is allowed to vary within a series; this variation is shown by mapping units that are a subdivision of the series, called soil types.

Soil types are subdivided into soil phases, which are the mapping units used in detailed soil surveys. Phase differentiation is based on slope steepness, physiographic position, thickness of soil profile or individual horizons, amount of erosion, stoniness, and salinity within the series limits.

A detailed soil map at the series, type, and phase level is useful in interpretating the geology of an area. In Iowa, for example, the soils mapped at the surface are a complex mosaic that can best, and perhaps only, be understood by mapping at the series and phase levels (Fig. 2–1). There, the different series and phases elucidate parent-material differentiation, the presence of buried soils and their effect on overlying soils, and provide a qualitative measure of the rate of hillslope erosion. In California, mapping at these same levels has helped differentiate Quaternary terrace deposits of different ages (Fig. 2–2). Once the potential usefulness and shortcomings of such maps is recognized, they can be compiled into moderately accurate, regional maps showing Quaternary stratigraphic units (Fig. 2–3).

Soil classification at the higher categories

Soil classification at the higher categories is in a state of flux. Many

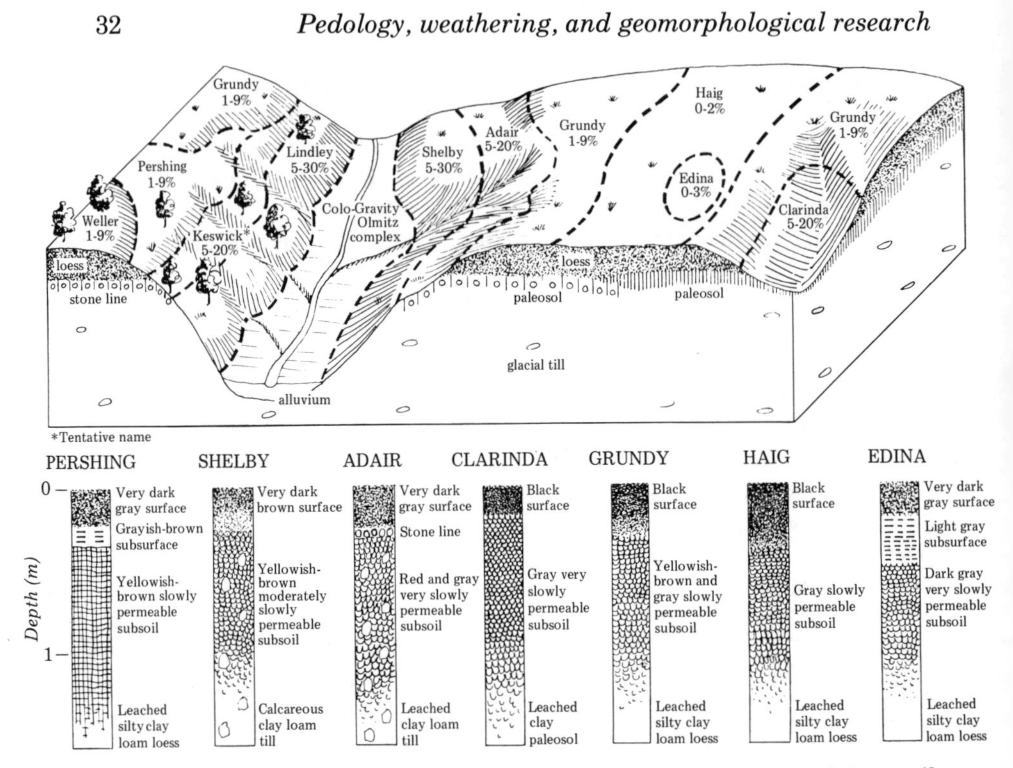

Fig. 2–1 Relationship of slope (%), vegetation, and-parent material to soil series mapped in southern Iowa. Upland soils have formed from Wisconsin loess, and valley side soils have formed either from re-exposed, pre-Wisconsin soils or from unweathered Kansan Till. Morphology of some of the diagnostic soil series is diagrammatic. (Taken from Oschwald and others,[12] Fig. 15.)

classification schemes have been proposed over the years, and the history of these is reviewed elsewhere.[5,13,16] There is still no world-wide agreement on soil classification; in general, each country has its own, although there are some features that are common to many of the systems now in use. At the present time an effort is being made to put together a soil classification scheme (FAO) for use with a soils map of the world.[7] Workers from many parts of the world have had a part in this study, which is continuing, and perhaps it will gain international recognition and use. Until that time, one must know several classifications.

Two classification systems currently are in use in the United States. One is the old system introduced by Marbut,[10,11] and recognized by such familiar terms as pedocal, pedalfer, Red Desert Soil, and Podzol. In 1960, a new classification was proposed[16,17,18] because the old system was quite difficult to use since definitions were imprecise, quantitative guidelines were lacking, and a genetic bias was present. The new

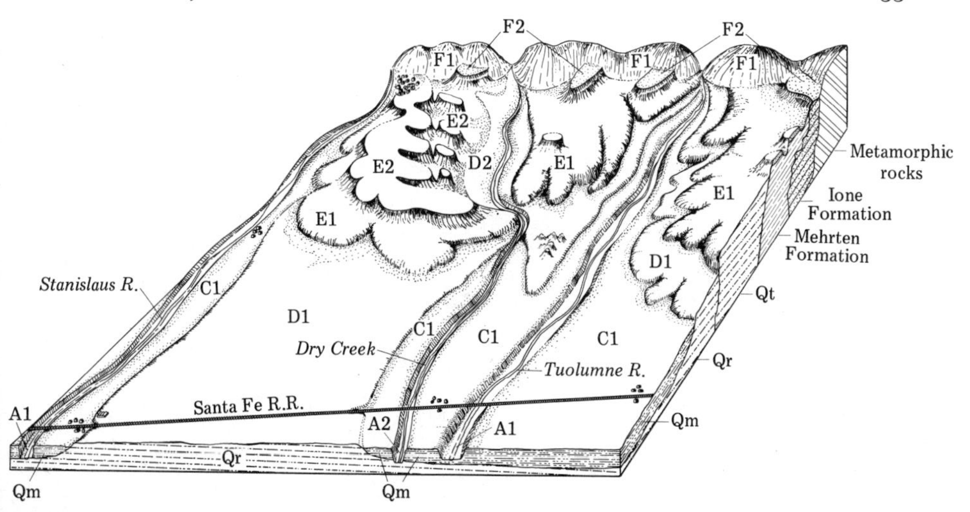

Fig. 2–2 Relationship of deposits, landscapes, and development of soils in the eastern Stanislaus area, California. (Taken from Arkley,[4] Fig. 21.) This diagram summarizes the more detailed soil survey of the area. The table below gives data for soils formed on Quaternary sedimentary deposits, mainly of glacial outwash origin. Although many soil series, types, and phases are mapped on each stream-terrace deposit, soil development is fairly consistent for any one deposit and is a strong clue to the geologic age of the deposit. F, soils formed on pre-Quaternary rocks (Mehrten Formation and older); F1, shallow soils formed from metamorphic rocks; F2, shallow soils formed from marine sediments.

Data for soils formed on stream terrace deposits

AGE	LANDSCAPE SYMBOL*	LANDSCAPE POSITION	SOIL DEVELOPMENT	GEOLOGIC FORMATION
Holocene	A	Present-day floodplain	None to weak	
Pleistocene	C	Low-level stream terraces	Weak; some moderate	Modesto (Qm)
	D	Intermediate-level stream terraces	Strong	Riverbank (Qr)
	E	High-level stream terraces	Strong	Turlock Lake (Qt)

*The number following the letter in the figure refers to the lithology of the parent material: 1 = granitic; 2 = andesitic. Parent-material texture of the terrace deposits is generally sand and sandy loam.

classification is based mainly on observable properties and not some genetic concept, is quantitative, runs over 200 pages, and carries such exotic combinations of Latin and Greek as Cryaqueptic Haplaquoll, Aquic Ustochrept, and Natraqualfic Mazaquerts. This classification is

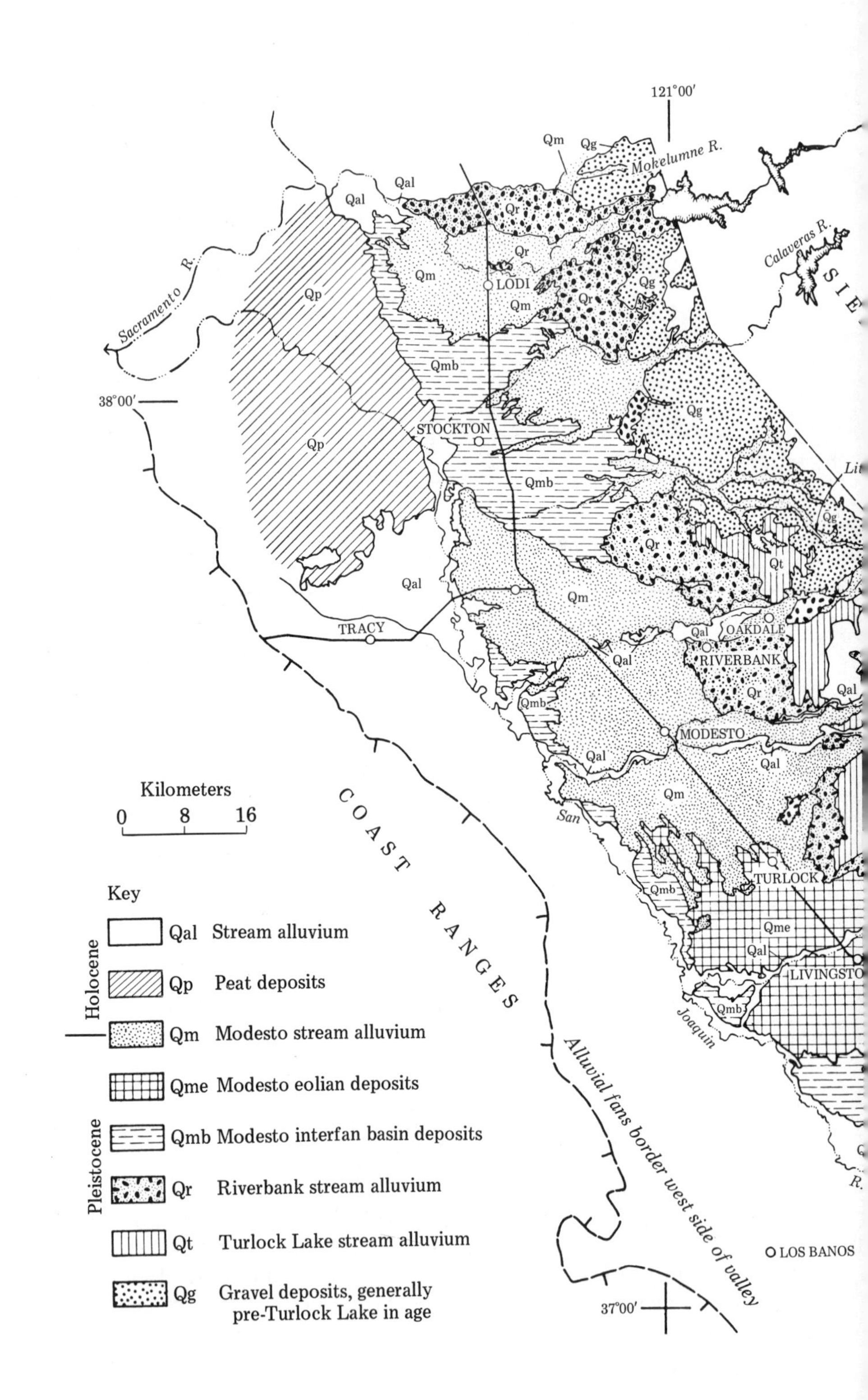
121°00′
Mokelumne R.
Calaveras R.
Sacramento R.
38°00′
LODI
STOCKTON
TRACY
OAKDALE
RIVERBANK
MODESTO
TURLOCK
LIVINGSTO
LOS BANOS
San Joaquin R.
COAST RANGES
Alluvial fans border west side of valley
37°00′
Kilometers
0 8 16
Key
Holocene
Pleistocene
Qal Stream alluvium
Qp Peat deposits
Qm Modesto stream alluvium
Qme Modesto eolian deposits
Qmb Modesto interfan basin deposits
Qr Riverbank stream alluvium
Qt Turlock Lake stream alluvium
Qg Gravel deposits, generally pre-Turlock Lake in age

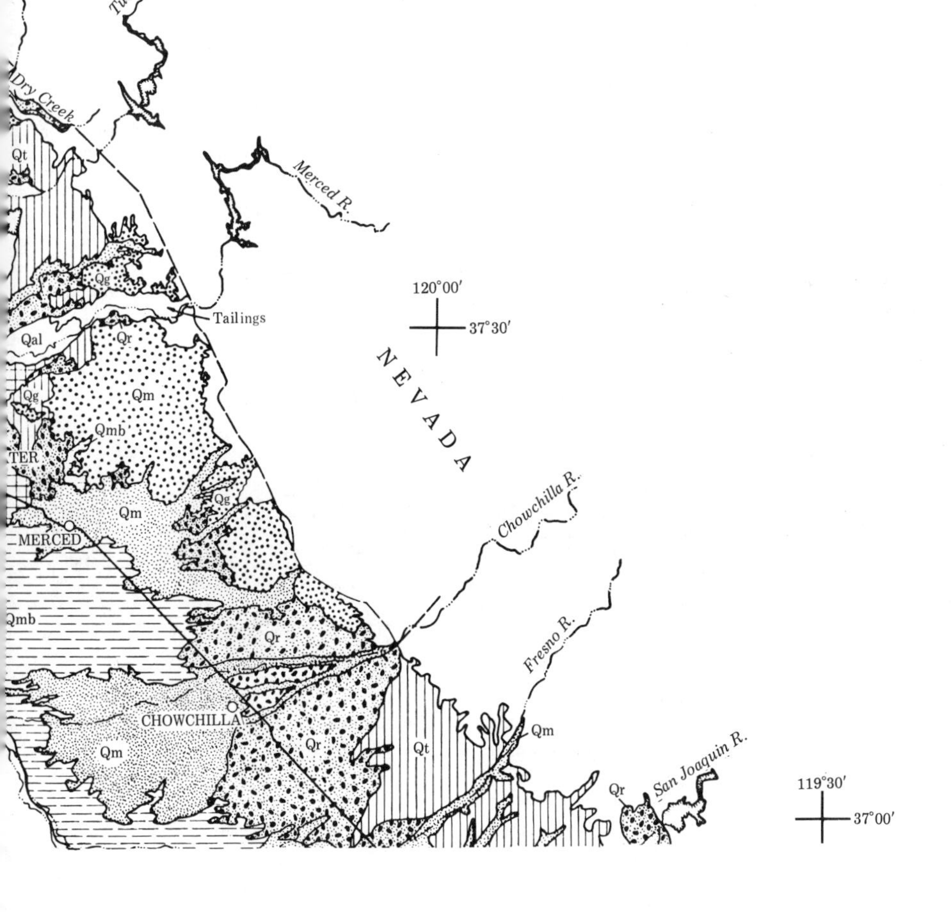

Fig. 2–3 Generalized geologic map of part of the San Joaquin Valley, California, compiled from soil surveys of the area,[3,4,21,22] along with some field checking by myself and R. J. Janda. Soils were mapped in extreme detail. Published surveys also give a generalized soil map, compiled from data of the detailed survey. The soils of the San Joaquin Valley are closely related to the age of the Quaternary deposits, so these maps can be generated from generalized soil maps. The topographic position of most of the deposits and the development of the soils are given in Fig. 2–2, a part of this area. Topographic position and sedimentary characteristics of the Modesto deposits, and the soil series mapped on them, allow for differentiation into stream alluvium, eolian deposits, and interfan basin deposits.

Table 2–1 Generalized key to soil orders and some

<table>
<tr><th>DIAGNOSTIC SOIL HORIZONS</th><th colspan="5">ORDER</th></tr>
<tr><th></th><th>ENTISOL</th><th>INCEPTISOL</th><th>ARIDISOL</th><th>MOLLISOL</th><th>ALFISOL</th></tr>
<tr><td>O</td><td></td><td>Aquept[1]*</td><td></td><td>Aquoll[1]*</td><td>Aqualf[1]*
Boralf*
Udalf*</td></tr>
<tr><td>Mollic A</td><td></td><td>Andept[2]*
Aquept[1]*
Tropept*</td><td></td><td>Rendoll[3]*
Boroll[4]*
Udoll[5]*
Ustoll[5,6]*
Xeroll[5,6]*</td><td></td></tr>
<tr><td>Umbric A</td><td></td><td>Aquept[1]*
Umbrept*
Tropept*</td><td></td><td></td><td>—*</td></tr>
<tr><td>Ochric A</td><td>—*</td><td>Aquept[1]*
Ochrept*
Tropept*</td><td>—*</td><td></td><td>—*</td></tr>
<tr><td>E</td><td>If at the surface</td><td></td><td></td><td>Alboll[7]*</td><td>Boralf*
Udalf[8]*</td></tr>
<tr><td>Cambic B or Cox</td><td></td><td>—*</td><td>Orthid*</td><td rowspan="3">Base saturation $>$ 50% in some part*</td><td rowspan="3">Base saturation $>$ 35%*</td></tr>
<tr><td>Argillic B</td><td></td><td></td><td>Argid*</td></tr>
<tr><td>Natric B</td><td></td><td></td><td>—*</td></tr>
<tr><td>Spodic B</td><td></td><td></td><td></td><td></td><td></td></tr>
<tr><td>Oxic B</td><td></td><td></td><td></td><td></td><td></td></tr>
<tr><td>K</td><td></td><td></td><td>—*</td><td rowspan="4">Ustoll[6]*
Xeroll[6]*</td><td rowspan="3">Ustalf[6]*
Xeralf[6]*</td></tr>
<tr><td>Cca</td><td>—*</td><td>—*</td><td>—*</td></tr>
<tr><td>Csi</td><td>—*</td><td>—*</td><td>—*</td></tr>
<tr><td>Csa</td><td>—*</td><td>—*</td><td>—*</td><td></td></tr>
</table>

Table notes:

[1]Soil shows evidence of gleying

[2]Contains high content of allophane, volcanic ash, or both

[3]Mollic A horizon rests on calcareous parent material

[4]Moist chroma < 1.5

[5]Moist chroma > 1.5

[6]Has properties associated with distinct dryness for varying parts of the year

suborders based on one or more diagnostic horizons

ORDER

SPODOSOL	ULTISOL	OXISOL	VERTISOL	HISTOSOL
Aquod[1]*	Aquult[1]*	Aquox[1]*	May or may not have diagnostic horizons; has > 30% clay and shrinks and swells with moisture variation to form cracks that extend to the surface; can have slickensides and gilgai microrelief	Organic soils, best described in 1968 suppl. to the new classification
	—*	—*		
	—*	—*		
	—*	Torrox[14]*		
—*	—*			
	Base saturation < 35%* Humult[12]* Udult[13]* Ustult[13,6]* Xerult[13,6]*			
Ferrod[9]* Orthod[10]* Humod[11]*				
		Humox[12] Orthox[13] Ustox[6]		

[7]If argillic or natric B horizon underlie the E, and there is some evidence of gleying
[8]E horizon is not continuous and chroma is 2 or less
[9]Fe:C > 6
[10]Fe:C < 6
[11]Fe < 0.7%
[12]High organic matter content (> 20 kg) in uppermost 1 m^3 of soil, exclusive of the O horizon
[13]Low organic matter content (< 20 kg) in uppermost 1 m^3 of soil, exclusive of the O horizon
[14]Including moist color value ≥ 4 in all subhorizons

*Indicates diagnostic horizons used in the soil classification. Some of the suborders with asterisks are shown opposite the horizon or horizons that best separate them from other suborders. Not all suborders are listed, however. Taken from Soil Survey Staff[16,17,18] and Smith[14].

so detailed and has such an abundance of names that most geologists have not abandoned the older classification for it. To use the new classification in detail requires a good working knowledge of it and long experience. In many instances, field observations are not enough to classify a soil; these must be supplemented by laboratory quantification of some properties, as well as such information as the number of wet or dry days per year or the mean annual soil temperature. The system is not universally accepted as it stands,[19,20] but much of it has been incorporated into the FAO system.

The new classification does have some appeal for geomorphologists and ecologists, at least down to the suborder level. The reason for this is that soil profile development is included in the classification, as well as base saturation, amount of organic matter, and properties indicative of relative wetness and dryness. Most geomorphological soils studies deal with climatic and time factors, and the above properties will later be shown to be related to both factors. Furthermore, the new classification has gained wide approval (or at least use) with U.S. pedologists, and it is the system used in many journal articles. Hence, it will be used here.

Ten orders are recognized in the new classification, and these are subdivided into forty-seven suborders. The orders are basically differentiated by the horizon or horizon combinations that occur in the soil profile. These usually can be recognized in the field without recourse to laboratory analysis. One criticism of the new classification, from a geomorphological point of view, is the overemphasis on soil classification by surface horizon. In contrast, horizons beneath the A horizon probably are more important to geomorphologists. Classification into suborders requires an increasingly quantitative knowledge of soil properties and soil-moisture and soil-temperature regimens. Many times, however, it is not necessary to take these measurements, because with experience they can be estimated from properties recognizable in the field. Classification below the suborder level is not important to many geomorphic studies and would introduce terms that could be understood only after a detailed study of the new classification.

I have tried, in Table 2–1, to list the horizons most diagnostic for classification of soils at the order and suborder level. Only those suborders thought to be most useful to Quaternary research are included. It should be stressed that because of the extremely complex nature of the new classification any such simplification is bound to contain some errors, especially at the suborder level. It is an attempt, anyway, to simplify the system so that it can be used by workers who have neither the necessary time nor desire to learn the new system in detail.

Entisols are soils in which pedologic processes have left only a faint imprint. Thus, only a weakly developed surface horizon is present, and there may be salt or silica accumulations at depth.

Inceptisols are better developed than Entisols in that a well-developed A horizon is present, and oxidation extends below the base of the A (Fig. 2–4, A). These soils would be termed weakly developed profiles in the soil-profile development classification given earlier. Almost all other orders, exclusive of the Vertisols and the Histisols, differ from the Inceptisols in having better developed B horizons and/or $CaCO_3$ accumulations.

Aridisols are characterized by a low content of organic matter in the A horizon and salt or silica accumulations at depth. The two suborders are based on the presence or absence of an argillic B horizon (Fig. 2–4, B). This is a significant differentiation for geomorphic studies because clay buildup in many soils generally is related to duration of soil formation.

Mollisols have a high content of organic matter in the A horizon as a major distinguishing property, and the base saturation throughout is usually higher than 50 per cent. At the suborder level, Mollisols can be differentiated on properties indicative of wetness or dryness, such as A horizon chroma, the presence of an E horizon, and the presence or absence of salts or silica at depth (Fig. 2–4, C and D).

Alfisols may have a content of organic matter in the A horizon similar to that of the Mollisols, but the A horizon properties do not meet the requirements of a mollic A horizon, and base saturation in some parts of the B horizon is lower than 50 but higher than 35 per cent (Fig. 2–4, E and F). Like the Mollisols, soils that are dry for various parts of the year can have $CaCO_3$ or silica accumulations at depth.

Spodosols differ from all other orders in the presence of a spodic B horizon. An E horizon is not mandatory, but it is usually present in uncultivated sites, along with an O horizon (Fig. 2–4, G). Various combinations of iron and organic matter in the spodic B serve to differentiate most of the suborders.

Ultisols may have any surface A horizon, but they differ from the other orders in having less than 35 per cent base saturation in the argillic B horizon (Fig. 2–4, H). Organic-matter content in the uppermost cubic meter of the mineral soil and properties associated with dryness are other criteria that serve to separate Ultisols at the suborder level.

Oxisols differ from the other orders by the presence of an oxic B horizon (Fig. 2–4, I). Again, organic-matter content and properties associated with dryness serve to separate Oxisols at the suborder level.

Suborder names are formed by combining two syllables to indicate the order and some distinguishing characteristic of the suborder (Table

(A) Inceptisol, Searles Lake, California (B) Argid, Las Cruces, New Mexico

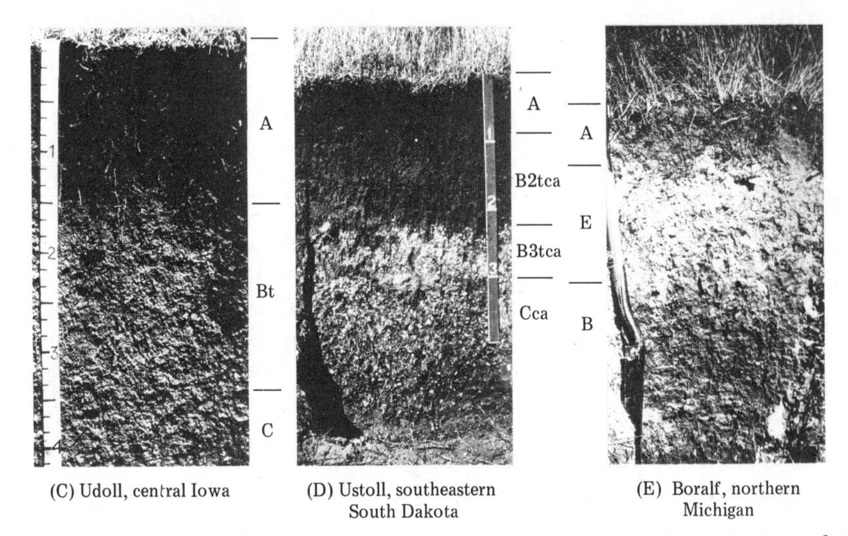

(C) Udoll, central Iowa (D) Ustoll, southeastern South Dakota (E) Boralf, northern Michigan

Fig. 2–4 Characteristic soil profiles of several soil suborders. All photographs but A and B are from the Marbut Memorial Slide Collection, prepared and published by the Soil Science Society of America (Madison, Wisconsin) in 1968. Originals are in color: scale feet and inches.

2–2). The last syllable of each suborder name is the clue to the order in which it is grouped. The first syllable indicates some diagnostic property or groups of properties common to the suborder. Thus, Aquolls are gleyed Mollisols. Argids are Aridisols with argillic B horizons, and Humox are Oxisols characterized by large amounts of organic matter to fairly great depth.

The old soil classification scheme used in the United States before 1960 is compared in Table 2–3 with the approximate equivalent orders and suborders of the new classification. Three main orders are recognized in the old system. Zonal soils are well-developed, well-drained,

(F) Xeralf, central California

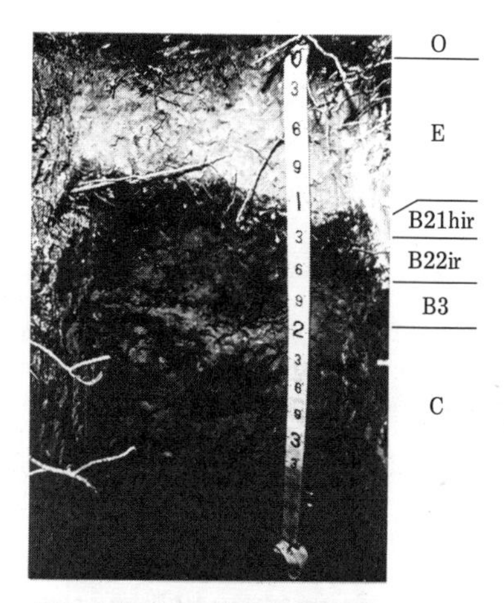

(G) Orthod, northern New York

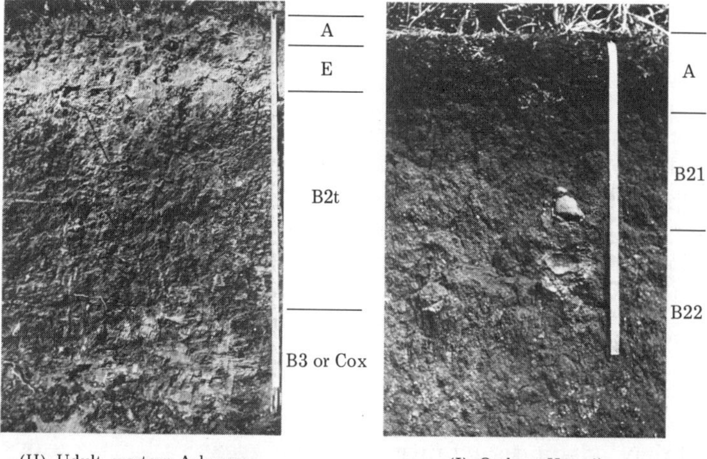

(H) Udult, western Arkansas

(I) Orthox, Hawaii

and on gently undulating uplands; their characteristics are determined mainly by the active soil-forming factors of climate and vegetation. Intrazonal soils vary in development, and their characteristics are strongly influenced by such local factors as a high water table or calcareous parent material. Azonal soils, for reasons of young age, rapid erosion, or parent material, are poorly developed and consist mainly of A horizons overlying bedrock or unconsolidated materials. Within the zonal soils, as well as in some other soils, one can recognize two main groups, the pedalfers and pedocals. The basic difference between the two is a horizon of pedogenic $CaCO_3$ accumulation; pedo-

Table 2–2

Suborder nomenclature key for new classification[16]

	*Prefix indicates some diagnostic property of the suborder**	*Suffix indicates the order*
alb-	presence of E horizon	-ent (Entisol)
aqu-	associated with prolonged wetness	-ept (Inceptisol)
arg-	presence of argillic B horizon	-id (Aridisol)
bor-	associated with cool environment; relatively high organic-matter content	-oll (Mollisol)
		-od (Spodosol)
ferr-	high iron content	-alf (Alfisol)
hum-	high organic-matter content	-ult (Ultisol)
ochr-	presence of an ochric A horizon	-ox (Oxisol)
orth-	the group of soils within the suborder that best typifies the order	-ert (Vertisol)
		-ist (Histosol)
torr- ust- xer-	associated with dryness lasting various lengths of time; usually low organic-matter content	
trop-	associated with continually warm climate	
ud-	associated with humid climates; moderate to low organic-matter content	
umbr-	presence of umbric A horizon	

*Only those prefixes that I consider most useful to the reader are here listed.

cals contain such accumulations, and pedalfers do not. The pedalfer-pedocal break has been a useful one for geomorphic studies, but it is not readily recognized in the new classification at the suborder level.

The great soil groups of the old soil classification have been widely used, but lately they have been criticized. The criticism is basically that genetic bias is built into the system and that the definitions are neither precise nor quantitative. Thus, one was allowed a fair amount of latitude in classifying a soil, and at times it even helped to know the vegetation. I will not define the great soil groups of the old classification here. For those interested, several recent publications have presented some fairly adequate definitions of the great soil groups of the old system, although ambiguities still exist.[1,2]

One major criticism of the new classification is that it groups together many soils that traditionally have been separated. The Mollisols, for example, encompass many of the Chestnut, Chernozem, and Prairie (Brunizem) soils. Furthermore, soils like the Humic Gleys are found in several of the new orders. Such changes reflect the fact that now there are more data on these soils, and they can be differentiated at a high category of classification. Perhaps the problem is not as acute as it seems. When one compares regional or U.S. soil maps based on both systems, one is impressed with the gross similarities in soil pattern and boundary placement, and so it seems that many traditional breaks of the old classification do show up at the suborder level of the new classification.

The geomorphologist has two choices for soil classification at this time—the old system or the new system. The suggestion here is that we give the new system a try as that seems to be the wave of the future in the United States. One can fairly easily classify a soil or group of soils at the suborder level and then describe other characteristic features in plain language without recourse to the great groups or subgroups of the new classification. One problem with the very detailed new classification is that it might take either the investigator or the reader longer to find the definition in the book than it would to classify at the suborder level and briefly spell out the pertinent characteristics of the soil. We can even use the terms pedalfer and pedocal if the suborder terms do not clearly convey information on the presence or absence of $CaCO_3$ accumulations. For example, a soil with an ochric A horizon, an argillic B horizon, and a K horizon simply could be called an Argid with a K horizon. If one looks into the new classification, he finds that it could be a petrocalcic Paleargid if it meets certain stringent requirements, among which are depth to top of the K horizon (<1 m) and a complex relationship between texture and organic-carbon content. Another reason for using the new classification is that it does set greater limits on individual interpretation than did the old system. It is hoped that this will result in more uniform classification by different investigators.

Distribution of soil orders and suborders in the United States

A soils map of the United States has been prepared, based on the new classification (Fig. 2–5). It shows some regional trends that can be roughly related to the soil-forming factors, mainly climatic and vegetation patterns and geology. One major trend is that, generally, the soil distribution east of the Rocky Mountains seems less complex than that to the west.

The soil pattern east of the Rocky Mountains generally follows the gradual regional climatic gradient, although there is some variation probably due in part to erosion in mountainous areas and to age of the landscape. Just east of the Rockies is a wide expanse of Ustolls. Entisols are interspersed with the Ustolls and are related to the erosive shales of eastern Montana and to the dune sands of Nebraska. Udolls lie east of the Ustolls, but in part of the glaciated region to the north these give way to Borolls and Aquolls, the latter being mainly associated with the relatively impermeable sediments of glacial Lake Agassiz. Proceeding eastward, there is a large area of Udults formed on fairly old landscapes south of the glacial boundary that extend almost to the Aquults of the

Table 2–3
Pre-1960 U.S. soil classification and the approximate equivalent orders and suborders of the new classification[6,16]

PRE-1960 U.S. CLASSIFICATION TO 1960 — THE NEW CLASSIFICATION

Order		*Suborder*	*Great Soil Groups*	*Suborder*	*Order*
ZONAL	*Pedalfer*	Soils of forested warm-temperate and tropical regions	Red-brown Lateritic	Humult, Udalf, Udult	Oxisol, Ultisol, Alfisol, Inceptisol
			Red-Yellow Podzolic	Udult	
			Low-humic Latosol	Tropept, Ustox	
			Latosol	Tropept, Humult Andept, Ustult	
			Humic Latosol	Humult, Humox, Andept	
			Laterite	Orthox	
		Soils of cold regions	Polar Desert		Inceptisol
			Arctic Brown		
			Tundra	Aquept, Ochrept Umbrept, Andept	
			Alpine Turf		
		Soils of forested cool-temperate regions	Podzol	Orthod, Humod	Spodosol, Ultisol, Alfisol, Inceptisol
			Brown Podzolic	Orthod, Andept, Ochrept	
			Gray-Brown Podzolic	Udalf, Udult	
			Gray Wooded	Boralf	
			Sols Bruns Acides	Ochrept, Umbrept	
			Western Brown Forest*		
		Soils of forest-grassland transition	Degraded Chernozem	Boralf, Boroll	Alfisol, Mollisol, Inceptisol
			Noncalcic Brown**	Xeralf, Ochrept	
		Dark-colored soils of semiarid, subhumid, and humid grasslands	Reddish Prairie	Ustoll	Mollisol, Alfisol
			Prairie (Brunizem)	Udoll, Boroll Xeroll, Ustoll	
	Pedocal		Chernozem	Boroll, Ustoll, Xeroll	
			Reddish Chestnut	Ustalf, Ustoll	
			Chestnut	Xeroll, Ustoll, Boroll	
		Light-colored soils of arid regions	Reddish Brown	Ustalf, Orthid, Argid	Aridisol, Mollisol, Alfisol
			Brown	Ustoll, Xeroll, Argid, Orthid, Boroll	
			Sierozem	Argid, Orthid	
			Red Desert	Argid, Orthid	
			Desert	Argid, Orthid	
			Polar Desert	Argid, Orthid	

PRE-1960 U.S. CLASSIFICATION TO 1960			THE NEW CLASSIFICATION	
Order	*Suborder* ·	*Great Soil Groups*	*Suborder*	*Order*
INTRAZONAL	Hydromorphic soils in areas of imperfect drainage or high water table	Humic Gley	Aquoll, Aquept, Aquult, Aqualf	Inceptisol, Mollisol, Alfisol, Spodosol, Ultisol, Entisol
		Low-Humic Gley	Aquult, Aquent, Aquept, Aqualf	
		Alpine Meadow	Aquod, Aquoll, Umbrept	
		Bog	Suborders of Histosol	
		Half Bog	Aquept, Aquoll, Aqualf	
		Planosol	Aqualf, Alboll	
		Ground-Water Podzol	Aquod	
		Ground-Water Laterite	Aquult, Udult, Usult	
	Halomorphic soils (saline and alkali) in areas of imperfect drainage in arid and coastal areas	Solonchak	Orthid, Aquept	Inceptisol, Aridisol, Mollisol, Alfisol
		Solonetz	Natric great groups of Alfisol, Mollisol, Aridisol	
		Soloth	Natric subgroups of Mollisol and Alfisol	
	Calcimorphic soils formed from calcareous parent materials	Brown Forest	Ochrept, Xeroll, Udoll	Inceptisol, Mollisol
		Rendzina	Rendoll	
AZONAL		Lithosol		Entisol, Inceptisol, Mollisol
		Regosol		
		Alluvial		

*Those without $CaCO_3$ accumulations should be included in the Pedalfers.

**Large tracts of Noncalcic Brown soils in California are in semiarid regions far removed from forests. They have properties common to the Brown Soils, but lack horizons of $CaCO_3$ accumulation.

east coast. Florida does not follow this regional trend and is covered mainly with Aquods, Entisols, and Histosols. Inceptisols are associated with the Ultisols, as shown by Aquepts on the Mississippi River floodplain deposits and Ochrepts in the Appalachian and Ouachita Mountains. Just east of the Mississippi River is a belt of Udalfs that seems to be associated with loess deposits of glacial age. In the glaciated region of the central and northeastern United States, Udalfs are common to the south, and these grade into Orthods to the north. The only major exceptions are Udolls over much of Illinois and Boralfs in northern Minnesota.

West of the eastern front of the Rocky Mountains there is an intricate mosaic of climatic, vegetation, topographic, and geologic patterns. The topography consists of a multitude of mountain ranges separated by intermontane valleys. Bedrock makes up most of the mountains, whereas unconsolidated alluvium of various ages underly the valleys. In almost all places the climate and vegetation are closely associated

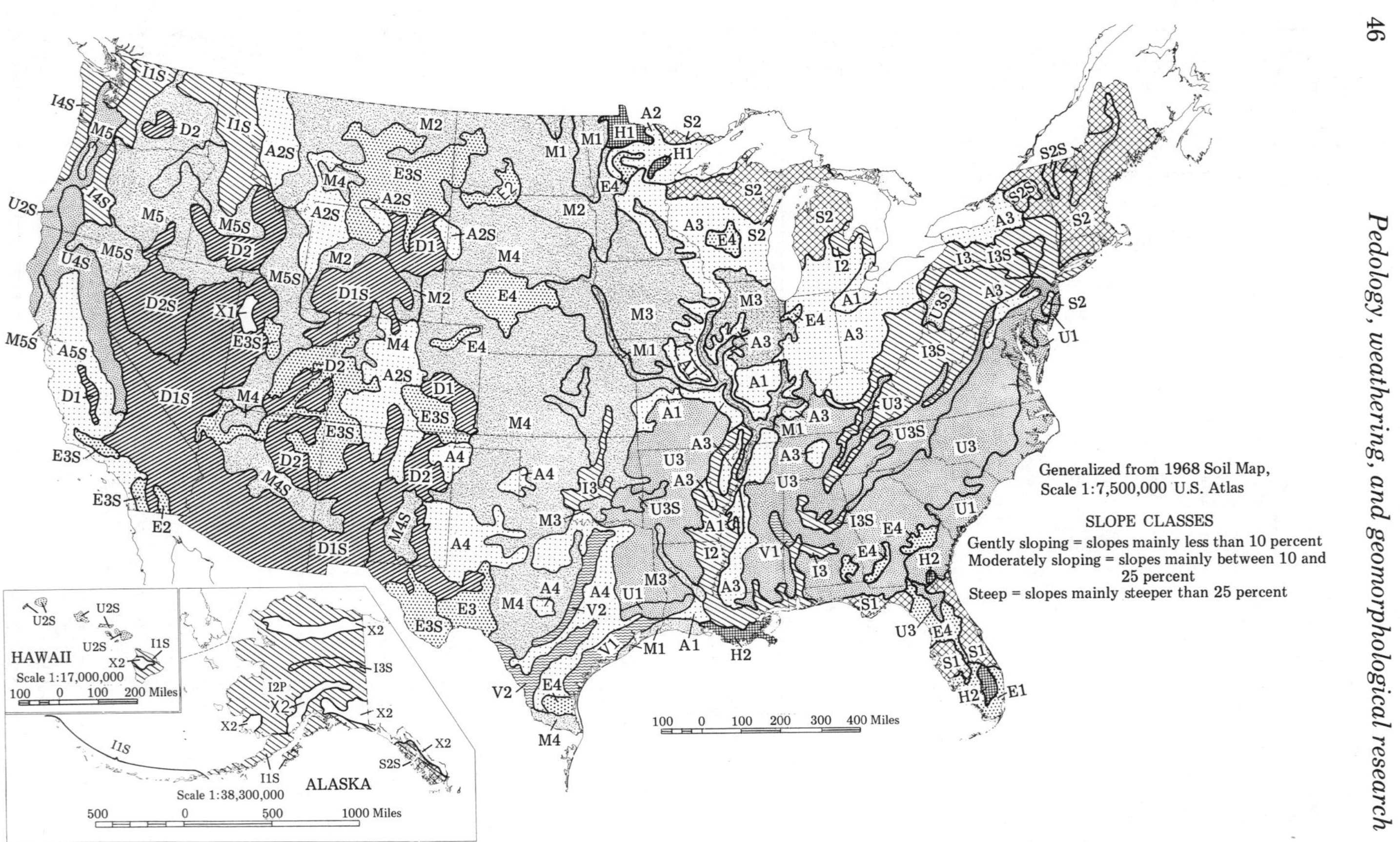
Generalized from 1968 Soil Map,
Scale 1:7,500,000 U.S. Atlas
SLOPE CLASSES
Gently sloping = slopes mainly less than 10 percent
Moderately sloping = slopes mainly between 10 and 25 percent
Steep = slopes mainly steeper than 25 percent
100 0 100 200 300 400 Miles
HAWAII
Scale 1:17,000,000
100 0 100 200 Miles
ALASKA
Scale 1:38,300,000
500 0 500 1000 Miles

LEGEND

Only the dominant orders and suborders are shown. Each delineation has many inclusions of other kinds of soil. General definitions for the orders and suborders follow. For complete definitions see Soil Survey Staff, *Soil Classification, A Comprehensive System, 7th Approximation*, Soil Conservation Service, U.S. Department of Agriculture, 1960 (for sale by U.S. Government Printing Office) and the March 1967 supplement (available from Soil Conservation Service, U.S. Department of Agriculture). Approximate equivalents in the modified 1938 soil classification system are indicated for each suborder.

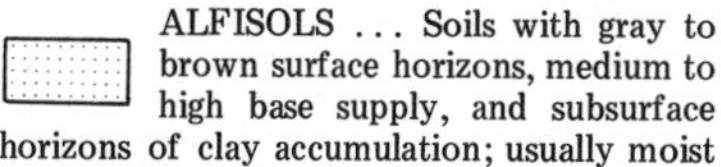

ALFISOLS . . . Soils with gray to brown surface horizons, medium to high base supply, and subsurface horizons of clay accumulation; usually moist but may be dry during warm season

A1 AQUALFS (seasonally saturated with water) gently sloping; general crops if drained, pasture and woodland if undrained (Some Low—Humic Gley soils and Planosols)

A2 BORALFS (cool or cold) gently sloping; mostly woodland, pasture, and some small grain (Gray Wooded soils)

A2S BORALFS steep; mostly woodland

A3 UDALFS (temperate, or warm, and moist) gently or moderately sloping; mostly farmed, corn, soybeans, small grain, and pasture (Gray—Brown Podzolic soils)

A4 USTALFS (warm and intermittently dry for long periods) gently or moderately sloping; range, small grain, and irrigated crops (Some Reddish Chestnut and Red—Yellow Podzolic soils)

A5S XERALFS (warm and continuously dry in summer for long periods, moist in winter) gently sloping to steep; mostly range, small grain, and irrigated crops (Noncalcic Brown soils)

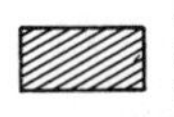

ARIDISOLS . . . Soils with pedogenic horizons, low in organic matter, and dry more than 6 months of the year in all horizons

D1 ARGIDS (with horizon of clay accumulation) gently or moderately sloping; mostly range, some irrigated crops (Some Desert, Reddish Desert, Reddish Brown, and Brown soils and associated Solonetz soils)

D1S ARGIDS gently sloping to steep

D2 ORTHIDS (without horizon of clay accumulation) gently or moderately sloping; mostly range and some irrigated crops (Some Desert, Reddish Desert, Sierozem, and Brown soils, and some Calcisols and Solonchak soils)

D2S ORTHIDS gently sloping to steep

ENTISOLS . . . Soils without pedogenic horizons

E1 AQUENTS (seasonally saturated with water) gently sloping; some grazing

E2 ORTHENTS (loamy or clayey textures) deep to hard rock; gently to moderately sloping; range or irrigated farming (Regosols)

E3 ORTHENTS shallow to hard rock; gently to moderately sloping; mostly range (Lithosols)

E3S ORTHENTS shallow to rock; steep; mostly range

E4 PSAMMENTS (sand or loamy sand textures) gently to moderately sloping; mostly range in dry climates, woodland or cropland in humid climates (Regosols)

HISTOSOLS . . . Organic soils

H1 FIBRISTS (fibrous or woody peats, largely undecomposed) mostly wooded or idle (Peats)

H2 SAPRISTS (decomposed mucks) truck crops if drained, idle if undrained (Mucks)

INCEPTISOLS . . . Soils that are usually moist, with pedogenic horizons of alteration of parent materials but not of accumulation

I1S ANDEPTS (with amorphous clay or vitric volcanic ash and pumice) gently sloping to steep; mostly woodland; in Hawaii mostly sugar cane, pineapple, and range (Ando soils, some Tundra soils)

I2 AQUEPTS (seasonally saturated with water) gently sloping; if drained, mostly row crops, corn, soybeans, and cotton; if undrained, mostly woodland or pasture (Some Low—Humic Gley soils and Alluvial soils)

(continued)

I2P AQUEPTS (with continuous or sporadic permafrost) gently sloping to steep; woodland or idle (Tundra soils)

I3 OCHREPTS (with thin or light-colored surface horizons and little organic matter) gently to moderately sloping; mostly pasture, small grain, and hay (Sols Bruns Acides and some Alluvial soils)

I3S OCHREPTS gently sloping to steep; woodland, pasture, small grains

I4S UMBREPTS (with thick dark-colored surface horizons rich in organic matter) moderately sloping to steep; mostly woodland (Some Regosols)

MOLLISOLS . . . Soils with nearly black, organic-rich surface horizons and high base supply

M1 AQUOLLS (seasonally saturated with water) gently sloping; mostly drained and farmed (Humic Gley soils)

M2 BOROLLS (cool or cold) gently or moderately sloping, some steep slopes in Utah; mostly small grain in North Central States, range and woodland in Western States (Some Chernozems)

M3 UDOLLS (temperate or warm, and moist) gently or moderately sloping; mostly corn, soybeans, and small grains (Some Brunizems)

M4 USTOLLS (intermittently dry for long periods during summer) gently to moderately sloping; mostly wheat and range in western part, wheat and corn or sorghum in eastern part, some irrigated crops (Chestnut soils and some Chernozems and Brown soils)

M4S USTOLLS moderately sloping to steep; mostly range or woodland

M5 XEROLLS (continuously dry in summer for long periods, moist in winter) gently to moderately sloping; mostly wheat, range, and irrigated crops (Some Brunizems, Chestnut, and Brown soils)

M5S XEROLLS moderately sloping to steep; mostly range

SPODOSOLS . . . Soils with accumulations of amorphous materials in subsurface horizons

S1 AQUODS (seasonally saturated with water) gently sloping; mostly range or woodland; where drained in Florida, citrus and special crops (Ground—Water Podzols)

S2 ORTHODS (with subsurface accumulations of iron, aluminum, and organic matter) gently to moderately sloping; woodland, posture, small grains, special crops (Podzols, Brown Podzolic soils)

S2S ORTHODS steep; mostly woodland

ULTISOLS . . . Soils that are usually moist, with horizon of clay accumulation and a low base supply

U1 AQUULTS (seasonally saturated with water) gently sloping; woodland and pasture if undrained, feed and truck crops if drained (Some Low—Humic Gley soils)

U2S HUMULTS (with high or very high organic-matter content) moderately sloping to steep; woodland and pasture if steep, sugar cane and pineapple in Hawaii, truck and seed crops in Western States (Some Reddish-Brown Lateritic soils)

U3 UDULTS (with low organic-matter content; temperate or warm, and moist) gently to moderately sloping; woodland, pasture, feed crops, tobacco, and cotton (Red-Yellow Podzolic soils, some Reddish-Brown Lateritic soils)

U3S UDULTS moderately sloping to steep. woodland, pasture

U4S XERULTS (with low to moderate organic-matter content, continuously dry for long periods in summer) range and woodland (Some Reddish-Brown Lateritic soils)

VERTISOLS. . . Soils with high content of swelling clays and wide deep cracks at some season

V1 UDERTS (cracks open for only short periods, less than 3 months in a year) gently sloping; cotton, corn, pasture, and some rice (Some Grumusols)

V2 USTERTS (cracks open and close twice a year and remain open more than 3 months); general crops, range, and some irrigated crops (Some Grumusols)

☐ AREAS with little soil . . .

X1 Salt flats X2

X2 Rock land (plus ice fields in Alaska)

NOMENCLATURE

The nomenclature is systematic. Names of soil orders end in *sol* (L. *solum*, soil), e.g., ALFISOL, and contain a formative element used as the final syllable in names of taxa in suborders, great groups, and subgroups.

Names of suborders consist of two syllables, e.g., AQUALF. Formative elements in the legend for this map and their connotations are as follows:

and — Modified from Ando soils; soils from vitreous parent materials

aqu — L. *aqua*, water; soils that are wet for long periods

arg — Modified from L. *argilla*, clay; soils with a horizon of clay accumulation

bor — Gr. *boreas*, northern; cool

fibr — L. *fibra*, fiber; least decomposed

hum — L. *humus*, earth; presence of organic matter

ochr — Gr. base of ochros, pale; soils with little organic matter

orth — Gr. *orthos*, true; the common or typical

psamm — Gr. *psammos*, sand; sandy soils

sapr — Gr. *sapros*, rotten; most decomposed

ud — L. *udus*, humid; of humid climates

umbr — L. *umbra*, shade; dark colors reflecting much organic matter

ust — L. *ustus*, burnt; of dry climates with summer rains

xer — Gr. *xeros*, dry; of dry climates with winter rains

Fig. 2–5 Patterns of soil orders and suborders of the United States (courtesy of the U.S. Dept. Agriculture, Soil Conservation Service).

with the topography, with the valleys being the driest and the mountain slopes receiving an increasing amount of moisture in proportion to altitude; vegetation follows these trends. The one major exception to these climatic trends is the lowland regions west of the crests of the northern Sierra Nevada and the Cascade Range; these areas receive abundant moisture from air masses moving inland from the Pacific Ocean. Still, even here the general climate-altitude relationship holds.

Soil patterns in the Cordilleran mountain ranges follow the overall climatic trends. Boralfs are common in the eastern parts of the Rocky Mountains, whereas the western parts are dominated by Ustolls to the south, as well as in the mountains of Arizona; Xerolls dominate to the north. Andepts are found in parts of the northern Rockies and in the Cascade Range, where they are most often associated with volcanic rock. The Sierra Nevada are dominated by Xerults, although soils south of Lake Tahoe might be closer to Xeralfs. Entisols occur in the ranges south of the Sierra Nevada. The ranges along the west coast grade from Umbrepts in the north to Humults in the central sector to Xeralfs in the south.

The basins show a soil variation from north to south. Those in eastern Oregon and Washington are dominated by Xerolls, with Orthids in the drier parts of the Columbia Plateau. Orthids are common also in the Snake River Plain of southern Idaho and in the northwestern Basin and Range Province, and they grade into Argids to the south in that province. This gradation can be explained, at least in part, by age of landscape, because the widespread alluvial fan and pediment deposits to the south seem to be older than those to the north. Entisols and Orthids are the major soils of the Colorado Plateau.

Humults dominate in the Hawaiian Islands. Oxisols are present there to a limited extent.

REFERENCES

1. Agricultural Experiment Stations of the North Central Region of the U.S., 1960, Soils of the north central region of the United States: Univ. Wisconsin Agri. Exp. Sta. Bull. 544, 192 p.
2. Agricultural Experiment Stations of the Western States Land-Grant Universities and Colleges, 1964, Soils of the western United States: Washington State University, Pullman, 69 p.
3. Arkley, R. J., 1962, Soil survey of Merced area, California: U.S. Dept. Agri., Soil Survey series 1950, no. 7, 131 p.
4. ____ 1964, Soil survey of the eastern Stanislaus area, California: U.S. Dept. Agri., Soil Survey series 1957, no. 20, 160 p.
5. Bunting, B. T., 1965, The geography of soil: Aldine Publ. Co., Chicago, 213 p.
6. Douglas, J. F., Austin, M. E., and Smith, G. D., 1969, General soil map of the United States: Soil Sci. Soc. Amer. Proc., v. 33, p. 746–749.
7. Dudal, R., 1968, Definitions of soil units for the soil map of the world: World soil resources reports no. 33, World soil resources office, FAO, Rome, 72 p.
8. Harris, S. A., 1968, Comments on the validity of the law of soil zonality: 9th Internat. Cong. Soil Sci., Trans. v. 4, p. 585–593.
9. Jenny, H., 1946, Arrangement of soil series and types according to functions of soil-forming factors: Soil Sci., v. 61, p. 375–391.
10. Marbut, C. F., 1927, A scheme for soil classification: 1st Internat. Cong. Soil Sci., Proc. v. 4, p. 1–31.
11. ____ 1935, Soils of the United States: U.S. Dept. Agri., Atlas of American Agriculture, part III, 98 p.
12. Oschwald, W. R., Riecken, F. F., Dideriksen, R. I., Scholtes, W. H., and Schaller, F. W., 1965, Principal soils of Iowa—Their formation and properties: Iowa State Univ. Coop. Ext. Serv. Spec. Rep. 42, 76 p.
13. Simonson, R. W., 1962, Soil Classification in the United States: Science, v. 137, p. 1027–1034.
14. Smith, G. D., 1965, Lectures on soil classification: Pedologie, Special issue 4, Belgium Soil Science Society, Ghent, 134 p.
15. Soil Survey Staff, 1951, Soil survey manual: U.S. Dept. Agri. Handbook no. 18, 503 p.

16. —— 1960, Soil classification, a comprehensive system (7th approximation): U.S. Dept. Agri., Soil Cons. Service, 265 p.
17. —— 1967, Supplement to soil classification system (7th approximation): U.S. Dept. Agri., Soil Cons. Service, 207 p.
18. —— in press, Soil taxonomy, a basic system of soil classification for making and interpreting soil surveys: U.S. Dept. Agri., Soil Cons. Service.
19. Stephens, C. G., 1963, The 7th approximation: its application in Australia: Soil Sci., v. 96, p. 40–48.
20. Tavernier, R., 1963, The 7th approximation: its application in western Europe: Soil Sci., v. 96, p. 35–39.
21. Ulrich, R., and Stromberg, L. K., 1962, Soil survey of Madera area, California: U.S. Dept. Agri., Soil Survey series 1951, no. 11, 155 p.
22. Weir, W. W., 1952, Soils of San Joaquin County, California: Univ. Calif. Agri. Exp. Sta., Soil Survey no. 9, 137 p.

CHAPTER 3

Weathering processes

Weathering is the physical and chemical alteration of rock or minerals at or near the earth's surface. Most rocks and minerals exposed at and immediately beneath the earth's surface are in an environment quite unlike that under which they formed. This is especially true for igneous or metamorphic rocks that formed under high temperatures and, with the exception of some volcanic rocks, at great confining pressures. Weathering can be defined as the process of rock and mineral alteration to more stable forms under the variable conditions of moisture, temperature, and biological activity that prevail at the surface.

Two main types of weathering are recognized.[23,26] One is physical weathering, in which the original rock disintegrates to smaller-sized material, with no appreciable change in chemical or mineralogical composition. The other is chemical weathering, in which the chemical and/or mineralogical composition of the original rock and minerals is changed. In nature, physical and chemical weathering occur together, and it may be difficult to separate the effects of one from the effects of the other.

Physical weathering

The mechanism common to all processes of physical weathering is the establishment of sufficient stress within the rock so that it breaks. If the rock is ruptured along fracture planes, blocks or sheets of varying size are produced. If, however, the lines of weakness are along mineral grain boundaries, physical weathering can produce materials whose

Fig. 3–1 Sheeting joints developed in granitic rock. (A) The Sierra Nevada, California. (Photo by W. C. Bradley.) (B) The Wind River Mountains, Wyoming.

Fig. 3–2 Weathered granitic stones in Baja California. In many places, the undersides of these stones are virtually unweathered. Salt crystallization in small cracks may be a major factor in the surface weathering of the stones.

size is determined by the size of the grains in the original rock; smaller sizes are possible if the minerals are cross-cut by small-scale fractures. The processes that are reported to be most common to physical weathering are unloading by erosion, expansion in cracks or along grain boundaries by freezing water or crystallizing salts, fire, and possibly thermal expansion and contraction of the constituent minerals.[26]

Rock bodies that are either homogeneous or layered can have numerous fracture planes or joints, nearly parallel to the ground surface, that divide the rock into a series of layers or sheets (Fig. 3–1). The spacing between the joints generally increases with depth, and they can be observed for several tens of meters below the surface. The origin of the fractures seems to be the release of stresses contained within the rock. While buried, the rock is under high confining pressures. With erosion of the overlying rock mass, however, the rock has less overburden pressure, so it can expand. If it is close enough to the surface, expansion can only be upward or toward the valley wall—in any direction in which the rock body is not confined. This expansion can lead to rupture of the rock along fracture planes oriented at right angles to the direction of the pressure release and, thus, to development of sheeting parallel to the surface.

Water, upon freezing, can set up pressures sufficient to disintegrate most rocks. At 0° C the increase in volume with the conversion from water to ice is 9 per cent. At localities in which there is a sufficient supply of moisture and a low enough temperature, the moisture contained in the rock can freeze, and the accompanying internal pressures are sufficiently great to exceed the strength of the rock, and the rock ruptures. Even though the water in the system might not be confined, pressures in an unconfined system are probably great enough to rupture most rock types. The direction of the fractures produced could be determined by minute, pre-existing planes of weakness, such as joints, or along fractures produced by unloading. In this case the result could be a large field of angular blocks. In some places this process might be responsible for granular disintegration, provided water has access to voids or cracks between the grains. This process is most effective in environments in which surface temperatures fluctuate across 0° C many times each year.

Saline solutions, if they have access to fractures in the rock or to the boundaries between grains, can bring about disintegration of rocks either into blocks or individual grains.[8,9,11] Several processes are recognized as important to the breakup of the rock by salts. One is the internal pressures set up during the growth of crystals from solution (Fig. 3–2). A common pedological example demonstrating this effect is the development of K horizons. In many places the K horizon contains over 70 per cent $CaCO_3$, and the original silicate grains are no longer in contact. Because there is little evidence that the original grains were dissolved, it appears that they were pushed aside during the crystallization of the $CaCO_3$ from the soil solution. Another important process is the thermal expansion of salts upon heating, which occurs because many common salts have thermal expansion coefficients higher than those of some common rocks.[9] This process might be important in many of the hot deserts that are characterized by large changes in daily temperatures. A final process that is important to rock disintegration is the stresses set up by volume increase that accompanies hydration of the various salts. Winkler,[38] for example, feels that the main cause of exfoliation of Cleopatra's Needle is hydration of salts that formed in the monument in Egypt; these salts were hydrated and expanded once the monument was moved to the humid climate of New York City. Hydration of clay minerals may have similar disruptive effects on rocks.

Fire is probably an important, but often overlooked, factor in the physical weathering of rock.[5] Because rock is a poor conductor of heat, a thin surface layer attains a high temperature during a fire, and there is a rapid decrease in temperature with depth. The heated surface layer will expand more rapidly than will rock at greater depth, and this expansion can lead to rupture of the rock into thin sheets that eventually fall to the ground. An example of the weathering effects of fire is seen on Mt. Sopris in western Colorado (Fig. 3–3). The rock-glacier deposits on Mt. Sopris do not have a forest cover. The stones on the surfaces of the rock-glacier deposits and of non-vegetated blockfields have thick oxidation weathering rinds, stone surfaces are oxidized to a reddish color, and the corners of some stones, even after several thousands of years of weathering, still are not too rounded. However, stones adjacent to or in the forest which was burned several decades ago have more rounded corners and thin or no weathering rinds, and they lack the pronounced surface oxidation. The differences in corner rounding and in weathering are attributed to fire, in both historic and prehistoric times.

These observations on the effect of fire on weathering are important to Quaternary stratigraphic studies, because commonly the weathered condition of the surface of a stone is one criterion used for age differentiation (Ch. 8). One finds in parts of the Colorado Front Range, for example, that stones in young tills above timber line have more highly pitted surfaces than have stones in much older tills within the present forest zone (Fig. 3–4). One suspects that fire may explain the difference in weathering of individual stones, because in the forest zone the stone surfaces can be continually renewed by spalling during a fire. The

Fig. 3–3 Comparison of stone weathering in (A) a forested and in (B) an adjacent non-forested environment, western Colorado. Note the abundance of fresh fracture faces in (A), due to expansion on heating by fire, and their absence in (B), which experiences no burning.

Fig. 3–4 Pitted stone on surface of Holocene till above timber line in the Colorado Front Range. Benedict[3] dates these boulders as early neoglacial, and thus the weathering has taken place over at least the last 3000 to 5000 years.

evidence for ancient fire is indirect, but surely lightning was an important cause. Blackwelder[5] emphasized that fire, in some places, might be the main weathering process in physical weathering, and I think the stratigraphic and boulder weathering data from some places would support him. If present stone weathering can indeed be related to the presence or absence of fire, and thus to the timber line, stone-weathering studies might be used as one criterion to help estimate the expansion of the upper and lower limit of past forests.

Diurnal fluctuations in surface temperatures, if great enough, are thought by some workers to bring about surface fracturing or granular disintegration of some rocks.[6] This idea is based on repeated observations of weathered rocks in deserts, weathering that seemed best ascribed to physical processes. In theory, the surface of a rock, where temperature fluctuations are greatest, should expand and contract the most. The effect is not unlike that of fire, that is, the rock is eventually weakened to the point where a part of the surface flakes off or mineral grains are dislodged. The process envisaged is that each different mineral will expand and contract a different amount at a different rate with surface-temperature fluctuations. With time, the stresses produced are sufficient to weaken the bonds along grain boundaries, and thus flaking of rock fragments or dislodging of grains occurs. The importance of this process in rock weathering has been debated over the years. Griggs,[13] in laboratory experiments in which rock samples were heated and cooled over large fluctuations in temperature to simulate 244 years of weathering in the absence of water, showed that little disaggregation of the rock occurred. Ollier[26] argues, however, that the time factor was not taken into account sufficiently in Griggs' study, and that small stresses applied over long periods of time might lead to permanent strain. This process might explain some weathering in arid regions. In most desert regions, however, some moisture is present, and therefore other weathering processes could be operative.

Plant growth contributes to some physical weathering. Pressures exerted by roots during growth are sometimes able to rupture the rock or force blocks apart. That such pressures can do this is shown by the common observation of cracked and heaved concrete sidewalks adjacent to tree roots. Lichens growing on rock surfaces also contribute to physical weathering. Loose mineral and rock fragments become attached to the undersides of the lichens and are pulled free of the surface when the lichen contracts during a dry spell. If the lichen is removed from the rock surface, it takes rock material with it. One would suspect, however, that the grains would have had to be loosened by some other process prior to removal from the rock surface with the lichen.

Chemical weathering

Chemical weathering occurs because rocks and minerals are seldom in equilibrium with near-surface waters, temperatures, and pressures. The products that form, however, are more stable in near-surface environments. If the soil environment does change with time, so too can the initial products of weathering. That some change occurs during chemical weathering is shown by field evidence for oxidation and for clay formation, by the different chemical and mineralogical composition of the weathered material relative to that of the assumed parent material, and by the chemistry of the waters that move through the soil. The change on weathering can be very slight and involve nothing more than the oxidation state of iron ions, or it can be quite intense and result in a product much different from the assumed parent material, such as the formation of an Oxisol from a mafic rock. There are several processes involved in chemical weathering of common rocks and minerals.

Hydrolysis

The most important process in the chemical weathering of the common silicate and aluminosilicate minerals is hydrolysis. Water, in which a variety of ions can be present, reacts with these minerals. A general equation for the reaction of a cation-bearing silicate is

$$\underset{c}{\text{Silicate}} + \underset{l}{H_2O} + \underset{aq}{H_2CO_3} = \underset{aq}{\text{cations}} + \underset{aq}{OH^-} + \underset{aq}{HCO_3^-} + \underset{aq}{H_4SiO_4}$$

where c is the crystalline species, l is the liquid, and aq denotes aqueous species. For an aluminosilicate a general reaction is

$$\underset{c}{\text{Aluminosilicate}} + \underset{l}{H_2O} + \underset{aq}{H_2CO_3}$$
$$= \underset{c}{\text{clay mineral}} + \underset{aq}{\text{cations}} + \underset{aq}{OH^-} + \underset{aq}{HCO_3^-} + \underset{aq}{H_4SiO_4}$$

The usual reaction is that of water and acid on the mineral. The acid shown here is H_2CO_3. Other acids, such as those resulting from the decay of organic matter, also are important H ion sources. The common by-products of hydrolysis are H_4SiO_4, HCO_3^-, and OH^-, along with clay minerals if aluminum is present in the decomposing minerals and if certain chemical conditions are met. More detailed weathering equations are given in Table 3–1.

The fate of the by-products of weathering varies, and this will be discussed in more detail later. Cations can remain in the soil either in the soil solution, as part of the crystal lattice of the clay mineral, or as exchangeable ions adsorbed to the surfaces of the colloidal particles. Some ions can be cycled through the biosphere from the soil and back again. Some cations can be removed from the system, along with HCO_3^-, with the percolating waters; indeed, one measure of the rate of chemical weathering of a region can be gained from the composition of the waters draining the region. Silica is quite soluble over the normal soil pH range (Fig. 4–3), and it is almost always present in the parent minerals in higher amounts than are necessary to form most clay minerals; therefore some is removed in solution. Aluminum is not very soluble over the normal soil pH range (Fig. 4–3), and so it generally remains near the site of release by weathering to form clay minerals or hydrous oxides. Iron also remains near the point of release in most soils and gives the soil or weathered rock the commonly observed oxidation colors.

Table 3–1
Equations representing the hydrolysis of orthoclase and albite with various clay minerals as a by-product

$$\underset{\text{(orthoclase)}}{2\,\underset{c}{KAlSi_3O_8}} + 2\,\underset{aq}{H^+} + 9\,\underset{l}{H_2O} = \underset{\text{(kaolinite)}}{\underset{c}{H_4Al_2Si_2O_9}} + 4\,\underset{aq}{H_4SiO_4} + 2\,\underset{aq}{K^+}$$

$$\underset{\text{(orthoclase)}}{3\,\underset{c}{KAlSi_3O_8}} + 2\,\underset{aq}{H^+} + 12\,\underset{l}{H_2O} = \underset{\text{(illite)}}{\underset{c}{KAl_3Si_3O_{10}(OH)_2}} + 6\,\underset{aq}{H_4SiO_4} + 2\,\underset{aq}{K^+}$$

$$\underset{\text{(albite)}}{2\,\underset{c}{NaAlSi_3O_8}} + 2\,\underset{aq}{H^+} + 9\,\underset{l}{H_2O} = \underset{\text{(kaolinite)}}{\underset{c}{H_4Al_2Si_2O_9}} + 4\,\underset{aq}{H_4SiO_4} + 2\,\underset{aq}{Na^+}$$

$$\underset{\text{(albite)}}{8\,\underset{c}{NaAlSi_3O_8}} + 6\,\underset{aq}{H^+} + 28\,\underset{l}{H_2O}$$

$$= \underset{\text{(montmorillonite)}}{3\,\underset{c}{Na_{0\cdot66}Al_{2\cdot66}Si_{3\cdot33}O_{10}(OH)_2}} + 14\,\underset{aq}{H_4SiO_4} + 6\,\underset{aq}{Na^+}$$

One effect of the hydrolysis reaction is that hydrogen ion is consumed, hydroxide is produced, and the solution becomes more basic. This effect is especially noticeable when the various silicate and aluminosilicate minerals are ground in distilled water, and the pH of the solution, called the abrasion pH, is taken.[33] The pH resulting from this grinding and initial hydrolysis is a function of the rapidity at which cations are released to the solution and the strengths of the bases formed (Table 3–2). In any weathering environment, the leaching of the cations and the production of hydrogen ion offsets this tendency for most reactions to become basic as weathering proceeds. Grant[12] has shown that the abrasion pH of weathered material that includes some clay is less than the pH of the original rock, because some cations have been removed, and abrasion pH's of clay minerals commonly are lower than those of the common rock-forming minerals.

Little is known of the details of the reactions that take place at the crystal surface during hydrolysis. Jenny,[19] however, has presented a useful model. He notes that there are unsatisfied bonds between the cations and the oxygens and hydroxides at the crystal surface. Water, being a polar substance, is attracted to the differently charged sites on the mineral surface (Fig. 3–5). The water dipoles may be attracted by silicon and aluminum with such force that they dissociate; the hydrogen ions then can combine with the oxygen ions of the crystal surface, and the hydroxide ions combine with either silicon or aluminum. Hydrogen ions also may replace cations in the crystal lattice. This exchange of hydrogen ions for cations has a disrupting effect on

Table 3–2
Abrasion pH values for some common minerals

FORMULA	FORMULA	ABRASION pH
Silicates		
Olivine	$(Mg,Fe)_2SiO_4$	10,11
Augite	$Ca(Mg,Fe,Al)(Al,Si)_2O_6$	10
Hornblende	$Ca_2Na(Mg,Fe)_4(Al,Fe,Ti)_3Si_6O_{22}(O,OH)_2$	10
Albite	$NaAlSi_3O_8$	9,10
Oligoclase*	$Ab_{90-70}An_{10-30}$	9
Labradorite*	$Ab_{50-30}An_{50-70}$	8,9
Biotite	$K(Mg,Fe)_3AlSi_3O_{10}(OH)_2$	8,9
Microcline	$KAlSi_3O_8$	8,9
Anorthite	$CaAl_2Si_2O_8$	8
Hypersthene	$(Mg,Fe)_2Si_2O_6$	8
Muscovite	$KAl_3Si_3O_{10}(OH)_2$	7,8
Orthoclase	$KAlSi_3O_8$	8
Montmorillonite	$(Al_2,Mg_3)Si_4O_{10}(OH)_2 \cdot n\,H_2O$	6,7
Halloysite	$Al_2Si_2O_5(OH)_4$	6
Kaolinite	$Al_2Si_2O_5(OH)_4$	5–7
Oxides		
Boehmite	$AlO(OH)$	6,7
Gibbsite	$Al(OH)_3$	6,7
Quartz	SiO_2	6,7
Hematite	Fe_2O_3	6
Carbonates		
Dolomite	$CaMg(CO_3)_2$	9,10
Calcite	$CaCO_3$	8

*Ab = albite; An = anorthite. (Taken from Stevens and Carron,[33] by permission from the Mineralogical Society of America.)

the crystal surface because of the high charge-to-size ratio of hydrogen. The polyhedra of aluminum and silicon also are no longer held tightly to the mineral, and they are able to move from the crystal surface into the soil solution.

Recent experimental work by Wollast[40] adds more detail to the above general scheme. He notes that in artificially weathering potassium-feldspar over the normal pH range, the increase of potassium ion in solution is accompanied by a decrease in hydrogen ion. This is expected from the general hydrolysis equation and abrasion pH data. He also found that the aluminum and silicon released by weathering formed a thin surface coating composed of amorphous $Al(OH)_3$ and SiO_2 or H_4SiO_4 (Fig. 3–5). The solution is saturated with respect to aluminum but not with respect to silicon. Silicon continues to be released by weathering but at a lower rate, because it must diffuse

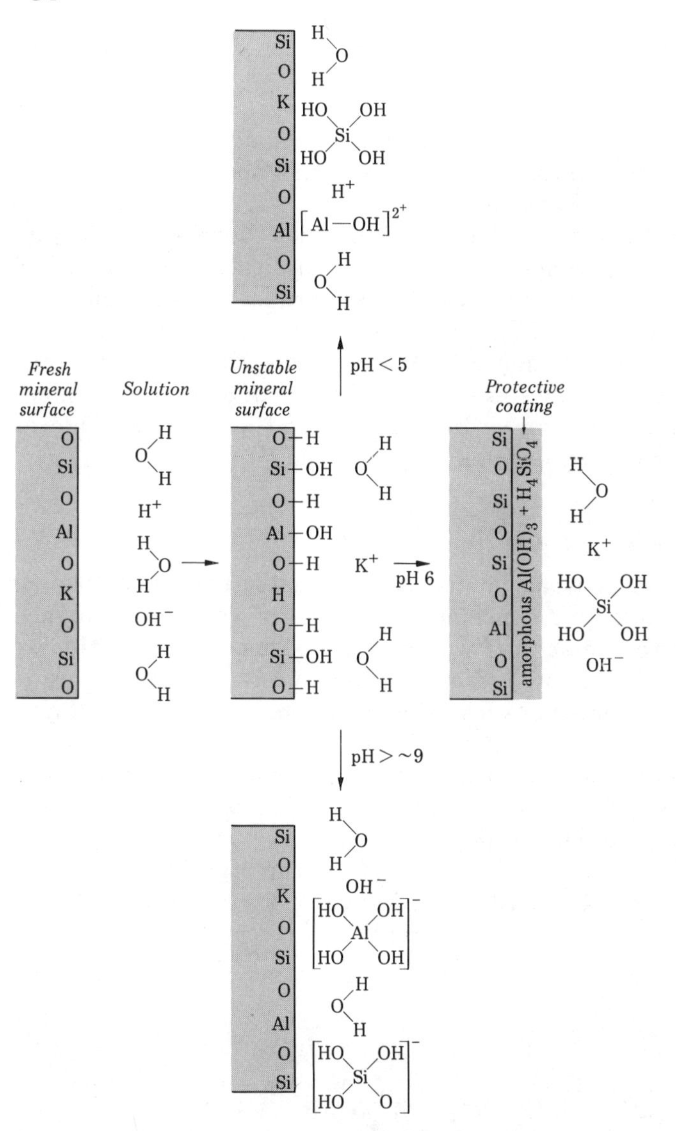

Fig. 3–5 Scheme of an orthoclase surface reacting with water at various pH's. Conditions in acid soils might approximate that depicted for pH less than 5, those for near-neutral soils by that for pH 6, and those for quite basic soils by that for pH near 9. Data partly from Jenny[19] and Wollast.[40]

through the surface amorphous coating. Thus, the initial weathering products, if they accumulate on the surface of the weathering mineral, can bring about a decrease in weathering rate with time. Below a pH of 5, aluminum is soluble as Al^{3+} or aluminum-hydroxy ion (Fig. 4–3), and the blocking effect is not seen (Fig. 3–5). Perhaps this explains, in part, the observation of more rapid weathering of minerals at low pH's.

Minerals also weather at high pH's, and one explanation for this may be seen in the model of Jenny and Wollast. At high pH there are abundant hydroxide ions to combine with the aluminum and silicon at the crystal surface, and the compounds thus formed are quite soluble (Fig. 4–3), perhaps as negatively charged hydroxy ions. Aluminum, therefore, would not have a blocking effect on weathering by forming a film on the mineral grain surfaces; thus, ions released by weathering would go into the solution. The rate of weathering, as under acid conditions, is governed somewhat by the rate of removal of cations from the site of weathering so that the mineral and the solution are not in equilibrium, a point well made by Todd.[34]

Chelation

Evidence is accumulating to suggest that chelating agents are responsible for a considerable amount of weathering; in fact, in some places the amount of weathering by this process might exceed that brought about by hydrolysis alone. Chelating agents are formed by biological processes in soil and excreted by lichens growing on rock surfaces. Their structure is varied and complex and can be described as ". . . the formation of more than one bond between the metal and a molecule of the complexing agent and resulting in the formation of a ring structure incorporating the metal ion" (Lehman,[22] p. 167; see Fig. 3–6). Hydrogen ion is released from the organic molecule during the reaction and can participate in hydrolysis reactions. Once in solution, the chelate may be stable at pH conditions under which the included cation would ordinarily precipitate out, a topic to be discussed later.

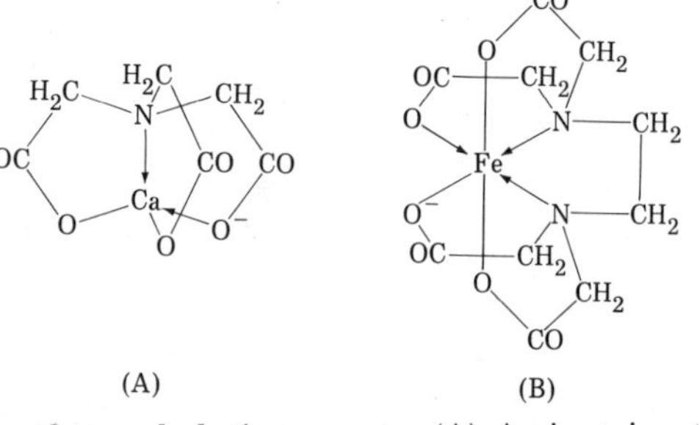

Fig. 3–6 Structure of two chelating agents. (A) Aminotriacetic acid binding calcium; (B) ethylenediaminetetracetic acid (EDTA) binding iron. Taken from Ponomareva.[27] (English translation by Israel Program for Scientific translations, Jerusalem.)

The problem here, however, is to assess the effect of chelating agents on rock and mineral weathering.

Recent laboratory and field work has demonstrated that chelating agents in contact with rocks or minerals can bring about a significant amount of weathering. Schalscha and others[29] ground up various minerals and granodiorite and allowed these materials to react with solutions containing chelating agents. Cations were released to the solution at rates greater than would be predicted by a hydrogen ion effect alone. In fact, these workers found little correlation between pH and weathering rate. They concluded that the weathering of these materials is a combination of the effect of chelating agents and hydrogen ion. This is a departure from much past thinking that has ascribed much weathering to the hydrogen ion only.

Lichens growing on rock surfaces or on soils can bring about substantial amounts of weathering. Lichens excrete chelating agents,[30] and, thus, are important to the understanding of weathering of rocks and minerals. Jackson and Keller[17] have studied the weathering of a recent basalt in Hawaii under a lichen cover and in the absence of such a cover (Table 3–3). They found good evidence that weathering rinds

Table 3–3
Comparison of weathering of Hawaiian basalt erupted in 1907 under two surface weathering environments: lichen-covered and lichen-free

		LICHEN-COVERED ROCK	LICHEN-FREE ROCK
Mean thickness of weathering rind (mm)		0.142 (*color:* 1OR 3/4–4/6)	<0.002
Concentration ratio of weathered crust: fresh rock for these elements	Fe	6.36	1.21
	Al	0.58	0.47
	Si	0.21	1.20
	Ti	0.27	0.965
	Ca	0.004	1.24

*The lichen is *Stereocaulon volcani*. (Taken from Jackson and Keller.[17])

are thicker and chemical alteration is more extensive beneath a lichen cover than in lichen-free areas of the same rock. Data for the lichen-free rock show slight enrichment in iron, silicon, and calcium, little change in titanium, and some depletion of aluminum. In contrast, the lichen-covered rock showed a sixfold enrichment of iron and depletion in varying amounts of all other elements. This study is unique in that it provides a comparison of weathering with and without biological input. These workers concluded that chelating agents in the presence of high hydrogen ion concentration, due to respiratory CO_2 and organic acids, increased weathering beneath lichens.

Little is known of the mechanism by which chelating agents bring about the weathering of a mineral grain surface. The agents might combine directly with exposed cations, much in the manner in which hydroxide ions become attached to aluminum at the grain surface (Fig. 3–5), and this could be followed by the movement of the complex into solution.

Chelating agents render substances more soluble as chelates under certain pH conditions. A common example is aluminum, which may be soluble as a chelate over pH's at which it is insoluble as an ion (Fig. 4–3). Hence, solubilizing aluminum with a chelating agent could offset the blocking action depicted in Fig. 3–5 and thus allow more rapid weathering of the mineral surface.

Oxidation

Oxidation is the process by which an element loses an electron. This loss results in an increase in positive valency for the element. Iron is the element most commonly oxidized in a soil or weathering environment, and the oxidation products give the altered material the characteristic yellowish brown to red colors. In soils, and in many other weathering environments, the common oxidizing agent is oxygen dissolved in the water involved in the weathering reactions. Two aspects of oxidation will be discussed here: oxidation as a process in mineral breakdown and oxidation of iron after its release by other weathering processes.

Oxidation of iron in a mineral can alter the mineral.[2] Iron exists as Fe^{2+} in the common rock-forming minerals. Oxidation to Fe^{3+} disrupts the electrostatic neutrality of the crystal such that other cations leave the crystal lattice to maintain neutrality. These latter cations leave vacancies in the crystal lattice, vacancies that either bring about the collapse of the lattice or render the mineral more susceptible to attack by other weathering processes. The alteration of biotite to vermiculite is one example of weathering primarily due to oxidation.

Weathering of iron-bearing minerals commonly releases Fe^{2+} which, if in contact with oxygenated waters, is oxidized and forms an oxide or hydrous oxide of iron. An example of this reaction for fayalite is given by Krauskopf[20]

$$\underset{c}{Fe_2SiO_4} + \underset{aq}{2\,H_2CO_3} + \underset{l}{2\,H_2O} \rightarrow \underset{aq}{2\,Fe^{2+}} + \underset{aq}{2\,HCO_3^-} + \underset{aq}{H_4SiO_4} + \underset{aq}{2\,OH^-} \text{ (hydrolysis)}$$

$$\underset{aq}{2\,Fe^{2+}} + \underset{aq}{4\,HCO_3^-} + \underset{g}{\tfrac{1}{2}\,O_2} + \underset{l}{2\,H_2O} \rightarrow \underset{c}{Fe_2O_3} + \underset{aq}{4\,H_2CO_3} \text{ (oxidation)}$$

where g denotes the gaseous phase. Here the Fe^{2+} is released by weathering, and it enters an oxidizing environment and forms a precipitate, in this case Fe_2O_3. Usually, other more hydrated iron compounds are formed (e.g., goethite).

The rate at which oxidation takes place and is noticeable in many soils seems to depend on the rate of release of iron upon weathering. Stratigraphic studies of unconsolidated materials indicate that the depth and intensity of oxidation increase with age, and that this increase is a fairly slow process on a geologic time scale. For example, chronosequences of soils in the western United States that span tens of thousands of years commonly show differences in the depths of oxidation that are a function of soil age. Although there are few data on oxygen content with depth in soils, those of Black[4] suggest that porous, well-drained soils can be well oxygenated below the zone of noticeable oxidation (Cox horizon) and into seemingly unoxidized materials (Cn horizon). It would seem, therefore, that the slow extension of oxidation with depth in the individual members of a soil chronosequence is directly related to the rate of release of iron during weathering, because oxygenated waters are passing through materials that appear to be unweathered by most field criteria.

Hydration and dehydration

Hydration and dehydration are processes by which water molecules are added to or removed from a mineral. The result is the formation of a new mineral. These processes probably are not too important in overall chemical weathering because few minerals are affected, and they are not too common. An example of hydration-dehydration is the formation of gypsum and anhydrite by adding or removing H_2O

$$\underset{\text{(gypsum)}}{\underset{c}{CaSO_4 \cdot 2\,H_2O}} \rightleftharpoons \underset{\text{(anhydrite)}}{\underset{c}{CaSO_4}} + \underset{l}{2\,H_2O}$$

A marked increase in volume accompanies the reaction anhydrite to gypsum, and, if this takes place within a rock, physical disintegration can occur. Perhaps a more common reaction is that involving iron oxides and hydrous oxides

$$\underset{\text{(hematite)}}{\underset{c}{Fe_2O_3}} + \underset{l}{H_2O} \rightleftharpoons \underset{\text{(goethite)}}{\underset{c}{2\,FeOOH}}$$

This reaction probably can go both ways in a weathering environment, although in chronosequences of well-drained soils the increased redness of the soils with age suggests that hematite probably is the stable end product. However, hematite has not been identified by X-ray analysis in some old desert soils that are quite red.[36] Hence, although hematite does give red colors, it does not follow that all red colors are imparted by hematite. The above equation suggests that hematite forms only on dehydration. While this may be true, it is not known what water contents favor the reverse reaction, or if other physical-chemical factors play a dominant role.[21,31,37] The fact remains, however, that red soils can form on old land surfaces in either hot-dry or hot-moist climates and that the red color is due either to hematite or to some other red iron oxide compound.

Ion exchange

Some weathering from one mineral to another can occur through the exchange of ions between the solution and the mineral. The most readily exchangeable cations are those between the layers of the phyllosilicates, such as sodium and calcium. During the exchange the basic structure of the mineral is unchanged, but interlayer spacing may vary with the specific cation. Because this mechanism is important in the alteration of one clay mineral to another, it is discussed more fully later.

Weathering of carbonate rock

Calcium carbonate is quite soluble under surface conditions and dissolves according to the equation

$$\underset{c}{CaCO_3} + \underset{g}{CO_2} + \underset{l}{H_2O} \rightleftharpoons \underset{aq}{Ca^{2+}} + \underset{aq}{2\,HCO_3^-}$$

Krauskopf[20] discusses carbonate equilibria and points out that the solubility of $CaCO_3$ varies with differences in CO_2 pressure and in H^+ concentration. Increases in either CO_2 pressure or in H^+ concentration will increase the rate at which $CaCO_3$ dissolves. The partial pressure of CO_2 in the soil atmosphere under vegetation is greater than atmospheric, and therefore $CaCO_3$ solubility is greater under vegetated surfaces than under surfaces that lack vegetation. Rode[28] gives $CaCO_3$-solubility values for various CO_2 pressures. It is also known that CO_2 partial pressure in water is temperature-dependent, with colder waters able to contain more CO_2 than warmer waters. Thus, $CaCO_3$ should dissolve more readily in cooler climates than in warmer climates. Arkley[1] gives two graphs showing the relationship of $CaCO_3$ solubility with pH and with temperature (Fig. 5–7), and these may be helpful in estimating the rate of weathering of carbonate rocks. In addition, chelating agents combine with Ca^{2+} and in this way increase the rate of solution of $CaCO_3$ and the mobility of Ca^{2+}.

Soils formed from limestone commonly consist only of the insoluble residue left behind as Ca^{2+} and HCO_3^- are leached from the system. The properties of the soil, therefore, are strongly dependent on the properties of the insoluble fraction.

Measurement of the amount of chemical weathering that has taken place

Total chemical analysis is the most widely used way of determining the amount of chemical weathering that has taken place in a rock. Table 3–4 compares analytical data for fresh rock with data for various stages of weathered rock (saprolite), rated from 1 (least weathered) to 4 (most weathered). Note that all data are in per cent, and therefore all one has are the relative changes that take place upon weathering, not absolute gains and losses. The obvious relative trends are the loss of silicon and gains of iron and aluminum upon weathering. It is commonly assumed that Al_2O_3 content does not change upon weathering because it is relatively insoluble at normal pH's, and much of it is tied up in the clay minerals that form. Thus, if one assumes a constant Al_2O_3 content, all constituents can be recalculated by multiplying by the factor

$$\%Al_2O_3 \text{ in fresh rock}/\% \; Al_2O_3 \text{ in weathered material}$$

This is done in Table 3–4 (columns 4a) for saprolites weathered to stage 4. Gains and losses can be determined by subtracting data in columns 4a from those of the fresh rock, and these are shown in columns 4a-R. The main error in this method is that the gains and losses depend upon Al_2O_3 content remaining constant on weathering, a condition not always attained.

A more accurate way of determining chemical weathering is by gains and losses in weight of material on a volume basis. Many times this cannot be done, but two examples from northern California serve to demonstrate the usefulness of the method (Table 3–5). Because rock structure is retained in saprolite, it could be shown that little volume change occurred in going from rock to saprolite. Hence, from data from the chemical analyses (Table 3–4) and the bulk densities for fresh material and for material of all weathering stages, the actual gains and losses in weight can be calculated (Table 3–5). It can be seen from this analysis that Al_2O_3 is depleted, and thus the assumption of a constant

Table 3–4
Chemical analyses, molar ratios of oxides, and abrasion pH values for two andesite rocks and their weathered products, southern Cascade Range, California

	Hypersthene andesite	*Saprolite samples (increasing weathering→)*			
		1	*2*	*3*	*4*
SiO_2	57.0	45.7	41.3	38.9	39.[illegible]
Al_2O_3	16.7	22.3	29.1	31.6	31.[illegible]
Fe_2O_3	2.0	5.2	9.8	11.1	11.[illegible]
FeO	4.7	4.2	1.1	0.50	0.[illegible]
TiO_2	1.0	1.0	1.2	1.3	1.[illegible]
MnO	0.12	0.24	0.26	0.27	0.[illegible]
P_2O_5	0.15	0.07	0.03	0.03	0.[illegible]
CaO	7.2	4.8	1.2	0.62	0.[illegible]
MgO	6.1	7.2	2.7	1.4	0.[illegible]
Na_2O	3.1	1.2	0.86	0.79	0.[illegible]
K_2O	1.2	0.24	0.28	0.11	0.[illegible]
H_2O^+	0.35	7.4	11.8	12.6	12.[illegible]
Total	99.6	99.9	99.6	99.2	99.[illegible]
Molar sa ratio†	5.79	3.47	2.40	2.09	2.[illegible]
Molar sa ratio saprolite / Molar sa ratio rock		0.60	0.41	0.36	0.[illegible]
Molar ba ratio‡ ($\times 10^{-3}$)	3.36	2.81	1.03	0.59	0.[illegible]
Molar ba ratio saprolite / Molar ba ratio rock		0.84	0.31	0.18	0.[illegible]
Abrasion pH	8.9	5.5	5.1	4.9	4.[illegible]

*Weight of each oxide (g) assuming that Al_2O_3 content remains constant on weather
**Gains and losses by weight (g) of each oxide obtained by subtracting the rock anal from that of column 4a. Total weight is an approximation of that that remains from weathering of 100 g of rock.
(Taken from Hendricks and Whittig,[16] Table 1)

a*	4a–R**	*Olivine andesite*	*Saprolite samples (increasing weathering →)* 1	2	3	4	4a*	4a–R**
.9	−36.1	53.8	42.6	38.3	36.3	36.8	20.2	−33.6
.7	0	16.9	21.7	30.3	30.5	30.7	16.9	0
.9	+3.9	2.3	6.7	15.2	16.0	15.6	8.6	+6.3
.21	−4.49	5.8	2.9	0.36	0.24	0.10	0.06	−5.74
.68	−0.32	1.2	1.1	1.5	1.5	1.5	0.83	−0.37
.14	+0.02	0.12	0.15	0.09	0.17	0.10	0.06	−0.06
.02	−0.13	0.15	0.25	0.04	0.03	0.04	0.02	−0.13
.23	−6.97	8.4	5.2	0.79	0.20	0.12	0.07	−8.33
.14	−5.96	7.4	9.7	0.22	0.15	0.15	0.08	−7.32
.33	−2.77	2.5	0.89	0.19	0.09	0.08	0.04	−2.46
.08	−1.12	0.60	0.31	0.27	0.15	0.15	0.08	−0.52
.8	+6.45	0.71	9.1	13.1	13.9	13.7	7.5	+6.79
.13	−47.48	99.9	100.6	100.3	99.3	99.0	54.44	−45.44
		5.40	2.88	2.15	2.03	2.04		
			0.53	0.40	0.38	0.38		
		3.72	3.44	0.25	0.11	0.09		
			0.92	0.07	0.03	0.02		
		8.6	6.3	5.6	5.4	5.5		

lar sa ratio = SiO_2/Al_2O_3
lar ba ratio = $(CaO + MgO + Na_2O + K_2O)/Al_2O_3$.

Al_2O_3 content must be re-evaluated. It does appear fairly certain, however, that by assuming constant Al_2O_3 one can at least show the minimum changes that have occurred upon weathering (compare relative losses in columns 4a-R of Table 3–4 with that in columns 4-R of Table 3–5).

In regional comparisons of many soils by chemical data, it is cumbersome to use total chemical analyses. Workers generally recalculate the data to a single number for these comparative studies. Molar ratios (per cent oxide divided by molecular weight) provide the best data. Common molar ratios are[18]

Table 3–5

Weight change in oxide content upon weathering for two andesite rocks of the southern Cascade Range, California*

	HYPERSTHENE ANDESITE	SAPROLITE (INCREASING WEATHERING→) 1	2	3	4	OLIVINE ANDESITE	SAPROLITE (INCREASING WEATHERING→) 1	2	3	4
SiO_2	157	77	48	43	42	143	82	51	47	46
Al_2O_3	46	38	34	35	34	45	42	40	39	39
Fe_2O_3	5.5	8.8	11	12	12	6.1	13	20	20	20
FeO	13	7.1	1.3	0.6	0.4	15	5.6	0.5	0.3	0.1
TiO_2	2.8	1.7	1.4	1.4	1.4	3.2	2.1	2.0	1.9	1.9
MnO	1.3	0.4	0.3	0.3	0.3	0.3	0.3	0.1	0.2	0.1
P_2O_5	0.4	0.1	0.03	0.03	0.03	0.4	0.5	0.05	0.05	0.05
CaO	20	8.1	1.4	0.7	0.5	22	20	1.1	0.3	0.2
MgO	17	12	3.1	1.6	1.0	20	19	0.2	0.3	0.1
Na_2O	8.5	2.0	1.0	0.9	0.4	6.6	1.7	0.3	0.1	0.1
K_2O	3.3	0.4	0.3	0.1	0.2	1.6	0.6	0.4	0.2	0.2
H_2O	1.0	13	14	14	14	1.6	18	17	18	17
Total Fe as Fe_2O_3	20	17	12	13	13	23	19	21	20	20

DIFFERENCES BETWEEN SAPROLITE (1, 2, 3, 4) AND PARENT ROCK (R)

	Hypersthene Andesite 1 − R	2 − R	3 − R	4 − R	*Olivine Andesite* 1 − R	2 − R	3 − R	4 − R
SiO_2	−80	−109	−114	−115	−61	−92	−96	−97
Al_2O_3	−8	−12	−12	−11	−3	−5	−6	−6
Fe_2O_3	+3.3	+5.5	+6.5	+6.5	+7	+14	+14	+14
FeO	−5.9	−11.7	−12.4	−12.6	−9	−15	−15	−15
TiO_2	−1.1	−1.4	−1.4	−1.4	−1.1	−1.2	−1.3	−1.3
CaO	−12	−19	−19	−19	−12	−21	−22	−22
MgO	−5	−14	−15	−16	−1	−20	−20	−20
Na_2O	−6.5	−7.5	−7.6	−8.1	−5	−6	−7	−7
K_2O	−2.9	−3.0	−3.2	−3.1	−1.0	−1.2	−1.4	−1.4
H_2O	+12	+13	+13	+13	+16	+15	+16	+15
Total Fe as Fe_2O_3	+3.2	−7.5	−7.6	−8.1	−3.4	−2.0	−2.3	−2.5

*All values in cg/cm^3 (centigrams per cubic centimeter). The volume of the rock was shown to not change upon weathering to the various stages of saprolite.

(Taken from Hendricks and Whittig,[16] Tables 2 and 3)

Silica:alumina	SiO_2/Al_2O_3
Silica:iron	SiO_2/Fe_2O_3
Silica:sesquioxides	$SiO_2/(Al_2O_3 + Fe_2O_3)$
Bases:alumina	$(K_2O + Na_2O + CaO + MgO)/Al_2O_3$

In general, these all decrease upon weathering, as shown by some examples in Table 3–4. One can use the above ratios for the parent material and for weathered materials or soil horizons to calculate ratios that express differences between the two. One such ratio is what Jenny[18] calls the leaching factor

$$\text{Leaching factor} = \frac{(K_2O + Na_2O)/SiO_2 \text{ of weathered horizon}}{(K_2O + Na_2O)/SiO_2 \text{ of parent material}}$$

Such ratios have the advantage of expressing six analytical values as a single number. To illustrate this method, the $SiO_2:Al_2O_3$ ratio and $(CaO + MgO + Na_2O + K_2O):Al_2O_3$ ratio for saprolite:rock are presented in Table 3–4.

The abrasion pH of weathered materials can be used as a rough measure of the weathering that has taken place. Grant[12] has shown that

$$\text{Abrasion pH} = f\,\frac{\text{Na} + \text{K} + \text{Ca} + \text{Mg}}{\text{Clay minerals}}$$

These are compared with various molar ratios in Table 3–4, and it is seen that as the various molar ratios decrease, so too does the abrasion pH. One could also combine abrasion pH's into ratios of weathered material:rock, in much the same way that various oxides can be combined.

Oxide ratios can be calculated on the basis of either weathered rock:parent material or weathering by-product:parent material. The former ratio gives the overall changes in the parent material. The latter, however, gives the direction in which the weathering reactions are going, as reflected in the composition of the by-product. This is especially true for the $SiO_2:Al_2O_3$ ratio because this ratio differs with the clay mineral that forms.

Chemical weathering disguised as physical weathering

In many field situations it is difficult to determine quantitatively the amount of weathering due to physical processes relative to that due to chemical processes. Indeed, if visual evidence for chemical weathering is lacking, one usually is tempted to look for evidence supporting some physical weathering process. The weathering may be chemical, however, and the evidence for it is either quite subtle, or it has been removed from the site of weathering.

One common example is the formation of granitic grus measuring meters or tens of meters thick. In some localities there is little visual evidence for chemical weathering, such as clay formation or iron oxidation, and the problem confronting many workers is just how is this deep weathering accomplished. The answer lies in the examination of the rock and constituent minerals under a petrographic microscope or by X-ray. Wahrhaftig[35] studied the origin of grus in the Sierra Nevada and was able to show that the granitic rock was shattered to considerable depth by the expansion of several minerals upon alteration. Some biotite and some plagioclase had altered to clay. Volume increase accompanying the alteration was sufficient to fracture the surrounding minerals, as seen in thin section, and this brought about the disaggregation of a great thickness of rock. Other work confirms this origin for grus. For example, biotite alteration was found to be important in the formation of grus in Wyoming.[10] In this case, however, the biotite was initially altered during Precambrian high-temperature oxidation. Later weathering in a near-surface environment exploited the already altered biotite grains, causing them to alter further, expand, and internally shatter the rock to grus. Grus formation in southern California also is attributed to biotite alteration.[25] Although biotite alteration was not always seen with the petrographic microscope, it could be deciphered by X-ray examination. As an example of the amount of expansion possible, it was pointed out that the complete alteration of biotite to vermiculite can be accompanied by a 40 per cent volume increase, and some data were presented to indicate that this had happened in places.

Biotite-induced shattering of rocks opens up a whole list of possibilities for investigation into weathering processes. One can no longer rely on field evidence alone to estimate the weathering process involved in rock breakdown. Thin-section study of the rock is mandatory, and this may lead to X-ray studies as well as polished-section study under high magnification.

Another common example of weathering that might be attributed to physical processes is that which goes on at high altitudes. For example, stones lying on the surface of Holocene deposits above timber line in the Colorado Front Range are deeply pitted (Fig. 3–4). Commonly, the felsic mineral bands of coarse-grained gneisses form depressions, and the mafic mineral bands stand in relief. The climate is very rigorous, with long cold winters and short cool summers. Because many of the rock surfaces appear fresh, and there is little visual evidence for chemical weathering, one might suspect that most of the

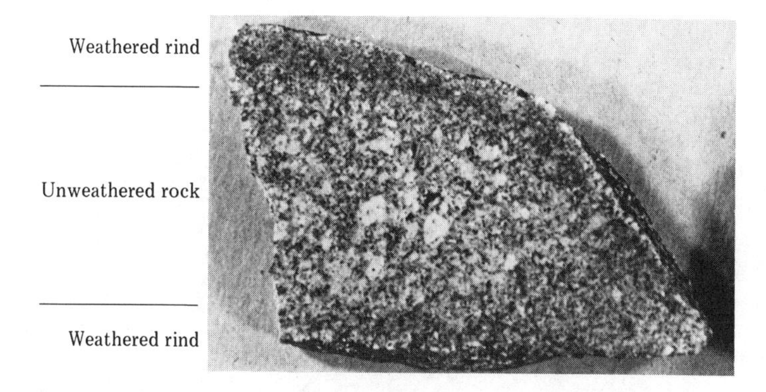

Fig. 3–7 Weathering rind, 1 cm thick, developed from a granitic rock on the surface of a late-Wisconsin rock glacier, western Colorado.

weathering is due to physical processes. However, fine-grained igneous and metamorphic rocks in the Colorado Front Range and in other parts of the Colorado Rockies, presumably of the same age or older, show considerable development of weathering rinds that result from chemical weathering (Fig. 3–7). Thus, chemical weathering of exposed rocks goes on under the rigorous climatic conditions of high altitudes. This relationship suggests that the coarse-grained, pitted rocks (Fig. 3–4) also are undergoing chemical weathering, that this weathering loosens the bonds between mineral grains, and that once the grains are loosened physical processes remove them from the rock surface. The evidence, therefore, for chemical weathering does not remain on the rock surface, which can always have a fresh appearance. In contrast, chemical weathering of fine-grained rocks results in a weathering rind that remains intact on the rock surface and thus provides the basic evidence for chemical weathering.

It is commonly stated that chemical weathering is nil in arid regions. Stratigraphic studies in the Basin and Range Province of the western United States, however, indicate considerable clay in soils on stable landscapes[24] (see Fig. 8–9). This indicates that the precipitation in these regions (less than 25 cm) is sufficient to bring about significant chemical weathering. Perhaps the reason many workers envisage little chemical weathering in these dry regions is because many of the landscapes are eroding, and therefore the products of chemical weathering are continually being swept from the site of weathering.

Rate of present-day chemical weathering

It would be helpful to have a measure of regional rates of chemical weathering in near-surface environments. Tombstone weathering

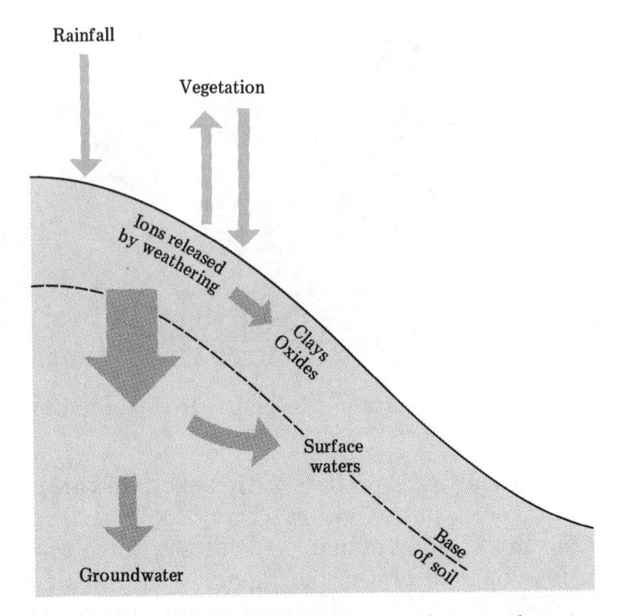

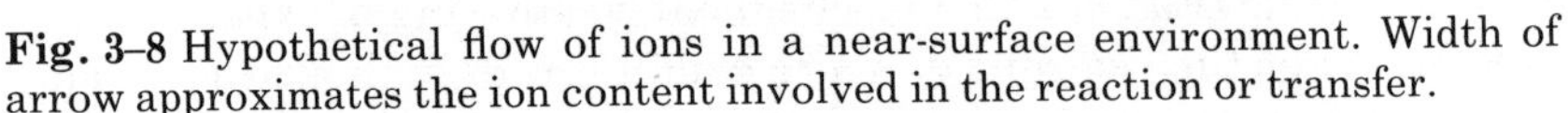
Fig. 3–8 Hypothetical flow of ions in a near-surface environment. Width of arrow approximates the ion content involved in the reaction or transfer.

rates could provide the necessary data, but the data are not too plentiful, and the rates obtained are for above-ground processes only. Clay formation in dated soils could also be used, but again we are hindered by the lack of data. Furthermore, clays form so slowly in some environments that thousands of years may have to elapse before clay formation is noticeable. Another complication is that the clays are a by-product of weathering and thus represent only part of the material released by weathering.

One approach to estimating the present-day rate of chemical weathering is to examine the chemistry of surface waters, because many of the ions in water come from weathering reactions. The system we are dealing with is an exceedingly complex one, as brought out by the detailed study of Cleaves and others.[7] A hypothetical flow of ions in a near-surface environment is depicted in Fig. 3–8. The main source of ions in the soil solution is release upon mineral weathering or organic-matter decomposition. Rainfall will add some ions, but this amount can be determined and subtracted from the overall ion content. Some ions will cycle through the biosphere, but it is probable that the amounts gained from and lost to the soil soon reach a steady state. Analyses of waters that have moved through the soil should provide the best data, but such analyses are few.[14,39] Of the water that moves through the soil, part goes to ground-water reservoirs and part becomes surface-water runoff. Assuming that the dissolved load in surface waters is proportional to the rate of chemical weathering, one might be able to use stream dissolved-load data to make rough estimates on chemical weathering in weight loss per unit area per year. Several examples of the kinds of data obtained from these analyses follow.

Hembree and Rainwater[15] have calculated what they termed the chemical degradation of the Wind River Mountains, Wyoming, by dissolved solids contained in surface runoff. They found that the NE flank of the range yields about twice the dissolved load than does the SW flank of the range, even though the stream runoff on the SW side is about 1.5 times that on the NE side. They calculate that 84,000 years are needed to remove 30.5 cm of material in solution on the SW flank

Table 3–6

Data on annual mechanical and chemical denudation for various rivers of the world

RIVER	MECHANICAL DENUDATION (TONS/KM²)	CHEMICAL DENUDATION (TONS/KM²)
I. *Northern rivers of temperate and cold climates*		
Mountain rivers		
Kolyma	7	5.5
Yana	10	3.9
Pechora	20	17.0
Indigirka	24	11.0
Amur	28	10.1
Yukon	103	22.0
Plains rivers		
Neva	3.9	10.0
Yenisei	4.0	11.4
Luga	4.0	17.0
Narova	4.0	17.7
Onega	4.0	20.0
Ob	6.0	12.2
Western Dvina	6.0	25.0
Mezen	10.0	16.0
Northern Dvina	16.5	48.0
II. *Rivers of temperate-hot, subtropical, and tropical climates*		
Plains rivers		
Dnieper	4.0	17.0
Southern Bug	11.5	14.0
Don	18.3	22.0
Ural	18.6	15.0
Volga	18.6	32.5
Dniester	31.5	39.5
Rivers starting in mountains or large uplands		
Kuma	33	9.0
Kalaus	40	5.0
Amazon	60	13.0
La Plata	75	18.0
Mississippi	118	28.4
Kuban	180	35.0
Mountain rivers		
Kura	213	23.4
Amu Darya	440	78.1
Rivers of south-eastern Asia (average)	390	93.0
Terek	587	125
Rion (Rioni)	2000	209
Samur	1700	270
Sulak	2000	290

(Taken from Strakhov[32])

compared with 45,000 years to remove the same amount of material on the NE flank. The more rapid rate of chemical degradation is explained by the distribution of rock types, the SW flank containing mostly granitic rock and the NE flank containing both granitic and sedimentary rocks.

Strakhov[32] has compiled dissolved-load data for many rivers of the world, and from these he has calculated chemical denudation. River basins differ markedly in their rate of chemical denudation (Table 3–6), and this variation no doubt is influenced by many factors. Some important factors are climate, susceptibility of the rocks to weathering, and length of time the water is in contact with the weathering materials. Although Strakhov[32] does not discuss these factors in detail, he does show that significant chemical denudation occurs in many diverse environments, including those characterized by low temperatures.

The point to this general discussion is to show that one might be able to make rough estimates on regional rates of chemical weathering with data such as those presented by Hembree and Rainwater, Strakhov, and Cleaves and others. The important factors of soil formation and of weathering will have to be taken into careful consideration, however, so that one is fairly certain of the influence of each. For example, the influence of rock type could be held constant by studying only those basins underlain with granitic rocks. The time factor could be held reasonably constant by restricting the study to areas characterized only by weakly or moderately developed soils. One could devise ways of keeping the other factors reasonably constant. Dissolved-load data then could be compared with variation in regional climates to determine if regions do differ in their rate of chemical weathering. Of course, this analysis carries with it the assumption that there is a strong correlation between the dissolved load of a stream and the rate of weathering in a near-surface environment, say, in a soil. Unfortunately, this ideal situation might not always obtain.

REFERENCES

1. Arkley, R. J., 1963, Calculation of carbonate and water movement in soil from climatic data: Soil Sci., v. 96, p. 239–248.
2. Barshad, I., 1964, Chemistry of soil development, p. 1–70 *in* F. E. Bear, ed., Chemistry of the soil: Reinhold Publ. Corp., New York, 515 p.
3. Benedict, J. B., 1968, Recent glacial history of an alpine area in the Colorado Front Range, U.S.A., II. Dating the glacial deposits: Jour. Glaciology, v. 7, p. 77–87.

4. Black, C. A., 1957, Soil-plant relationships: John Wiley and Sons, New York, 332 p.
5. Blackwelder, E. B., 1927, Fire as an agent in rock weathering: Jour. Geol., v. 35, p. 134–140.
6. ——1933, The insolation hypothesis of rock weathering: Amer. Jour. Sci., v. 226, p. 97–113.
7. Cleaves, E. T., Godfrey, A. E., and Bricker, O. P., 1970, Geochemical balance of a small watershed and its geomorphic implications: Geol. Soc. Amer. Bull., v. 81, p. 3015–3032.
8. Coleman, J. M., Gagliano, S. M., and Smith, W. G., 1966, Chemical and physical weathering on saline tidal flats, Northern Queensland, Australia: Geol. Soc. Amer. Bull., v. 77, p. 205–206.
9. Cooke, R. U., and Smalley, I. J., 1968, Salt weathering in deserts: Nature, v. 220, p. 1226–1227.
10. Eggler, D. H., Larson, E. E., and Bradley, W. C., 1969, Granites, grusses, and the Sherman erosion surface, southern Laramie Range, Wyoming: Amer. Jour. Sci., v. 267, p. 510–522.
11. Goudie, A., Cooke, R., and Evans, I., 1970, Experimental investigation of rock weathering by salts: Inst. British Geographers (London) Area no. 4, p. 42–48.
12. Grant, W. H., 1969, Abrasion pH, an index of weathering: Clays and Clay Minerals, v. 17, p. 151–155.
13. Griggs, D. T., 1936, The factor of fatigue in rock exfoliation: Jour. Geol., v. 44, p. 781–796.
14. Hay, R. L., and Jones, B. F., 1972, Weathering of basaltic tephra on the island of Hawaii: Geol. Soc. Amer. Bull., v. 83, p. 317–332.
15. Hembree, C. H., and Rainwater, F. H., 1961, Chemical degradation on opposite flanks of the Wind River Range, Wyoming: U.S. Geol. Surv. Water Supply Pap. 1535-E, 9 p.
16. Hendricks, D. M., and Whittig, L. D., 1968, Andesite weathering II. Geochemical changes from andesite to saprolite: Jour. Soil Sci., v. 19, p. 147–153.
17. Jackson, T. A., and Keller, W. D., 1970, A comparative study of the role of lichens and "inorganic" processes in the chemical weathering of recent Hawaiian lava flows: Amer. Jour. Sci., v. 269, 446–466.
18. Jenny, H., 1941, Factors of soil formation: McGraw-Hill, New York, 281 p.
19. —— 1950, Origin of soils, p. 41–61 *in* P. D. Trask, ed., Applied sedimentation: John Wiley and Sons, New York, 707 p.
20. Krauskopf, K. B., 1967, Introduction to geochemistry: McGraw-Hill, New York, 721 p.
21. Langmuir, D., 1971, Particle size effect on the reaction goethite = hematite + water; Amer. Jour. Sci., v. 271, p. 147–156.
22. Lehman, D. S., 1963, Some principles of chelation chemistry: Soil Sci. Soc. Amer. Proc., v. 27, p. 167–170.
23. Loughnan, F. C., 1969, Chemical weathering of the silicate minerals: American Elsevier Publ. Co., Inc., New York, 154 p.

24. Morrison, R. B., 1964, Lake Lahontan: geology of the southern Carson Desert, Nevada: U.S. Geol. Surv. Prof. Pap. 401, 156 p.
25. Nettleton, W. D., Flach, K. W., and Nelson, R. E., 1970, Pedogenic weathering of tonalite in southern California: Geoderma, v. 4, p. 387–402.
26. Ollier, C. D., 1969, Weathering: Oliver and Boyd, Edinburgh, 304 p.
27. Ponomareva, V. V., 1969, Theory of podzolization: Israel Program for Scientific Translations, Jerusalem, 309 p.
28. Rode, A. A., 1962, Soil science: Israel Program for Scientific Translations, Jerusalem, 517 p.
29. Schalascha, E. B., Appelt, H., and Schatz, A., 1967, Chelation as a weathering mechanism—I. Effect of complexing agents on the solubilization of iron from minerals and granodiorite: Geochim. et Cosmochim. Acta., v. 31, p. 587–596.
30. Schatz, A., 1963, Chelation in nutrition, soil microrganisms and soil chelation. The pedogenic action of lichens and lichen acids: Jour. Agri. and Food Chem., v. 11, p. 112–118.
31. Schmalz, R. F., 1968, Formation of red beds in modern and ancient deserts: Discussion: Geol. Soc. Amer. Bull., v. 79, p. 277–280.
32. Strakhov, N. M., 1967, Principles of lithogenesis: Oliver and Boyd, London, 245 p.
33. Stevens, R. E. and Carron, M. K., 1948, Simple field test for distinguishing minerals by abrasion pH: Amer. Mineralogist, v. 33, p. 31–49.
34. Todd, T. W., 1968, Paleoclimatology and the relative stability of feldspar minerals under atmospheric conditions: Jour. Sed. Petrol., v. 38, p. 832–844.
35. Wahrhaftig, C., 1965, Stepped topography of the southern Sierra Nevada, California: Geol. Soc. Amer. Bull., v. 76, p. 1165–1190.
36. Walker, T. R., 1967, Formation of red beds in modern and ancient deserts: Geol. Soc. Amer. Bull., v. 78, p. 353–368.
37. ____ 1968, Formation of red beds in modern and ancient deserts: Reply: Geol. Soc. Amer. Bull., v. 79, p. 281–282.
38. Winkler, E. M., 1965, Weathering rates as exemplified by Cleopatra's Needle in New York City: Jour. Geol. Educ., v. 13, p. 50–52.
39. Wolff, R. G., 1967, Weathering of Woodstock granite near Baltimore Maryland: Amer. Jour. Sci., v. 265, p. 106–117.
40. Wollast, R., 1967, Kinetics of the alteration of K-feldspar in buffered solutions at low temperature: Geochim. et Cosmochim. Acta, v. 31, p. 635–648.

CHAPTER 4

The products of weathering

Materials released during weathering either are removed from the system in leaching water or react in the system to form a variety of crystalline and amorphous products. The most commonly observed reaction products are the clay minerals and hydrous oxides of aluminum and iron. These products can occur alone or in combination, and their distribution with depth can be uniform or highly variable. Characterization of these products is important because, of all the properties of a soil, they probably best reflect the long-term effect of the chemical and leaching environment of the soil. Their genesis, however, is varied; it can range from relatively simple ion-exchange reactions to the more complex combination of aluminum and silicon from solution or a gel to form a crystalline clay mineral. Within recent years, it has been possible to construct phase diagrams to test for mineral-water equilibria. This approach may prove to be very useful for predicting stable products of weathering, but it has yet to be tested fully in soils.

These minerals are generally so finely divided that their identification is difficult; it can be accomplished, however, by a variety of chemical, thermal, X-ray, and electron-microscope techniques.[4,11,43] Only those products most commonly found in soils will be dealt with here.

Clay minerals, allophane, and oxides of aluminum and iron

The common clay minerals are hydrated silicates of aluminum, iron, and magnesium arranged in various combinations of layers.[20,25,32] They are called layer silicates or phyllosilicates. Two kinds of sheet structures, the tetrahedral and the octahedral, make up the clay minerals, and variations in combinations of these structures and in their chemical makeup give rise to the multitude of clay minerals. The basic difference between these two sheets is in the geometrical arrangement of Si, Al, Fe, Mg, O, and OH. The arrangements differ with the cation because it is the size of the cation that determines how many O or OH ions surround it.

The silicon tetrahedron is the basic structural unit of the tetrahedral sheet. It consists of one Si^{4+} surrounded by four O ions (Fig. 4–1). A sheet is formed when the basal O ions are shared between adjacent tetrahedra and the remaining unshared O ions all point in the same direction. These sheets have a net negative charge.

In the octahedral sheet, the basic unit consists of the octahedral arrangement of six O or OH ions about a central Al, Mg, or Fe ion (Fig. 4–1). In the octahedral sheet, the O and OH ions are shared between adjacent octahedra. If a trivalent ion is the central cation, two-thirds of the available cation sites are filled and the structure is said to be dioctahedral; in contrast, divalent cations fill all available sites and exhibit trioctahedral structure.

The variety of clay minerals results in part from different arrangements of the tetrahedral and octahedral sheets. The two adjacent sheets are bound together tightly by the mutual sharing of the O ions

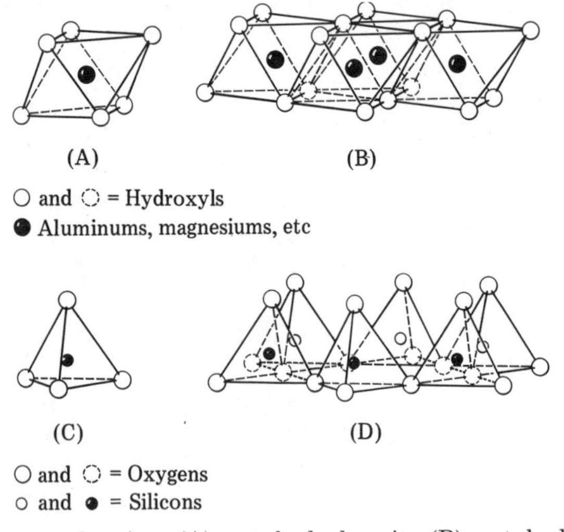

Fig. 4–1 Diagram showing (A) octahedral unit, (B) octahedral sheet, (C) tetrahedral unit, and (D) tetrahedral sheet. (Taken from Grim,[20] Figs. 4–1 and 4–2.)

of the tetrahedral sheet. Thus, in the octahedral sheet of a clay mineral the six ions surrounding the central cation include both O and OH ions. Three structural groupings of the two basic sheets are recognized. The first consists of a tetrahedral sheet attached to one side of an octahedral sheet to form a 1:1 layer phyllosilicate. The second is the symmetrical arrangement of two tetrahedral sheets about a central octahedral sheet to give a 2:1 layer phyllosilicate. A third arrangement is the presence of an octahedral sheet between adjacent 2:1 layers, and these are known as 2:1:1 layer phyllosilicates. In a clay mineral each layer combination, hereafter called a layer, extends for varying distances in the direction of the *a* and *b* crystallographic axes.

The 1:1 and 2:1 layers are stacked in the *c*-axis direction, and bonds between the layers are formed by more than one mechanism. One mechanism involves a bonding of the external O ions of one layer with the external OH ions of the adjacent layer. Bonding probably is due to the mutual sharing of the H ion between the O ions of both sheets to form a hydrogen bond. Bonding can also occur with polar water molecules forming a hydrogen bond between the two layers. The bonding here is as follows: OH ion (one sheet) to H ion (positive pole of water molecule) to OH ion (negative pole of water molecule) to H ion (adjacent sheet). Another mechanism for bonding involves isomorphous substitution within the clay lattice. Although silicon and aluminum are the more common cations, other cations of near similar size can replace them. The common substitutions are Al^{3+} for Si^{4+} in the tetrahedral sheet, and Fe^{3+}, Fe^{2+}, and Mg^{2+} for Al^{3+} in the octahedral sheet. These substitutions, with the exception of Fe^{3+} for Al^{3+}, are marked by an ion of lower valency substituting for one of higher valency, and the net result is a negative charge in the layer. This charge is balanced by some combination of substituting OH^- for O^{2-}, filling vacant cation sites in the octahedral sheets, and adsorbing hydrated cations at the edges of layers and between layers. Cations between layers can bond the layers together, especially if the opposing layers both have external O ions. The kind of interlayer bonding varies from clay mineral to clay mineral and can be correlated with some properties of the clay minerals.

Kaolinite and halloysite are the common 1:1 layer clay minerals. They are structurally similar (Fig. 4–2), the only difference being that halloysite commonly has a layer of water between successive layers; it is then called hydrated halloysite. Hydrogen bonding with a water interlayer (hydrated halloysite), or without a water interlayer (kaolinite, halloysite), binds the adjacent layers together. Isomorphous substitution is negligible.

The montmorillonite, or smectite, group of clay minerals make up a wide variety of 2:1 layer clay minerals (Fig. 4–2). Isomorphous sub-

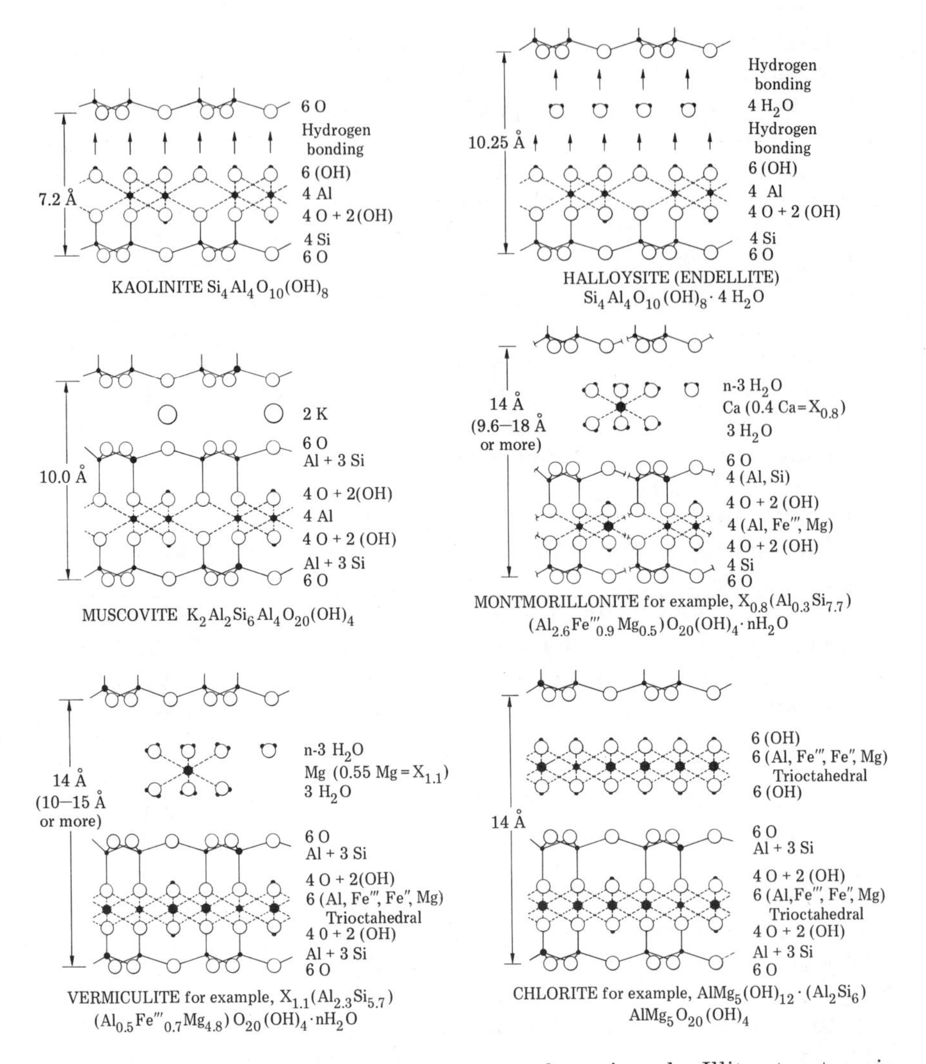

Fig. 4–2 Crystal structure of the common clay minerals. Illite structure is similar to that of muscovite. The *b* crystallographic axis is horizontal and the *c* crystallographic axis is vertical in these diagrams. (Taken from Jackson,[25] Figs. 2.3, 2.8, 2.9, 2.10, 2.14, and 2.15 *in* Chemistry of the Soil by F. E. Bear, ed., © 1964 by Litton Educational Publishing, Inc. Reprinted by permission of Van Nostrand Reinhold Company.)

stitution is common, and this gives rise to many minerals that differ in chemical composition. Although Al^{3+} does substitute for Si^{4+}, the more common substitution is Fe^{2+} and Mg^{2+} for Al^{3+} in the octahedral sheet. The resulting net negative charge is partly balanced by interlayer hydrated cations that bond adjacent layers. Since the negative charge is due mainly to substitution in the octahedral layer, and since that layer is some distance from the interlayer-bonding cation, bonding is relatively loose. Therefore, exchange of cations and water layers readily occurs. These minerals expand or contract as water layers are added or removed, and the amount of such change depends, in part, upon the cation present.

The illite group of 2:1 layer clay minerals are intermediate in composition between ideal montmorillonite and muscovite (Fig. 4–2). Some illites seem to be micas intermixed with layers of montmorillonite, vermiculite, and chlorite. Although the basic structure is similar to that of montmorillonite, most of the isomorphous substitution is Al^{3+} for Si^{4+} in the tetrahedral layer. The interlayer bonding cation is K^+, and because it is close to the site of the negative charge, bonds are especially strong, and interlayer expansion does not occur.

Vermiculite is also a 2:1 layer clay mineral (Fig. 4–2). Like montmorillonite, both dioctahedral and trioctahedral types are recognized and, like illite, the major isomorphous substitution is in the tetrahedral layer. This results in a fairly tight bond between the adjacent layers and the interlayer hydrated cations, and thus interlayer expansion is limited.

Chlorite is the 2:1:1 clay mineral common to soils (Fig. 4–2). It consists of alternating 2:1 layers and octahedral layers. The 2:1 layers are commonly trioctrahedral, and substitutions are mainly Al^{3+} for Si^{4+} in the tetrahedral sheet, which results in a negative charge. The octahedral interlayer contains Mg^{2+} and Fe^{2+} in addition to Al^{3+}, which results in a positive charge. Bonding between the two oppositely charged layers is strong, it is enhanced by hydrogen bonding, and expansion does not take place.

Mixed-layer minerals result from the interstratification of 2:1 and 2:1:1 layer clay minerals and aluminum hydroxide and magnesium hydroxide (gibbsite and brucite layers, respectively). Interlayering is possible because many of the clay-mineral species are structurally similar, and because some expand readily in water or upon slight weathering. Mixing varies from a regular repetition of the components in the direction of the *c* axis to a wholly random arrangement. Such mixed-layer clays are quite common, and their detailed identification is difficult.[25]

Some materials in soil, called allophane, have a chemical composition somewhat similar to those of the clay minerals, but they are not crystalline. Allophane is difficult to characterize but it appears to be an alumino-silicate gel containing varying amounts of Al_2O_3 and SiO_2, perhaps as octahedra and tetrahedra, respectively, H_2O, and perhaps

Fe_2O_3. Because it gives no X-ray diffraction pattern, it is commonly overlooked in many soils.

Aluminum and iron minerals are also formed in soils. The common aluminum species are gibbsite [$Al(OH)_3$ or $Al_2O_3 \cdot 3\,H_2O$] and boehmite (AlOOH), and iron commonly occurs as hematite (Fe_2O_3) and goethite [FeOOH or $Fe_2O_3 \cdot H_2O$]. The goethite that imparts a yellow-brown color to soil is a fine-grained hydrous variety commonly called limonite ($2\,Fe_2O_3 \cdot 3\,H_2O$).

Various quantitative methods using X-ray diffraction data have been employed to determine the content of clay minerals in soils and sediments. Unfortunately, many commonly used methods do not give comparable results.[34] Attainment of meaningful results within 1 per cent seems unlikely, and parts in ten might be a more reasonable goal. An alternate procedure would be to report results in peak heights or peak areas, or in qualitative terms such as abundant and moderately abundant, accompanied by sketches of representative diffractograms.

Cation-exchange capacity of inorganic and organic colloidal particles

Colloidal material in soils carries an electrical charge. Although both negative- and positive-charge sites exist, the origin of the negative sites has been studied most extensively.

The negative charge is approximated by the cation-exchange

Table 4–1
Representative cation-exchange capacities for various materials

Material	*Approximate cation-exchange capacity (me/100 g dry weight)*
Organic matter	150–500
Kaolinite	3–15
Halloysite	5–10
Hydrated halloysite	40–50
Illite	10–40
Chlorite	10–40
Montmorillonite	80–150
Vermiculite	100–150 +
Allophane	25–70
Hydrous oxides of aluminum and iron	4
Feldspars	1–2
Quartz	1–2
Basalt	1–3
Zeolites	230–620

(Taken from Carroll[12] and Grim[20])

capacity (CEC) of the material, and most of the cations attracted to these sites are exchangeable. The CEC varies in amount and origin with the soil material. In humus, the charge originates from the dissociation of H^+ from carboxyl and phenolic groups at the particle surface and within the particle. In clay particles, charge originates at the clay edge or along the surfaces parallel to the *ab* crystallographic plane.[12] Charges along the clay edge originate from unsatisfied (broken) bonds at the edge of the particle, say, between Si—O or Al—OH, or from the dissociation of H^+ from OH^-. Charge originating in this manner is common in the 1:1 clays and is dependent on particle size, because the number of exposed edges increases as particle size decreases. Ionic substitution in the clay lattice commonly results in a negative charge, mainly along the interlayer surfaces. This is probably the main origin of charge in the 2:1 clays, although some charge originates at the clay edge.

The CEC varies with the clay mineral, it is expressed in milli-equivalents (me)* per unit dry weight of the material, and the highest values are for organic matter (Table 4–1). The variation with clay mineral is due to a combination of ionic substitution and its extent, the degree of hydration, and the number of exchange sites at the edges of particles. Because organic matter can have such a high CEC, the presence of a small amount of organic matter can greatly affect the CEC of the soil. In contrast, nonclay minerals and rock fragments have a negligible effect on the CEC; zeolites, because of their structure, are an exception. Thus, one can obtain a rough estimate of the CEC of the clay fraction, and hence of the possible clay minerals present, by knowing the CEC of the soil and the amount of clay present (in per cent), and by sampling deep enough to avoid sizable amounts of organic matter. For example, if a soil has a CEC of 20 me/100 g soil and is 20 per cent clay, the CEC for the clay fraction would be approximately 20 me/20 g clay, or 100 me/100 g clay. The clay mineral could be montmorillonite or vermiculite (Table 4–1).

Anion exchange seems to be important to the overall charge of 1:1 clays mainly because it is thought to occur at clay edges. Grim[20] reviews the subject and cites data suggesting that the average ratio of cation- to anion-exchange capacity is 0.5 for kaolinite, 2.3 for illite, and 6.7 for montmorillonite. Anion-exchange sites are believed to originate

*One milliequivalent is defined as 1 mg of H^+ or the amount of any other cation that will displace it. If, for example, the CEC is 1 me/100 g oven-dry soil, 1 mg of H^+ is adsorbed. If Ca^{2+} displaces the H^+, the amount of Ca^{2+} has to be equivalent to 1 me of H^+; this amount can be calculated. Ca has an atomic weight of 40, compared to 1 for H, with a positive valency of 2; one Ca^{2+} ion, therefore, is equivalent to two H^+ ions, because the latter ion has only one charge. The amount of Ca^{2+} required to displace 1 me of H^+ is 40/2, or 20 mg, the weight of 1 me of Ca^{2+}.

from OH^- dissociation and from the linkage of complex ions, like phosphate, to silica sheet edges. Such complex ions would exchange for the silica tetrahedra because both have a similar geometry.

Ion mobilities

The behavior of ions, once released by weathering to the soil solution, varies; some ions participate in reactions involving mineral synthesis, some are adsorbed to colloid surfaces, and some are removed with the downward-moving water. Only when specific ions in the soil solution occur in the right proportions can mineral synthesis take place. Here we look into the factors governing the mobility of the more common ions.

Two approaches are used to rank ions on their mobility. One is to calculate the ratio of the per cent of an element in stream or spring water to the per cent in the parent rock. When this is done,[15,36] the ranking usually obtained is $Ca^{2+} > Na^+ > Mg^{2+} > K^+ > Si^{4+} > Fe^{3+} > Al^{3+}$, and the values for the latter two under the normal soil pH's are so low that their mobilities are almost negligible. Other studies indicate that this ranking can vary with rock type.[13,35] The other approach[31] is to calculate the ionic potential of the ions, which is the ratio of the charge in valency units to the ionic radius in angstrom units (Å). Ionic potential, therefore, is a measure of the intensity of the positive charge. Ions with a ratio of 3 or less ($K^+ = 0.75$: $Na^+ = 1.0$; $Ca^{2+} = 2.0$; $Fe^{2+} = 2.7$; $Mg^{2+} = 3.0$) can remain in solution as ions, whereas those between 3 and 9.5 precipitate out as hydroxides ($Fe^{3+} = 4.7$; $Al^{3+} = 5.9$). This behavior can be explained as a result of the attraction of the ion in solution for the O ion of the H_2O molecule. If the attraction of the ion for O^{2-} is weak, the ion remains in solution surrounded by water molecules. If, however, the attraction of the ion for the O^{2-} is comparable to that of the H^+ for the O^{2-}, one H^+ from the water molecule is expelled, and the ion is precipitated as an hydroxide. Much of the silicon in natural waters forms non-ionized silicic acid, H_4SiO_4, and so the abundance of silicon is due to other factors. Thus, the data indicate that iron and aluminum can remain close to the site of weathering and partake in synthesis reactions, whereas other ions can be transported away. This precipitation of iron and aluminum hydroxides takes place under oxidizing conditions over the pH range of most normal soils (Fig. 4–3).

When the data for the relative mobilities are compared with those for the ionic potential, the ions are not ranked in the same order. This apparent discrepancy is no doubt due to many factors. One factor is that minerals weather at different rates and, because minerals vary in their

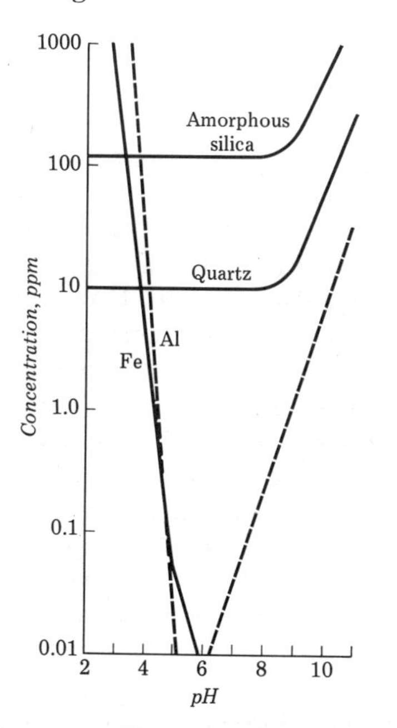

Fig. 4–3 Relationship between pH and solubility of aluminium, iron, amorphous silica, and quartz. Data for amorphous silica and quartz are from Krauskopf[29] (Fig. 6–3), and data for iron and aluminum are from Black.[8] (Figs. 1 and 2, © 1967, John Wiley and Sons.)

composition, so too will leaching waters. Another factor would be the fixation of certain ions into the crystal lattices of the clay minerals. A familiar example is the fixation of potassium between layers of illite where it is not readily exchangeable. Another example is the introduction of magnesium into the octahedral layer of montmorillonite. Ion exchange is also an important factor. Certain ions can replace others on clay-mineral surfaces, and, although the usual replacement series is $M^{2+} > M^{+}$, where M refers to any cation, replacement rank may be modified by a number of factors.[9,12,43] Another important factor is the selective uptake of ions by plants. In short, the difference in ranking by the two methods is explained mostly by processes operative once the ions enter the soil solution.

Genesis of crystalline products

The genesis of minerals in the soil is complex due to the variety of weathering environments, the possibility of unlike micro-environments in close proximity to one another, and the fact that some minerals are derived by precipitation from solution, whereas others are derived from solid-state alteration of pre-existing phyllosilicates. It is not surprising, therefore, that any soil sample commonly is characterized by several clay-mineral species. This is compounded by the facts that (a) reaction rates are very slow at surface conditions, (b) duplication of natural conditions in the laboratory is difficult, and (c) some minerals may not be part of the current mineral-water equilibrium but may have formed in the past under different conditions or even have been derived from the parent material. However, some general statements can be made regarding the chemical conditions favoring the formation of certain clay minerals, and progress in this area is being made on both the laboratory and theoretical fronts.

Formation of secondary iron-bearing minerals in the soil depends on the prevailing pH-Eh conditions.[18] Eh is a measure of the ability of a natural environment to bring about either oxidation or reduction.[29] Under oxidizing conditions, the Fe^{2+} released during weathering is oxidized to Fe^{3+} upon contact with soil water. The ferric oxide or hydrous oxide formed is insoluble in oxygenated soil environments over the normal pH range (Fig. 4–3), and so it precipitates as hematite or geothite. Iron compounds thus formed would be close to the source of iron. If conditions are reducing, Fe^{2+}, being more soluble, can migrate far from the site of release by weathering, until a change in conditions, such as a change in Eh, brings about precipitation. As will be discussed later, iron is also soluble as a chelate and as such can migrate in oxygenated water.

Clay minerals can form in several ways.[16,37] If, for example, the primary minerals undergoing weathering are not phyllosilicates, the formation of the clay mineral involves weathering with the release of cations and silica and alumina, followed by their recombination into phyllosilicate clay minerals. If, on the other hand, the weathering primary mineral is a phyllosilicate from the original igneous rock or from a soil or sediment, the alteration may take place essentially in the solid state. Both conditions will be explored here.

The type of clay mineral that forms in a soil solution mostly depends on the content of silica, the kind and concentration of cations present, the soil pH, and the amount of leaching.[3,27,32,33] Both iron and aluminum released by weathering precipitate as oxides and hydrous oxides over the normal soil pH range and so generally remain within the soil for possible reaction to form clay minerals. As mentioned earlier, many of the other common cations in the soil have relative mobilities much greater than iron and aluminum, and they can be leached from the soil environment and thus will not react. Because silicon and aluminum form the basic framework for many clay minerals, their presence in the correct proportions is essential. Davis[14] has shown that silicon is common in most natural waters. Thus the amount of silicon available for reaction depends less on its rate of release during weathering and more on the rate of leaching from the soil environment. The same can be said for many other cations, but their presence probably depends more on rock type. If the soil solution is characterized by a very low Si/Al molar ratio, gibbsite would form. Leaching conditions that would keep silicon content low would also result in low cation concentration. With increasing amounts of silicon, aluminosilicates form. A Si/Al ratio near 2, with a low cation content and a pH less than about 7, favors the formation of kaolinite and halloysite. The Si/Al ratio necessary to form the 2:1 clay minerals is more difficult to establish because of substitution of both Al^{3+} for Si^{4+} in tetrahedral layers and Mg^{2+}, Fe^{2+}, and Fe^{3+} for Al^{3+} in octahedral layers. For most of these clays, however, the ratio is greater than 2 and may go to 5 or more. Under these conditions montmorillonite forms in the presence of near neutral to alkaline pH and relatively high concentrations of Ca^{2+}, Mg^{2+}, and Na^{+}. Illite forms under these same conditions when the concentration of K^{+} is high. If the solution is high in Na_2CO_3 and $NaHCO_3$, however, extremely alkaline conditions prevail and zeolites form.[2,21,22] Vermiculite formation is favored by a high concentration of Mg^{2+} and a pH less than 7. Chlorite with a gibbsite-layer interlayer would probably form under these same conditions; if, however, conditions are more alkaline and Mg^{2+} is abundant, the interlayer probably would be brucite-like.

There is much conjecture regarding the manner in which clay minerals form from a solution,[3,25,26] because the reactions are very difficult to reproduce under laboratory conditions that even approach those in the field (for laboratory studies see Siffert,[39] Grim,[20] and Millot[32]). One way in which crystallization might proceed is by the mutual attraction of colloidal SiO_2 and $Al(OH)_3$, because the former has a negative charge and the latter a positive charge.[29] One problem here is the orientation of the combining constituents as tetrahedral and octahedral sheets. One way in which this could take place is by adsorption on both nonclay and clay-mineral surfaces. Orientation into sheets may be facilitated with substrates with an atomic lattice structure similar to that of the forming clay minerals. Because clay mineral surfaces carry a negative charge, one might visualize colloidal $Al(OH)_3$ first being attracted to the mineral surface followed by colloidal SiO_2 to build up a particular clay mineral.

Because most silica in natural water occurs as non-ionized silicic acid, however, one probably has to look for other possible mechanisms. Siffert[39] suggests that H_4SiO_4 might combine with Al-hydroxy ions to form two kinds of monomers according to the following equations

$$\begin{matrix}HO\diagdown \\ HO-Si-OH \\ HO\diagup\end{matrix} + [Al(OH)]^{2+} \rightarrow \left[\begin{matrix}HO\diagdown \\ HO-Si-O-Al-OH \\ HO\diagup\end{matrix}\right]^{+} + H^{+} \quad (1)$$

followed by

$$\left[\begin{matrix}HO\diagdown \\ HO-Si-O-Al-OH \\ HO\diagup\end{matrix}\right]^{+} + HO-Si\begin{matrix}\diagup OH \\ -OH \\ \diagdown OH\end{matrix} \rightarrow \begin{matrix}HO\diagdown \\ HO-Si-O-\underset{\displaystyle OH}{\underset{|}{Al}}-O-Si \\ HO\diagup\end{matrix}\begin{matrix}\diagup OH \\ -OH \\ \diagdown OH\end{matrix} + H^{+} \quad (2)$$

$$\begin{matrix}HO\diagdown \\ HO-Si-OH \\ HO\diagup\end{matrix} + [Al(OH)_2]^{+} \rightarrow \begin{matrix}HO\diagdown \\ HO-Si-O-Al \\ HO\diagup\end{matrix}\begin{matrix}\diagup OH \\ \\ \diagdown OH\end{matrix} + H^{+} \quad (3)$$

From the above equations it is seen that the Si/Al ratio would determine which monomer forms. The H^+ ions are neutralized in solution, but twice as many have to be neutralized in eqs. (1) and (2) as in eq. (3); thus eqs. (1) and (2) are favored by a higher pH. Sharma[38] has shown that $CaCO_3$ can neutralize the H^+ by the reaction

$$\underset{c}{CaCO_3} + \underset{aq}{H^+} \rightleftharpoons \underset{aq}{Ca^{2+}} + \underset{aq}{HCO_3^-}$$

and this surely could happen in soils in dry regions. These monomers can then link to other similar monomers to form the tetrahedral and octahedral sheets (Fig. 4–4). Formation of the octahedral sheet involves the rearrangement of the OH ions about the Al ions, but the formation of the tetrahedral sheet requires the removal of water from OH groups linked to Si ions. A somewhat similar scheme is depicted for the formation of a trioctahedral sheet. As in the other model, adsorption onto pre-existing mineral surfaces may facilitate the formation of the octahedral sheet.

Periodic dehydration of the soil would favor both clay–mineral-formation reactions. With dehydration the reacting constituents are brought into close proximity to each other, and this would foster clay formation. On the other hand, if a soil were moist most of the time with water moving through it, the soil solution would be dilute with respect to the reacting constituents, and this would slow the rate at which clays would form.

For clay minerals to form from solution, the aluminum has to be in sixfold coordination and form a gibbsitic layer. This generally has been a problem in the laboratory synthesis of clay minerals. Recently, however, Linares and Huertas[30] have shown that fulvic acid, extracted from peat, forms a soluble aluminum-fulvic acid complex in which the Al^{3+} is in sixfold coordination. With a change in pH, aluminum hydroxide is formed and oriented into a gibbsitic sheet, and this is

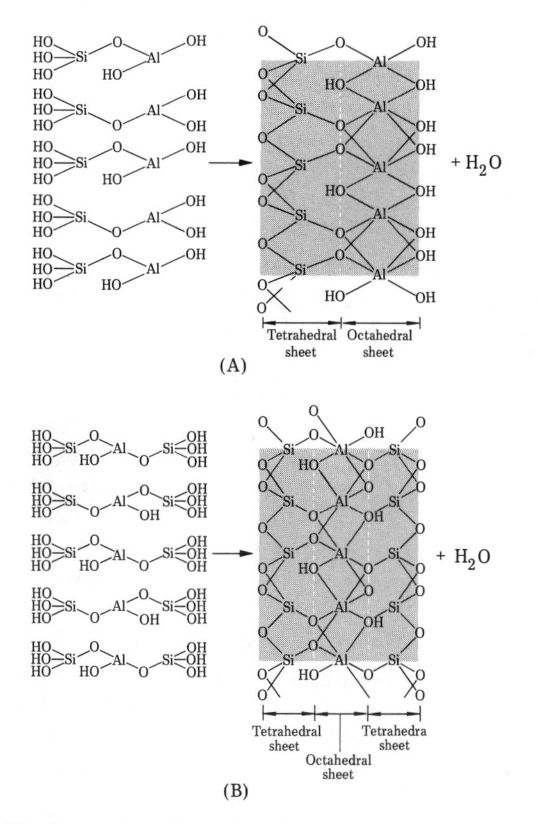

Fig. 4–4 Possible formation of octahedral and tetrahedral sheets by the polymerization of two different kinds of monomers. (A) Formation of a 1:1 clay; (B) formation of a 2:1 clay. (Taken from Siffert,[39] Fig. III_2, Published by Israel Universities Press, Jerusalem.)

followed by adsorption of a silica tetrahedral sheet to form kaolinite. Thus, it now looks as though the sixfold coordination of Al^{3+} is possible in many soil environments, and this could lead to the synthesis of the many different clay minerals.

This discussion does not, of course, exhaust the possible routes that precipitation of clay minerals from a solution can take; I have presented just two plausible mechanisms that seem consistent with existing data.

Various processes have been proposed for the alteration of micas derived from rock to clay minerals or for alteration from one clay mineral to another.[25] These changes almost always involve the interlayer areas because that is where ions can be exchanged, hydroxy ions introduced, or a silica sheet removed. Reactions usually begin at the crystal edge and proceed inward.

Alteration among the 2:1 and 2:1:1 groups of clay minerals probably takes place quite readily because of their similar structures and interlayer-bonding mechanisms. Ion exchange probably is the simplest reaction to envisage (Fig. 4–5, A). By this process ions from the soil solution replace interlayer ions, and there may be some replacement within the crystal lattice. The resulting product is mostly a function of the host mineral and the replacing ion. For example, replacement of interlayer K^+ with Mg^{2+} could alter muscovite or illite to montmorillonite or biotite to vermiculite. If, however, hydroxy ion groups (Al) are present, they could replace the interlayer ions, because they too carry a positive charge, and form chlorite (Fig. 4–5, B). In a similar manner, if the soil solution is rich in Mg^{2+}, brucite layers could form and become interlayered with 2:1 clays to form chlorite. Thus, trioctahedral montmorillonite and vermiculite could be altered to chlorite by these mechanisms. Such an alteration might proceed more rapidly for montmorillonite than for vermiculite, because the former is characterized by an expandable lattice. This expansion would allow for more rapid interchange of interlayer ion and hydroxy-ion or hydroxide sheet.

The alteration from 2:1 to 1:1 clay minerals involves a greater amount of structural reorganization. Two mechanisms proposed for this transformation are (a) gibbsite interlayering to form a chlorite-montmorillonite intergrade, followed by kaolinite formation and (b) the removal of silica tetrahedral sheets from montmorillonite layers (Fig. 4–5, C and D).

It is also known that gibbsite and kaolinite can form from one another. The formation of gibbsite from kaolinite involves the removal of a silicon sheet, perhaps in the manner shown in Fig. 4–5, D. The reverse reaction, the formation of kaolinite from gibbsite, is thought to involve partial dehydration of the gibbsite structure, the entry of silicon-enriched solutions between gibbsite layers, with the oxygen of the silicon tetrahedra occupying the lattice positions vacated by hydroxide ions during dehydration.[41]

The reactions between the 2:1 clay minerals, between 2:1 and 2:1:1 clay minerals, and between kaolinite and gibbsite seem to be reversible. The reaction 2:1 to 1:1 clay, however, may not be readily reversible under surface conditions. The reverse reaction would involve addition of a silicon sheet between each kaolinite layer, and this might be difficult because the hydrogen bond between the kaolinite layers is a fairly strong bond, the layers are held closely together, and there are no interlayer cations. Perhaps the reverse reaction proceeds by solution of the original kaolinite followed by precipitation of the constituents as montmorillonite.

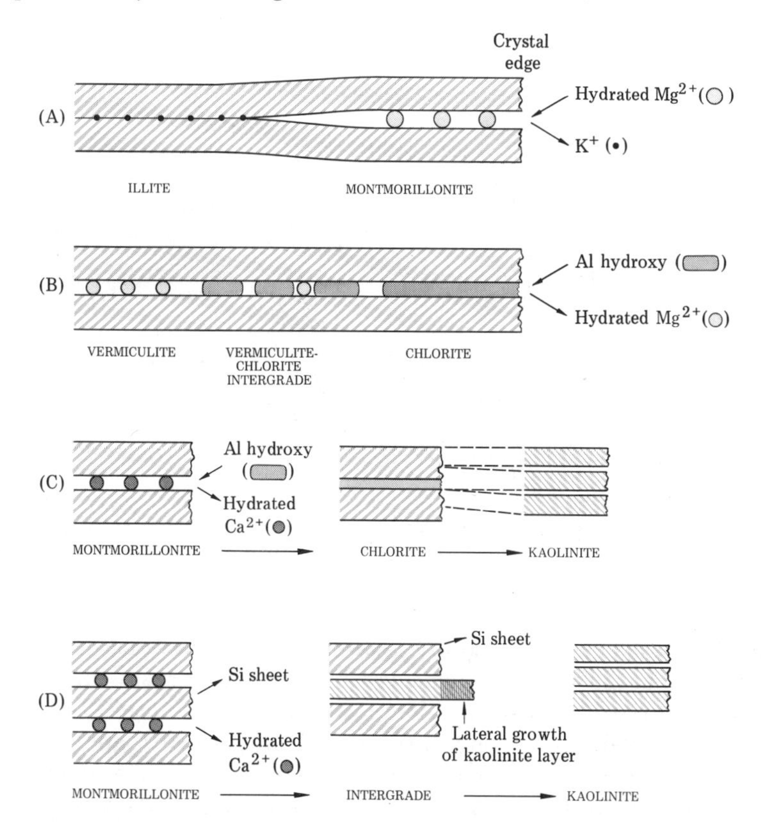

Fig. 4–5 Several clay-mineral-alteration schemes. (A) Alteration of illite to montmorillonite by ion exchange along the edges of the layers. (B) Alteration of vermiculite to chlorite by replacement of interlayer ions by aluminum hydroxy groups to form a gibbsite layer. (Modified from Jackson,[25] Fig. 2.12 *in* Chemistry of the Soil by F. E. Bear, ed., © 1964 by Litton Educational Publishing, Inc. Reprinted by permission of Van Nostrand Reinhold Company.) (C) Alteration of montmorillonite to chlorite to kaolinite. (Modified from Brindley and Gillery,[10] Fig. 1; see also Glenn and others.[19]) The silica tetrahedra in the middle kaolinite layer depicted must invert so that the interlinked bases of the tetrahedra all face upward. (D) Alteration of montmorillonite to kaolinite by stripping silicon sheets from montmorillonite. (Modified from Altschuler and others,[1] © 1963 by the American Association for the Advancement of Science.) Kaolinite growth can take place laterally from a newly formed kaolinite layer, and additional layers in turn can become oriented on these layers, Hydroxyls occupy the oxygen sites in the octahedral layers left vacant by silicon sheet removal.

Phase diagrams have been constructed to help quantify the relationship between the ionic concentration (or activity) of the solution and the mineral or minerals present.[15,17,18,23,28] Data on free energies of formation and on the chemistry of the species are used to construct

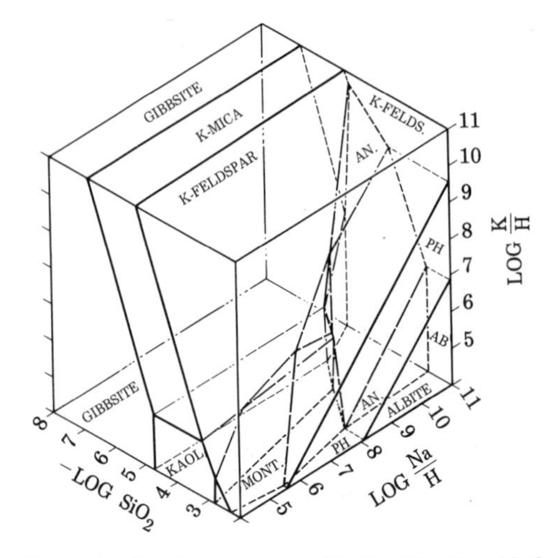

Fig. 4–6 Phase diagram for the system $K_2O—Na_2O—Al_2O_3—SiO_2—H_2O$ at 25°C and 1 atmos. KAOL, kaolinite; MONT, montmorillonite; PH, phillipsite; AN, analcite; and AB, albite. (Taken from Hess,[23] Fig. 1.)

such diagrams; Fig. 4–6 is one example. The diagrams can be considered to be only approximations because the crystallinity of the minerals varies, because some data on the activity-activity ratios are not available for low temperatures and must be extrapolated from higher temperature data, and because the precision of free energy data used to calculate the equilibrium constants varies. At any rate, the diagrams are useful in that they show the approximate stability fields for clay minerals precipitated from solutions of varying chemical composition. They also indicate the stability relationships of one or more minerals; minerals adjacent to one another in the diagram can be in equilibrium (e.g., kaolinite-montmorillonite), but those not adjacent to one another are not in equilibrium (e.g., gibbsite-montmorillonite). Furthermore, the use of such diagrams helps to predict possible successive alteration products with gradual variation in ion-activity ratios, and activity of H_4SiO_4. Kittrick[28] discusses some possible applications and relates them to field conditions. In any study involving equilibrium diagrams, it is important to compare the mineralogical data with the water data from the same horizon; river water collected several miles away should never be considered representative.

A usual extension of this mineral-equilibria work is to plot the water data on the same diagram to determine if the minerals are in equilibrium with the solution. This has been done for perennial and ephemeral springs in granitic terrain of the Sierra Nevada,[15] the Great Lakes,[40] and other systems.[18] Figure 4–7 shows the data of Feth and others.[15] If the water-chemistry data plot in a particular mineral field and the X-ray data confirm the presence of that particular mineral, the mineral probably is in equilibrium with the solution. This is the standard interpretation. However, equilibrium conditions may not exist for samples in which the water and X-ray mineralogical data do not coincide. In the Sierra Nevada data, for example, K-mica and montmorillonite are present in some samples, but the water data all plot in the kaolinite field. It is suggested that this means that the water composition may later shift to activity values at which all three minerals are in equilibrium, or perhaps the minerals present will in time convert to more stable phases.[15] Garrels and Mackenzie[18] plot data from many rock types and environments, and most fall within the kaolinite field. It is suggested that kaolinite is the stable end product in the chemical weathering of silicate minerals. This is not always borne out by field studies, however; perhaps more time is needed to

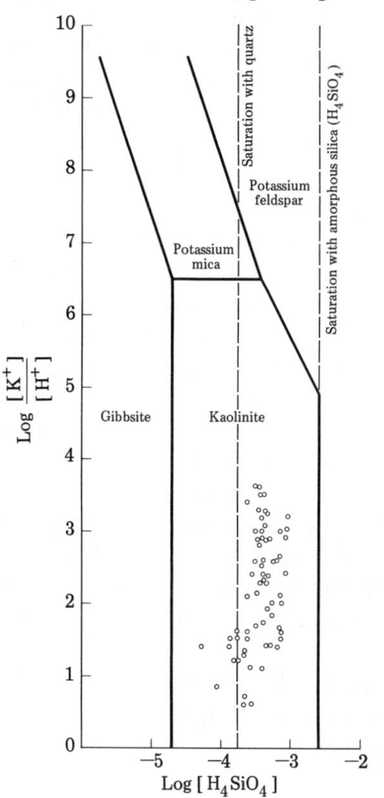

Fig. 4–7 Phase diagram for the system $K_2O—Al_2O_3—SiO_2—H_2O$ at 25°C and 1 atmos. Open circles are analytical data for water from springs or seeps in a granitic terrain in the Sierra Nevada. (Taken from Feth and others,[15] Fig. 13.)

reach equilibrium, or some of the basic data on free energies are not too accurate.

The applicability of these water-mineral equilibrium data has yet to be fully tested for soils. In any such application, it will be extremely important to know which, if any, minerals were inherited from the parent material and the direction in which any mineral alteration schemes might be proceeding. A consideration of the data available indicates that these applications are going to be difficult.

Within any one soil or soil horizon there is usually a variety of clay minerals.[5,6,7,24,25] Various hypotheses have been advanced to explain this diversity. One explanation is that the clay minerals in the assemblage are not in equilibrium with present conditions and are altering to other more stable clay minerals. Some of the clay minerals, for example, could have been inherited from the parent material. Another possible explanation is that some of the clay minerals could have formed in the past under different environmental conditions, and, reaction rates being so slow at surface conditions, these clay minerals are metastable in the present environment. If one considers the diverse microenvironmental conditions within a soil, perhaps diversity of clay minerals should be expected. For example, ionic concentration might vary from place to place due to the variation in charge of the particles. Furthermore, the primary minerals weather at slightly different rates, releasing a variety of substances into the soil. Thus, there will be times when the ionic species do not have a uniform distribution. Furthermore, there could be slight variations in pH that could be important to any ongoing reactions involving mineral synthesis. Such variation could be caused by proximity to a root, as they commonly have adsorbed H^+, or root CO_2 respiration and the local concentration of H_2CO_3. Data on abrasion pH of various primary minerals (Table 3–2), as well as their composition, also might lead one to suspect that similar local variations in the soil exist during mineral weathering and synthesis. Barshad[3] suggests that plant-nutrient cycling might control cation content and pH of the solution in some soils to such an extent that montmorillonite might form during periods of inactive plant growth and kaolinite might form during periods of active plant growth and nutrient uptake.

The above examples do not exhaust the list of possible chemical conditions at the microscopic level. They are mentioned to point out that a soil may have a variety of clay minerals due to a variety of local conditions. If this is the case, one may have to look carefully at the use of clay-mineral stability diagrams to depict whether a clay mineral or a clay-mineral assemblage is stable or not. The soil-solution chemical data needed to test for equilibrium on any diagram represent only the average conditions within the soil. On a microscopic level, the average may be difficult to find, controlled as it is by the rate of ion release by weathering, the chemical conditions at the site, and the rate of leaching.

Clay-mineral distribution with depth in soil profiles

Clay minerals vary in amount with depth in some soil profiles, but they remain relatively constant throughout in others. Such relationships seem to be closely associated with the leaching conditions within the soil. Three leaching conditions can be examined. With extensive leaching, many cations and silica may be carried to great depth. This can result in relatively uniform chemical conditions with depth in the soil and therefore a rather uniform clay-mineral distribution. The same uniformity involving different clay minerals would be expected in arid

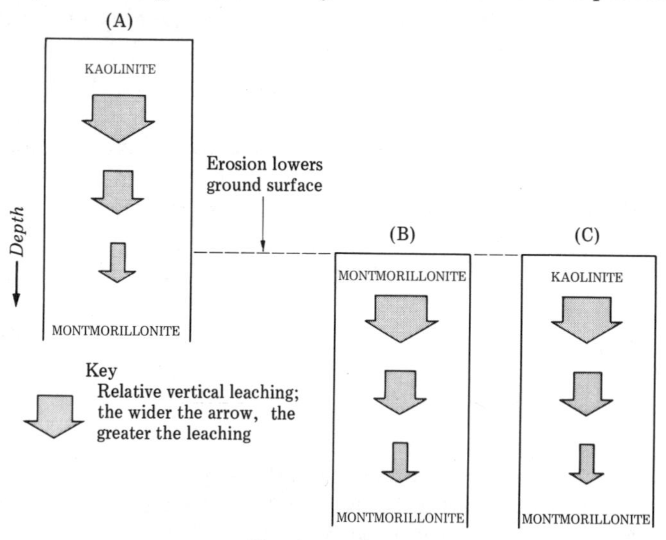

Fig. 4–8 Vertical distribution of clay minerals as a function of leaching conditions within the soil. (A) Leaching conditions favor the formation of kaolinite and montmorillonite; each forms at different levels in the soil profile. (B) Rapid lowering of the surface by erosion results in the montmorillonite residing in a soil environment of high leaching. (C) With time, the montmorillonite in the surface becomes desilicated and alters to kaolinite. Rates of ground-surface lowering, mineral alteration, and amount of leaching are important to stable end-product clay mineral formation. With slow ground-surface lowering, alteration may proceed at the same rate, and the clay-mineral distribution may always appear as in (A). If the rate of erosion far exceeds the rate at which the clay mineral can change, a distribution like (B) occurs temporarily.

regions, because what little leaching there is may not alter the chemical conditions enough to favor differences in clay mineralogy with depth. Between these two extremes, however, differences in leaching with depth might produce enough variation in chemistry to favor the formation of different clay minerals. Put another way, the ions and colloids released by weathering in the higher levels of the profile become reactants in mineral synthesis at depth.

The most common vertical variation is one in which the amount of silica and bases in the weathering product increases with depth. Thus, gibbsite at the surface grading downward to kaolinite or kaolinite at the surface grading downward to montmorillonite are common trends. Although it is possible that one or the other mineral may no longer be in equilibrium with the present environmental conditions, this need not be the case. Both minerals in the vertical sequence can be in equilibrium with the environmental conditions, and because the latter changes with depth, this change is reflected in the clay mineralogy. If conditions change, for example, due to erosional lowering of the surface with time, the montmorillonite at depth eventually may be in that part of the profile most conducive to kaolinite formation, and it may alter to kaolinite (Fig. 4–8). The variation of clay mineral with depth does not by itself constitute proof that one mineral is changing over to another. In this case, the lowering of the ground surface so changed the conditions within the soil that the original clays were no longer stable; they therefore changed to equilibrate with the environment.

REFERENCES

1. Altschuler, Z. S., Dwornik, E. J., and Kramer, H., 1963, Transformation of montmorillonite to kaolinite during weathering: Science, v. 141, p. 148–152.
2. Baldar, N. A., and Whittig, L. D., 1968, Occurrence and synthesis of soil zeolites: Soil Sci. Soc. Amer. Proc., v. 32, p. 235–238.
3. Barshad, I., 1964, Chemistry of soil development, p. 1–70 *in* F. E. Bear, ed., Chemistry of the soil: Reinhold Publ. Corp., New York, 515 p.
4. ——— 1965, Thermal analysis techniques for mineral identification and mineralogical composition, p. 699–742 *in* C. A. Black, ed., Methods of soil analysis (part I): Amer. Soc. Agron., Madison, Series in Agron., no. 9, 770 p.
5. ——— 1966, The effect of a variation in precipitation on the nature of clay mineral formation in soils from acid and basic igneous rocks: Proc. Internat. Clay Conf., Israel, 1966, v. 1, p. 167–173.
6. Birkeland, P. W., 1969, Quaternary paleoclimatic implications of soil clay mineral distribution in a Sierra Nevada–Great Basin transect: Jour. Geol., v. 77, p. 289–302.

7. —— and Janda, R. J., 1971, Clay mineralogy of soils developed from Quaternary deposits of the eastern Sierra Nevada: Geol. Soc. Amer. Bull., v. 82, p. 2495–2514.
8. Black, A. B., 1967, Applications: Electrokinetic characteristics of hydrous oxides of aluminum and iron, p. 247–300 *in* S. D. Faust and J. V. Hunter, eds. Principles and applications of water chemistry: John Wiley and Sons, New York, 643 p.
9. Black, C. A., 1957, Soil-plant relationships: John Wiley and Sons, New York, 332 p.
10. Brindley, G. W., and Gillery, F. H., 1954, A mixed-layer kaolin-chlorite structure: Proc., 2nd National Conf., Clays and clay minerals (Columbia, Mo., Oct. 15–17, 1953), Publ. 327 of the committee on clay minerals of the National Acad. Sci., National Res. Coun., Washington, D.C., p. 349–353.
11. Brown, G., ed., 1961, The x-ray identification and crystal structures of clay minerals: Mineralogical Society, London, 544 p.
12. Carroll, D., 1959, Ion exchange in clays and other minerals: Geol. Soc. Amer. Bull., v. 70, p. 749–780.
13. —— 1970, Rock weathering: Plenum Press, New York, 203 p.
14. Davis, S. N., 1964, Silica in streams and ground water: Amer. Jour. Sci., v. 262, p. 870–891.
15. Feth, J. H., Roberson, C. E., and Polzer, W. L., 1964, Sources of mineral constituents in water from granitic rocks, Sierra Nevada, California and Nevada: U.S. Geol. Surv. Water Supply Pap. 1535-I, 70 p.
16. Fieldes, M., and Swindale, L. D., 1954, Chemical weathering of silicates in soil formation: New Zealand Jour. Sci. and Tech., v. 36, p. 140–154.
17. Garrels, R. M., and Christ, C. L., 1965, Solutions, minerals, and equilibria: Harper & Row, New York, 450 p.
18. —— and Mackenzie, F. T., 1971, Evolution of sedimentary rocks: W. W. Norton and Co., Inc., New York, 397 p.
19. Glenn, R. C., Jackson, M. L., Hole, F. D., and Lee, G. B., 1960, Chemical weathering of layer silicate clays in loess-derived Tama silt loam of southwestern Wisconsin: Clays and Clay Minerals, v. 8, p. 63–83.
20. Grim, R. E., 1968, Clay mineralogy: McGraw-Hill, New York, 596 p.
21. Hay, R. L., 1963, Zeolitic weathering in Olduvai Gorge, Tanganyika: Geol. Soc. Amer. Bull., v. 74, p. 1281–1286.
22. —— 1964, Phillipsite of saline lakes and soils: Amer. Mineralogist, v. 49, p. 1366–1387.
23. Hess, P. C., 1966, Phase equilibria of some minerals in the K_2O—Na_2O—Al_2O_3—SiO_2—H_2O system at 25°C and 1 atmosphere: Amer. Jour. Sci., v. 264, p. 289–309.
24. Jackson, M. L., Tyler, S. A. Willis, A. L., Bourbeau, G. A., and Pennington, R. P., 1948, Weathering sequence of clay-size minerals in soils and sediments—I. Fundamental generalizations: Jour. Phys. Colloid. Chem., v. 52, p. 1237–1260.
25. —— 1964, Chemical composition of soils, p. 71–141 *in* F. E. Bear, ed., Chemistry of the soil: Reinhold Publ. Corp., New York, 515 p.

26. ____ 1965, Clay transformations in soil genesis during the Quaternary: Soil Sci., v. 99, p. 15–22.
27. Keller, W. D., 1964, Processes of origin and alteration of clay minerals, p. 3–76 *in* C. I. Rich, and G. W. Kunze, eds., Soil clay mineralogy: Univ. No. Carolina Press, Chapel Hill.
28. Kittrick, J. A., 1969, Soil minerals in the Al_2O_3—SiO_2—H_2O system and a theory of their formation: Clays and Clay Minerals, v. 17, p. 157–167.
29. Krauskopf, K. B., 1967, Introduction to geochemistry: McGraw-Hill, New York, 721 p.
30. Linares, J., and Huertas, F., 1970, Kaolinite: synthesis at room temperature: Science, v. 171, p. 896–897.
31. Mason, B. H., 1966, Principles of geochemistry: John Wiley and Sons, New York, 329 p.
32. Millot, G., 1970, Geology of clays: Springer-Verlag, New York, 429 p.
33. Pedro, G., Jamagne, M., and Begon, J. C., 1969, Mineral interactions and transformations in relation to pedogenesis during the Quaternary: Soil Sci., v. 107, p. 462–469.
34. Pierce, J. W., and Siegel, F. R., 1969, Quantification in clay mineral studies of sediments and sedimentary rocks: Jour. Sed. Petrol, v. 39, p. 187–193.
35. Reiche, P., 1950, A survey of weathering processes and products: Univ. New Mex. Publ. in Geol., no. 3, 95 p.
36. Rode, A. A., 1962, Soil Science: Israel Program for Scientific Translations, Jerusalem, 517 p.
37. Rich, C. I., and Thomas, G. W., 1960, The clay fraction of soils: Advances in Agron., v. 12, p. 1–39.
38. Sharma, G. D., 1970, Influence of CO_2 on silica in solution: Geochem. Jour., v. 3, p. 213–223.
39. Siffert, B., 1967, Some reactions of silica in solution: Formation of clay: Israel Program for Scientific Translations, Jerusalem, 100 p.
40. Sutherland, J. C., 1970, Silicate mineral stability and mineral equilibria in the Great Lakes: Environmental Science and Technology, v. 4, p. 826–833.
41. Tamura, T., and Jackson, M. L., 1953, Structural and energy relationships in the formation of iron and aluminum oxides, hydroxides, and silicates: Science, v. 117, p. 381–383.
42. Whittig, L., 1965, X-ray diffraction techniques for mineral identification and mineralogical composition, p. 671–698 *in* C. A. Black, ed., Methods of soil analysis (part I): Amer. Soc. Agron., Madison, Series in Agron., no. 9, 770 p.
43. Wiklander, L., 1964, Cation and anion exchange phenomena, p. 163–205 *in* F. E. Bear, ed., Chemistry of the soil: Reinhold Publ. Corp., New York, 515 p.

CHAPTER 5

Processes responsible for the development of soil profiles

Since many processes act together to form any one soil profile, it is difficult to discuss soil formation as a function of a specific process. The formation of a soil profile is viewed by Simonson[46] as the combined effect of additions to the ground surface, transformations within the soil, vertical transfers (up or down) within the soil, and removals from the soil (Fig. 5–1). For any one soil the relative importance of these processes varies, and the result is the variety of profiles seen in any landscape. The main additions to most soils are organic matter from the surface vegetation and their contained elements, ions and solid particles introduced with rainfall, and particles carried by the wind. Transformations include the multitude of organic compounds that form during organic-matter decomposition, the weathering of primary minerals, and the formation of secondary minerals and other products. Transfers generally involve the movement of ions and substances with the moving soil water. Soluble substances move with the percolating water unless changing chemical conditions or dehydration cause them to precipitate out of solution. In places where capillary rise of water is important, ions can be transferred upward and be precipitated high in the soil profile. In addition, biological activity is an important agent in the upward movement of materials. Ions can move upward through plants and be returned to the surface with litterfall. Soil-dwelling fauna can actively move solid particles in any direction. Finally, when water moves through the profile, it removes substances still in solution; these substances then become part of the dissolved constituents of the ground or surface waters. Because transformations have been dis-

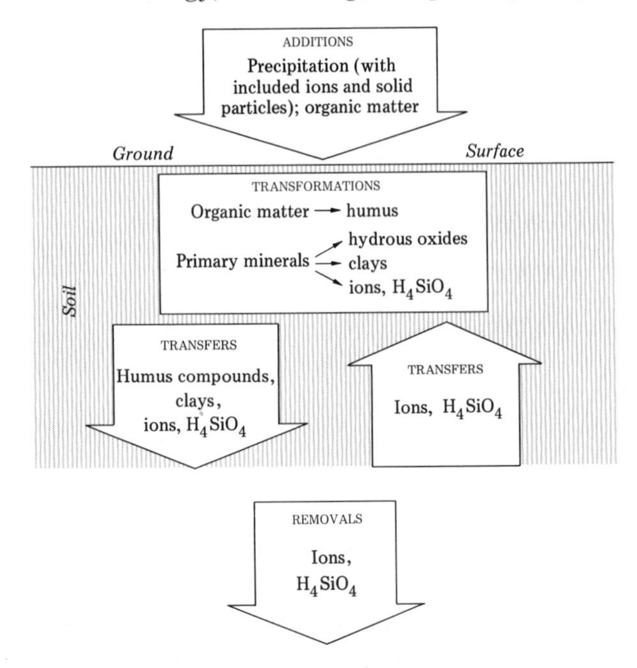

Fig. 5–1 A flow chart of major processes in soil-profile development (after Simonson[46]). Surface erosion might be added here, because erosion rates must be less than soil-formation rates for profiles to form.

cussed elsewhere in this book, the emphasis here will be on other processes that form the various horizons diagnostic of soil profiles.

Formation of the A horizon

The organic-matter content of a soil is a function of gains in organic matter to the soil surface and upper layers of the soil and losses that accompany decomposition. In the ideal case a newly formed surface has no organic matter. Once vegetation is established, an A horizon can begin to form. Early in the formation of a soil the gains exceed the losses, organic matter gradually accumulates, and the A or A and O horizons thicken. With time a steady-state condition is reached, and the gains equal the losses. Thereafter, even though the system is a dynamic one with continuing gains and losses, the amount of organic matter in the soil and its distribution with depth remain essentially constant. The length of time needed to reach the steady-state condition and the amount of organic matter in the soil at that time vary with the soil-forming factors. The maximum time to reach the steady state probably is about 3000 years, but in many environments it is reached much sooner.

Because of the dynamic nature of the A horizon, its age is difficult to define. However, in some stratigraphic studies, workers have tried to differentiate the A horizon from underlying horizons by age. Thus, one might see a description of a soil in which a modern A horizon overlies a Sangamon B horizon. Such differentiation is possible if the modern A horizon has formed from a younger parent material that was deposited on the Sangamon B. If this is not the case, such age differentiation is difficult. For example, the primary mineral fraction of both the A and B horizons are the same age in a soil formed from one parent material in a stable landscape. Clays and other weathering products are younger than the primary minerals, and the organic constituents are youngest of all. The organic constituents, however, vary from vegetative matter just deposited and in the initial stages of decomposition, to humus that may have resided in the soil several thousand years. In this sense, then, all A horizons at the land surface are modern or have a considerable modern component. If the interpretation is made that the A horizon is modern because the older A horizon was removed at some time in the past by erosion, evidence for the erosion that removed the A horizon may be preserved in the stratigraphic record as unconformities.

Translocation of iron and aluminum

Many soils show evidence of podzolization, a process that involves a pronounced downward translocation of iron, aluminum, and organic matter to form an eluvial E horizon overlying an illuvial spodic B horizon (Figs. 1–2 and 2–4). In many podzolized soils the processes behind such translocation are not readily apparent because the pH, although acid, can be in the range in which Fe^{3+} and Al^{3+} are essentially insoluble (Fig. 4–3). Furthermore, many of these soils are characterized by an oxidizing environment, which means that iron cannot move as the more soluble ionic species, Fe^{2+}. A fair amount of work has gone into the study of the processes by which iron and aluminum move in Spodosols or Spodosol-like soils.[32,39,50]

Present data indicate that iron and aluminum probably move in the soil as soluble metallo-organic chelating complexes (Fig. 3–7). Fulvic acids are thought to be the common chelating agents for a number of soils.[28,43,54] The stability of the chelate that forms is a function of size and valency of the metal ion, with increasing stability associated with smaller sizes and higher valencies. The process of translocation envisaged begins with the production of fulvic acids in the O or A horizon. A stable fulvic acid chelate is formed with Al^{3+} and Fe^{3+}, or with aluminum and iron hydroxy ions, and because these chelates are water-soluble they move downward with the percolating soil water.

Precipitation of these complexes can be brought about by a number of conditions at some depth in the soil profile. In some cases quite small changes in ionic content can bring about precipitation. For example, Wright and Schnitzer[54] report experimentally determined values of 0.13 ppm Ca^{2+} and 4.5 ppm Mg^{2+} to flocculate a complex of iron and organic matter and 10 and 45 ppm of the same cations, respectively, to flocculate a complex of aluminum and organic matter. Because of these relationships, it is possible that complexes of aluminum and organic matter could move deeper in the profile before flocculation than complexes of iron and organic matter. The position of the spodic B horizon, therefore, marks the position of flocculation of the chelating compounds and associated metal ions. Furthermore, this process explains the loss of appreciable iron, aluminum, and organic matter from the E horizons. Another way to precipitate aluminum and iron from the chelating complex at depth is through microbial action. Schuylenborgh,[44] for example, suggests that the organic matter portion of the chelating complex is decomposed by microorganisms at depth, leaving the metal ions free to precipitate as oxides or hydroxides.

Chelates are also formed with other cations, such as Ca^{2+} and Mg^{2+}. However, because calcium and magnesium also exist as free cations in soil solution chelate formation is less essential for their movement than it is for the movement of Fe^{3+} and Al^{3+}.

Conditions for chelate formation and movement of iron and aluminum are best for the Spodosols, but movement can also take place in other soils, as shown by some with E horizons (Table 2–1). In general, the degree of podzolization declines in a transect going from boreal forest to grassland,[28] that is, in a transect from Spodosols to Alfisols to Mollisols. In such a transect, the humic acids:fulvic acids ratio increases from about 0.5 to 1.5–2.0. Furthermore, the humic acids in the Spodosols are similar to the fulvic acids in that they are dispersed and mobile. Calcium-humic acid complexes that are both relatively stable and immobile form in Mollisols and thus are not capable of translocating iron and aluminum. As will be demonstrated later, evidence for iron, aluminum, and organic matter translocation can be used to reconstruct past vegetation and possible climatic effects on both relict and buried soils.

Translocation of clay-size particles

The distribution of clay-size particles in many moderately to strongly developed soils is marked by relatively low contents in the A and C horizons with the maximum amount in the B horizon, generally in the upper part of the B. Several processes may account for this distribution. One such process is that the constituents of clay are derived by weathering higher in the profile and that they move downward in solution with the percolating water and precipitate as clay minerals in the B horizon. A second process is that the clays have formed in place from mineral weathering in the B horizon. A third is that the clays have moved as particles in suspension in the downward-percolating water to accumulate in the B horizon because of flocculation, constrictions in the pores through which the water moves, or because the base of the B horizon marks the lower limit of most water movement.[32] No doubt, in most soils, clays form in the B horizon by all three processes, but the relative importance of each may vary from soil to soil. It should be possible to differentiate clay particles translocated from those formed in place, but I know of no criteria that can be used to identify clay precipitated from downward moving solutions.

Clay films lining ped surfaces or voids in both field and thin-section studies generally are taken as evidence for clay particle translocation (Fig. 1–4). This translocation can be verified by comparing detailed chemical analyses of the horizon with those of the film; the two usually differ because the film has been emplaced by downward movement more recently, and therefore the film has properties that more closely resemble those of the A horizon than of the B.[14] Thin-section analysis indicates that films of oriented clay along voids and ped surfaces that have sharp boundaries with the soil mass probably result from translocation.[8] Care must be taken, however, to eliminate the possibility of stress as the cause of oriented clay particles. Brewer[8] lists several criteria useful in identifying the origin of clay particle orientation. Thin-section evidence for origin of clay formation in place might be suggested by the lack of clay films or by textural relations suggesting that the clays are pseudomorphous after the parent mineral grains. Subsequent mixing might destroy the evidence, however.

Many kinds of data have been used to assess the relationship between the amount of clay formed in place from that introduced by translocation.[36] One common analysis involves the calculation of the volume of clay films in the soil, by horizon, to determine if the horizon of maximum clay content coincides with the horizon exhibiting the greatest amount of clay films. In many cases the match is not good because the films are commonly best expressed below the zone of maximum clay content.[8,15,34] In other cases, however, the match between the argillic B horizon and maximum expression of clay films is good.[24] Furthermore, in some soils the amount of illuviated clay, as evidenced by per cent of clay films in thin section, fall short of the total clay in the B horizon (Fig. 5–2), thus suggesting that the B-horizon clay is not derived entirely by illuvial processes.

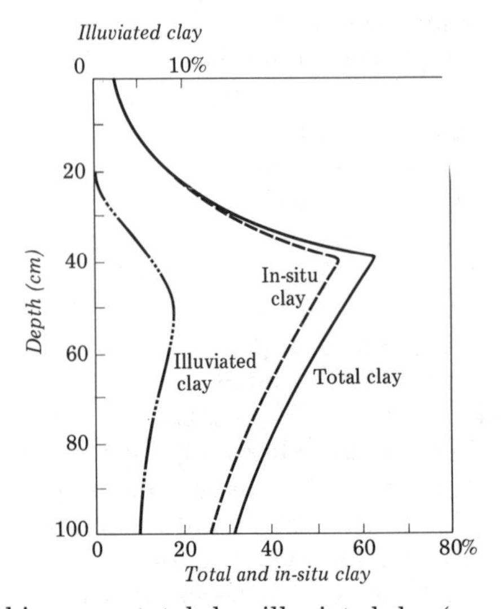

Fig. 5–2 Relationship among total clay, illuviated clay (recognized by oriented clay films), and, by difference, clay formed in place. (Taken from Brewer,[9] Fig. 3.) Because only some clay distribution is accounted for by illuviation, the clay in place represents that formed by weathering and/or that originally in the parent material.

One problem with using clay films as the only basis for demonstrating clay illuviation is that they are sometimes destroyed as soon as they form, or later. Nettleton and others[35] report clay-film destruction due to shrinkage and swelling in soils formed in the desert and mediterranean climates of the southwestern United States. They found, for example, that soils with a low shrink-swell potential and a lower than 40 per cent clay content can retain clay films and that soils with a higher shrink-swell potential and a higher than 40 per cent clay content cannot retain films. Thus, many soils may have translocated clay, but the thin-section evidence for it has been destroyed as the films become incorporated into the B horizon matrix. Gile and Grossman[24] note that clay films in the desert region are most stable on sand and pebble surfaces, but they can be destroyed during the accumulation of $CaCO_3$ in the soil. Clay-film destruction may also occur through mixing of the soil by roots and fauna.

Quantitative mineral-analysis techniques have been developed by various workers to determine weight gains and losses of both clay and nonclay (or primary) minerals within the profile; these techniques can be used to assess clay formation in place from a clay distribution due to translocation,[3,8] but they can only be used for soils in which the parent material is uniform throughout the soil profile (Ch. 7). Unfortunately, it turns out that few soils have a uniform parent material, and this limits the usefulness of these techniques. All changes within the soil are based on comparisons with the unweathered parent material, which might be either rock or unconsolidated deposits at depth. A resistant immobile primary mineral, such as zircon or quartz in the sand fraction, is used as an index mineral. If one assumes that the amount of weathering of the index mineral is negligible, the ratios of all other primary minerals to the index mineral can be calculated. These ratios will decrease from the parent material upward into the soil as the less resistant primary minerals weather, and they should be lowest where the weathering has been most intense. There should be some relationship between the loss in primary minerals and the formation of clay-size particles. Once the relationship is established, one can estimate the amount of clay formed per horizon and the amount of clay now present per horizon; the difference is the amount of clay that has been gained or lost by translocation (Fig. 5–3).

Brewer[8] has raised some objections to the above approach and

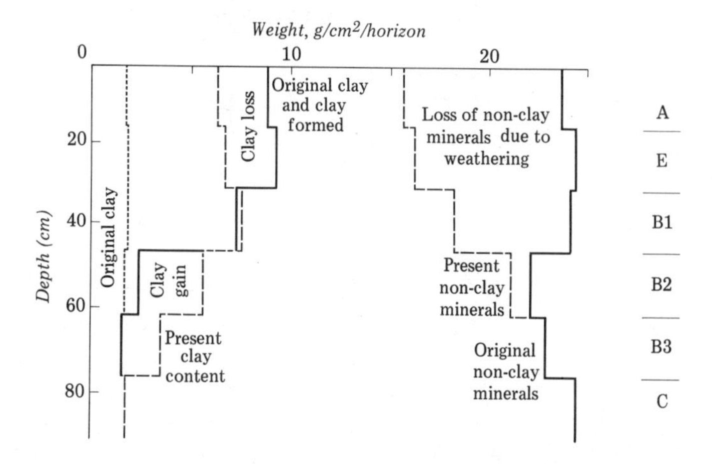

Fig. 5–3 Comparison between nonclay (or primary) mineral loss due to weathering and clay formation from constituents released during weathering, by the index-mineral method. (Data from Barshad,[3] Table 1.13 *in* Chemistry of the Soil by F. E. Bear, ed., © 1964 by Litton Educational Publishing, Inc. Reprinted by permission of Van Nostrand Reinhold Company.) Nonclay mineral losses and clay-mineral formation are estimated to be restricted to the upper 61 cm; this thickness would depend upon parent-material choice. Differences between the distribution of the original clay plus the clay formed and the present distribution give approximate values for overall clay migration. Although Brewer[8] has some doubts about the uniformity of the parent material for this profile and the mineral content in the parent material, these are the kind of results one can obtain from analysis of the clay and nonclay fractions on a volume basis.

called for the integration of both mineralogical and thin-section data to determine the formation and migration of clay-size material. Such a study was made on an Ultisol in Australia.[7] Knowing the approximate amounts of the rock-forming minerals that have weathered, as well as the clay mineralogy, one can calculate the maximum amount of clay that could form due to weathering in place (Fig. 5–4); this amount of clay will vary with the chemical composition of the clay fraction. Calculations were made for both kaolinite and for "complex" clay minerals (a variety of 2:1 minerals considered end products of weathering and found to be present on X-ray analysis). A comparison of the weight of kaolinite or of the "complex" clay mineral that could form from weathering in place with the actual clay present showed that much of the material released by weathering was removed from the soil and that only a small portion remained to form the clays. It is, therefore, not necessary to hypothesize clay illuviation to explain present clay distribution. Thin-section study revealed a lack of oriented clay and, thus, no evidence of clay illuviation. In this study, therefore, the weight of the evidence fails to support much clay illuviation as an explanation

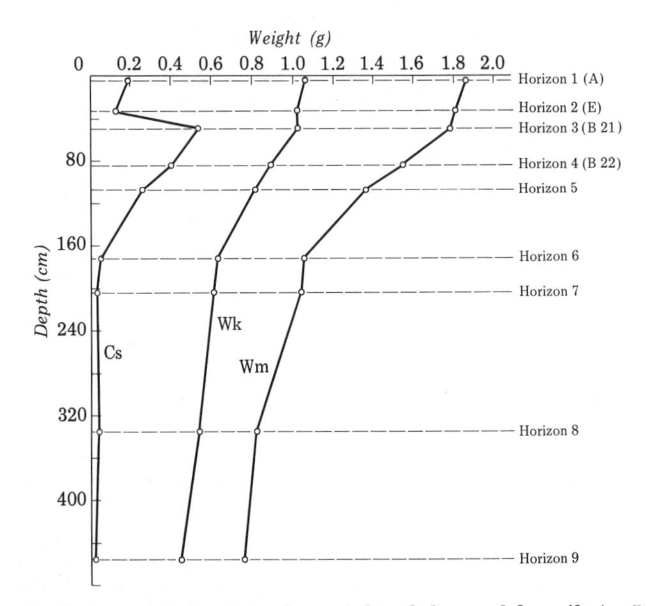

Fig. 5–4 Variation with depth in the weight of clay and fine silt (< 5 μ) present in the soil, in the maximum weights of <5μ material that could form from the amounts of mineral weathering and the specific clay minerals that have formed. (Taken from Brewer,[7] Fig. 4.) Here *Cs* is the weight of < 5 μ particles present in the soil derived from 1 cm^3 of parent rock, and *Wk* and *Wm* are the weights of kaolinite and "complex" clay minerals, respectively, that could have formed from the observed weathering of 1 cm^3 of parent rock.

of the present clay distribution with depth. There is the possibility, however, that the evidence for clay movement has been destroyed, because the B horizons contain 30 to 43 per cent clay.

The examples given here demonstrate how difficult it is to find evidence that translocation of clay-size particles in soil profiles has occurred. Oriented clays in thin section generally are accepted as evidence for such movement. However, several studies show that subsequent pedogenesis can destroy this evidence. Clay and nonclay mineral relationships also can be used, but it is extremely difficult to calculate the gains and losses accurately. I would suspect that, as shown by the study of Brewer,[7] the losses of material upon weathering in most soils will far outweigh the gains due to clay formation. To prove clay translocation by these methods, one would have to demonstrate that the clay content in a particular horizon exceeds the maximum amount that could form only by weathering in place.

Clay migration requires that the clay be dispersed so that it can remain in suspension and be transported by water moving slowly through pores or cracks in the soil. Dispersion is favored by several factors, among them a low electrolyte content in the soil solution and the absence of positively charged colloids, such as iron and aluminum hydroxides.[3] Flocculation is induced by high electrolyte content in the soil solution and the presence of positively charged colloids; clays under these conditions cannot migrate.

The dispersion or flocculation of clay-size particles also depends on the thickness of the ion layer that satisfies the negative charge of the particle, and this thickness can vary with the ion present.[37] Ions attracted to particle surfaces are distributed so that their concentration is highest close to the surface, and concentration diminishes away from the surface (Fig. 5–5, A). When two particles, each with positive ions attracted to their surfaces, move toward one another, the initial reaction is one of repulsion (Fig. 5–5, B). Thus the clays are dispersed. If, however, the clay particles can move closer together, van der Waals force takes over, and attraction and flocculation occur. The particles also can come close together if the thickness of the ionic layer is reduced (Fig. 5–5, C). This reduction can be brought about in two ways. One is to increase the electrolyte concentration of the soil solution, and the other is to replace the ion layer with ions of higher valency.

These flocculation-dispersion effects can be seen in soils. Calcium clays have a thin ion layer, they commonly are flocculated and, thus, they do not migrate. Soils high in $CaCO_3$, for example, show little evidence of clay migration; migration can take place only after the carbonate has been leached from the soil and some of the Ca^{2+} in the

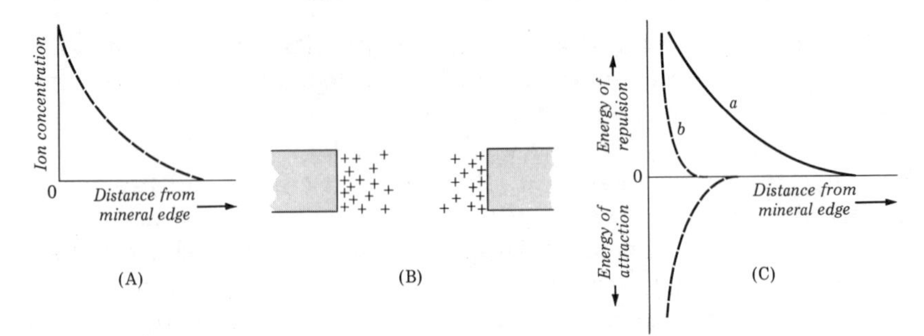

Fig. 5–5 Conditions for clay dispersion and flocculation. (Modified from Olphen,[37] Figs. 9 and 12 © 1963, John Wiley and Sons.) (A) Concentration of ions adjacent to a charged surface. Concentration is highest near the surface and decreases away from the surface. The total positive ion charge satisfies the negative clay particle charge. (B) Distribution of positive ions attracted to clay surfaces. If the two clays are brought close together, and the ion layers are thick enough, flocculation cannot take place, and the system remains dispersed. (C) Energy relations for repulsive and attractive forces between two clay particles with positive surface ions. In plot a the ion layer is thick; flocculation does not occur because the surfaces are too far apart for short-range van der Waals forces of attraction to operate. In plot b the ion layer is thin; because of either higher electrolyte content or higher valency cations, van der Waals forces operate once the clays come in close contact, and flocculation occurs.

exchange complex replaced. In contrast, clays with appreciable amounts of Na^+ in the exchange complex remain dispersed because the ionic layer is thick. Natric B horizons, for example, commonly show evidence of clay translocation. At high amounts of Na^+, however, flocculation results because the high electrolyte content in the soil solution compresses the adsorbed ion layer.

Some experimental work has been performed to determine the factors responsible for clay migration.[10,26] Besides verifying to some extent the cation relationships discussed above, the work indicates an upper limit of 20 to 40 per cent clay, depending on the clay mineral, above which migration in the pore spaces virtually ceases. At these clay contents, the pore spaces may be small enough to limit movement, or the swelling of clays upon hydration may close off some pores. Clay migration at greater clay contents would have to take place along soil-structure discontinuities, and, as mentioned earlier, soils with such high clay contents tend to lose the clay-film evidence for migration quite rapidly.

Conditions for clay migration can vary with depth in the profile or even seasonally. Quite commonly the electrolyte content will increase with depth because the upper parts of the profile have more water moving through them. Thus, clays could be dispersed near the surface but be flocculated at depth. In soils with high clay content, high shrink-swell potential, and distinct dry and wet periods, clay movement might take place only at the onset of the wet season, when an open prismatic soil structure with wide cracks reaches the surface and thus provides avenues for rapid clay translocation. In time, however, the clays could swell, the cracks seal closed, and further clay migration would cease.

Origin of oxic B horizons

The origin of oxic B horizons has been reviewed by several workers recently.[16,30,33,47,48,49] The oxic B horizon is characterized by the extreme chemical alteration of the original parent material, and in places this alteration is so complete that the parent material cannot be identified. Chemical analyses commonly show the near total depletion of cations and silica that originally had been combined in the primary aluminosilicate minerals (Fig. 5–6). The silica that remains in the soil generally is seen as the original quartz grains or is combined in secondary clay minerals. Oxides and hydrous oxides of iron and aluminum, alone or in combination, are common; in many places they constitute most of the material present. Clay minerals are usually of the 1:1 lattice variety. Weathering of primary minerals often has been so

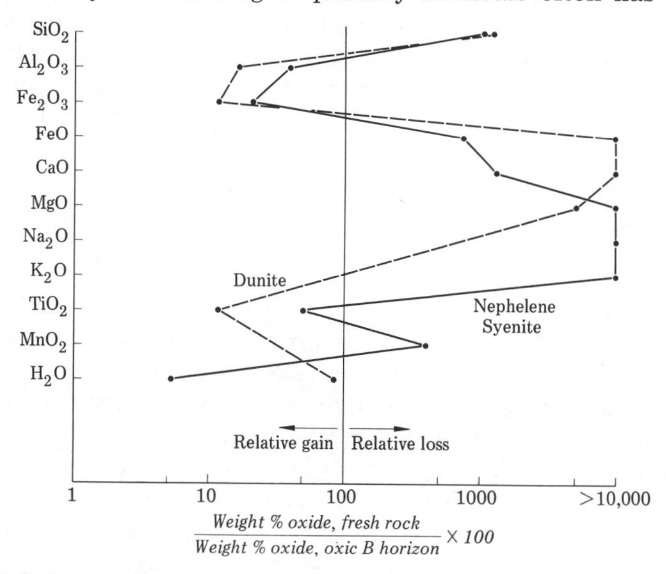

Fig. 5–6 Gain-loss diagram for oxic B horizons formed from two different parent materials. Water gain is due to its incoporation into the crystal lattices of the sesquioxides and clay minerals that form; FeO loss could be real or due in part to oxidation to Fe_2O_3. (Data from Maignien,[30] Tables 3 and 4, *Review of Research on Laterites, Natural Resources Research IV,* 1966, reproduced by permission of UNESCO.)

intense that there is very little cation reserve from further weathering of the primary minerals. What cations there are occur at exchangeable sites of the colloidal fraction or in the plant tissue. Plant nutrition, therefore, is dependent upon the cycling of nutrients between the upper part of the soil and the plant itself. If this cycling is disturbed, the cations can be leached and the soil rendered essentially devoid of nutrients.

Most of the processes that produce oxic B horizons operate in other soil horizons. Most cations and silica are soluble under the usual soil conditions and so can be leached from the soil. In many areas where these horizons occur, the leaching conditions are so intensive and the landscapes so old, dating perhaps to the Tertiary, that near total removal of cations and silica is not unexpected. However, under usual biochemical conditions the sesquioxides released from the parent rock are relatively insoluble and can accumulate. In some soils, however, the sesquioxide content, especially that of iron, is greater than would be expected from the weathering of the parent material. Under these circumstances, iron may be derived from overlying horizons, perhaps now partly eroded away, or be transported in with laterally moving ground waters from higher parts of the landscape (Fig. 9–9). Combination with organic compounds, perhaps as chelates, could solubilize iron, as could low acidity and reducing conditions. A high or fluctuating water table seems to be important mainly in producing conditions for more intensive alteration and the production of colorful mottling. The position of the top of the water table does not seem to be important in the accumulation of iron, however. Where clay minerals are not too plentiful, leaching conditions may have been too intense to provide the appropriate chemical environment for their formation. Once clays do form they do not seem to migrate very far, even though most oxic B horizons are relatively porous and can translocate large quantities of rainfall. Perhaps the mutual attraction of the negatively charged clay and the positively charged sesquioxides to give stable aggregates prevents much clay translocation, since this seems to be the reason why dispersion of laboratory samples is so difficult.

Some oxic B horizons are hard or are able to harden upon exposure to drying conditions.[1] The term "laterite" is restricted to these materials, and this hardened material is included in the plinthite of the new U.S. soil classification. Sufficient iron and dehydration seem to be necessary for hardening to occur, concomitant with increased crystallinity and continuity of the crystalline phase of already existing iron compounds. This results in the formation of goethite and hematite in a rigid network of crystals that cement the material together. Dehydration necessary for the hardening may come about by a natural change in vegetation from forest to savanna, by the clearing of forests for agricultural and other purposes, or by exposure by erosion.

Origin of $CaCO_3$-rich horizons

Soils in semiarid and arid regions commonly have Cca and K horizons at some depth below the surface, or, if the climate is dry enough or the surface erosion intensive enough, these horizons may extend to the surface. Although several origins have been presented for some of these horizons,[6,12,40] our concern here is with $CaCO_3$-rich horizons of pedogenic origin. Some of these horizons are the caliche of present and past geologic literature. Pedologists call these accumulations Cca and K horizons, and because they have recognized and defined stages in the buildup of $CaCO_3$-bearing horizons[22,23] (see Appendix I), this more precise terminology seems preferable to the general term caliche. Although both calcium and magnesium carbonate occur in soils, I will discuss only calcium carbonate here to demonstrate the processes involved.

The origin of carbonate horizons involves carbonate-bicarbonate equilibria, as discussed by Jenny[27] and Krauskopf,[29] and shown by the following

$$\underset{g}{CO_2} + \underset{l}{H_2O}$$

$$\downarrow\uparrow$$

$$\underset{c}{CaCO_3} + \underset{aq}{H_2CO_3} \rightleftharpoons \underset{aq}{Ca^{2+}} + 2\,\underset{aq}{HCO_3^-}$$

An increase in CO_2 content in the soil air or a decrease in pH will drive the reaction to the right; carbonate will dissolve and move as Ca^{2+} and HCO_3^- with the soil water. Dissolution is also favored by increasing the amount of water moving through the soil, as long as the water is not already saturated with respect to $CaCO_3$. Precipitation of carbonate occurs under conditions that drive the reaction to the left, that is, a lowering of CO_2 pressure, a rise in pH, or an increase in ion concentration to the point where saturation is reached and precipitation takes place.

All of the above conditions are found in soils in which $CaCO_3$ has accumulated. Carbon dioxide partial pressures in soil air are 10 to more than 100 times that in the atmosphere[4,13]; this decreases the pH which, in turn, increases $CaCO_3$ solubility (Fig. 5–7, A). The partial pressure of CO_2 is high as a result of CO_2 produced by root and microorganism respiration and organic matter decomposition. Thus, one would expect the highest CO_2 partial pressure to be associated with the

A horizon, with values diminishing down to the base of the zone of roots. The amount of water leaching through the soil also is greater near the surface than at depth, so, as the water moves vertically through the soil, the Ca^{2+} and HCO_3^- content might increase to the point of saturation after which further dissolution of $CaCO_3$ is not possible. Combining the effects of high CO_2 partial pressure and downward percolating water, we might visualize the formation of a $CaCO_3$-rich horizon as follows. In the upper parts of the soil, Ca^{2+} may already be present or may be derived by weathering of calcium-bearing minerals. Due to plant growth and biological activity, CO_2 partial pressure is high and forms HCO_3^- upon contact with water. Water leaching through the profile can carry the Ca^{2+} and HCO_3^- downward in the profile. Precipitation as a $CaCO_3$-rich horizon would take place by a combination of decreasing CO_2 partial pressure below the zone of rooting and major biological activity, and the progressive increase in concentration with depth in Ca^{2+} and HCO_3^- in the soil solution as (a) the water percolates downward, and (b) water is lost by evapotranspiration. The position of the $CaCO_3$-bearing horizon is therefore related to depth of leaching, which, in turn, is related to the climate.

Temperature also affects $CaCO_3$ equilibria. Because CO_2 is less soluble in warm water than in cold water, $CaCO_3$ solubility decreases with rising temperature (Fig. 5–7, B). This temperature effect may not be too great in one profile, but it is important in comparing the depths to the tops of Cca or K horizons between regions of contrasting temperature.

Several stages in the buildup of carbonate horizons are recognized.[23] In gravelly material the first stage is the appearance of carbonate coatings on the under sides of gravel particles; in nongravelly material the first stage is the occurrence of thin filaments (Fig. A–1, Appendix 1). The under sides of gravel probably are favored sites initially because downward-moving water would tend to collect there. With time, in both parent materials, the horizon is increasingly impregnated by carbonate deposition on solids until the voids become plugged and water percolation through the horizon is greatly restricted. At this point water tends to collect periodically over the plugged horizon; the resulting solution and re-precipitation produces the laminated part of the upper K horizon. In these later stages of K-horizon development the pores become progressively plugged, the vertical movement of water is restricted, and the horizon builds upward and so is younger in that direction. During the buildup of a K horizon, the carbonate, upon crystallization, forces the silicate grains and gravel apart. At the maximum development of the K horizon, $CaCO_3$ content may approach 90 per cent. The buildup of $CaCO_3$ in the diagnostic

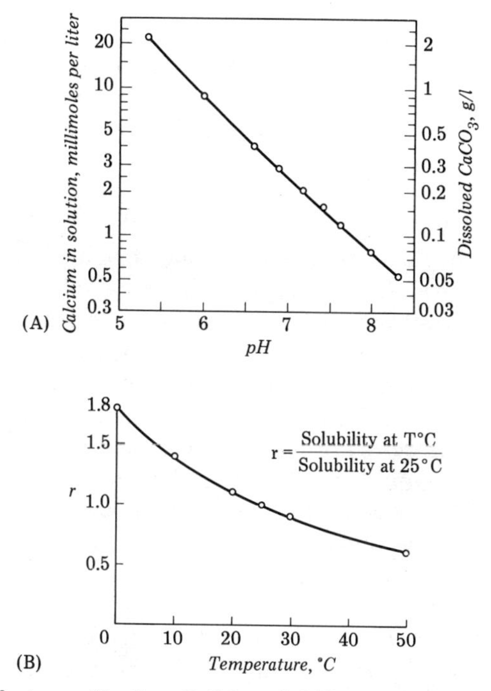

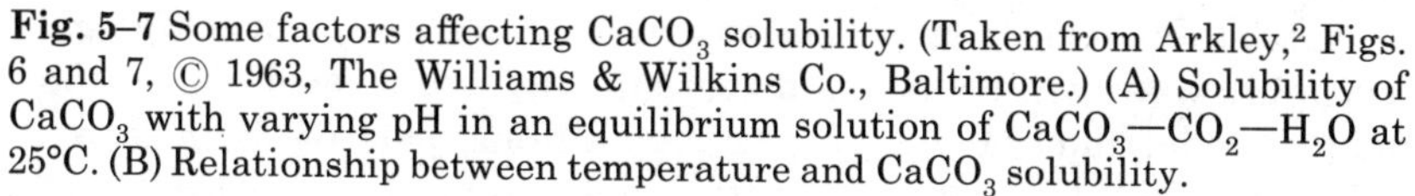

Fig. 5–7 Some factors affecting $CaCO_3$ solubility. (Taken from Arkley,[2] Figs. 6 and 7, © 1963, The Williams & Wilkins Co., Baltimore.) (A) Solubility of $CaCO_3$ with varying pH in an equilibrium solution of $CaCO_3$—CO_2—H_2O at 25°C. (B) Relationship between temperature and $CaCO_3$ solubility.

stages (Appendix I) is more rapid in gravel than in nongravelly material because gravel has less pore space.[20]

The $CaCO_3$ of the carbonate horizons may come from several sources: Ca^{2+} released by weathering could combine with HCO_3^- deeper in the profile, and, if this is the origin, there should be a close relationship between the CaO content of the parent material, the amount of weathering in the upper part of the soil to release Ca^{2+}, and the amount of $CaCO_3$ in the Cca or K horizon. In many places an external source of Ca^{2+} seems likely because the data indicate far too much $CaCO_3$ for the amount of weathering of the assumed parent material.[21] Detailed study in the Las Cruces region of New Mexico indicates that airfall is an important source for Ca^{2+} and $CaCO_3$,[23,25,41] and indeed this probably is true for the $CaCO_3$ in many carbonate horizons throughout the semiarid Basin and Range Province.[21] Accumulation of $CaCO_3$ by capillary rise from a perched high water table may explain some $CaCO_3$ occurrences,[31] but such occurrences are not thought to be widespread. This is because capillary rise in coarse alluvium is nil, and impermeable beds necessary to perch the water table are uncommon. Also, in many areas streams have downcut following deposition of alluvium, which has lowered the water table far below the level that could possibly be reached by the capillary rise of water with its dissolved carbonate salts.

Horizons associated with high water content or nearness to water table

Several other horizons will be mentioned briefly, because they constitute important pedological features in some soils. These horizons usually are associated with high water contents within the soil due to high regional or local water tables or to horizons within the soil that impede the downward movement of water. Quite commonly, soils with these horizons occur in the lower parts of the landscape, juxtaposed with soils exhibiting more normal drainage conditions. In some places, however, they can be quite extensive. In the old U.S. soil classification system, these horizons would be diagnostic of some intrazonal soils.

Poor water drainage and the accompanying low oxygen content in a soil leads to reducing conditions. The result is that iron and manganese are in the reduced state, and the compounds formed give the characteristic gray and bluish colors of gleyed horizons. If conditions are part oxidizing and part reducing, perhaps due to fluctuating water content or water table level, some of the iron will be oxidized, and compounds with yellow-brown, brown, and red color will be formed. Quite commonly, under fluctuating moisture conditions, part of the matrix is reduced and part is oxidized, and the characteristic colors are mixed; these intermixed colors are described as mottled (Fig. 5–8). Gleying conditions can occur throughout a soil if the water table is high or be restricted to only a part of the profile if downward movement of water is impeded.

Soils characterized by poor internal drainage, by the influence of a high water table, or by slight leaching can contain appreciable concentrations of salts. Common cations are Ca^{2+}, Na^{+}, Mg^{2+}, and K^{+}, and common anions are Cl^{-}, SO_4^{2-}, S^{2-}, and HCO_3^{-}. If the water table is close to the surface, salt accumulations can result from precipitation from water that has risen by capillarity and then evaporated. If the water table is low, however, salt accumulations can result from airfall or from weathering of the parent material.[55] Accumulation under conditions of a low water table can only come about if downward-moving water is impeded or if the climate is dry enough and leaching so slight that salts as soluble as these cannot be removed. Concentrations then will build up to levels required for precipitation. Although common to warm desert regions, salt accumulations are not restricted to them. Appreciable salt concentrations in soils have been reported

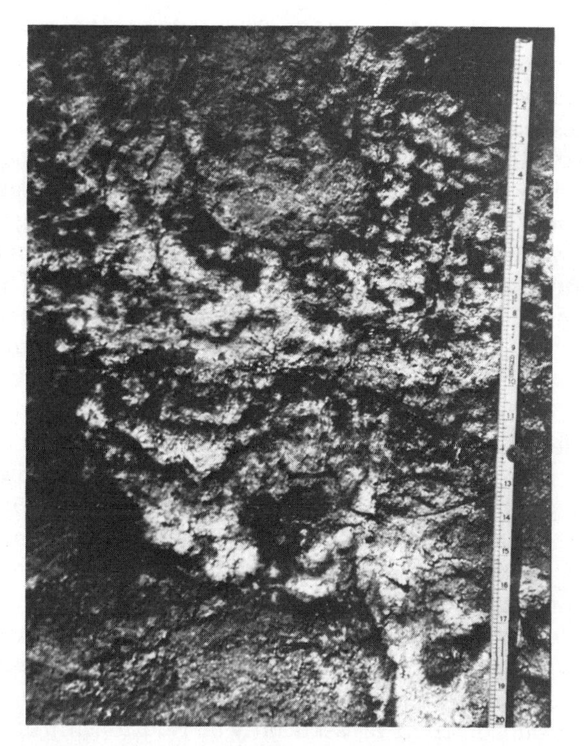

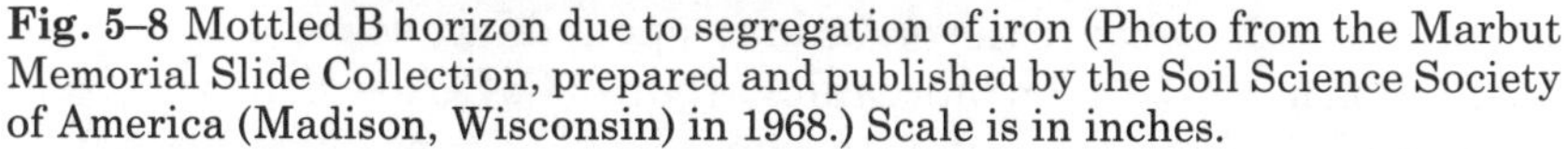

Fig. 5–8 Mottled B horizon due to segregation of iron (Photo from the Marbut Memorial Slide Collection, prepared and published by the Soil Science Society of America (Madison, Wisconsin) in 1968.) Scale is in inches.

from the far north[51] and Antarctica.[19,52] If ions for several salts are present, and their solubilities differ, the more soluble salts will be found at greater depth. For example, in soils with both calcium carbonate and sulpate (gypsum), the latter, being more soluble, will occur at a greater depth if the water from which it was precipitated was moving downward. If the gypsum accumulation layer lies above the $CaCO_3$ accumulation layer, then the salts might have formed by capillary rise of water from a high water table, because the more soluble salts would precipitate out closer to the surface.

Radiocarbon dating of soil horizons

Because soils contain both organic and inorganic carbon, radiocarbon dates can be obtained in an attempt to date the soil or some of its features (see papers in Yaalon[53]). Such dates should not be accepted at face value, however, because the systems are very complex, and both new and old carbon can be introduced or exchanged in the soil. Graphs have been prepared to estimate the error due to contamination.[17,42]

A radiocarbon date of organic carbon from the A horizon of a surface soil includes a mixture of organic matter varying from that fraction being added daily to that synthesized and resynthesized over several thousand years. The dates that reflect this dynamic system are mean residence times (MRT). The MRT for western Canada range from about 200 to 2000 years,[38] and values of 200 to 400 years are reported for the central United States.[11] Dates for various fractions of the organic matter yield different values, a reflection of the relative stability of the fractions.[38,45] The MRT have little meaning with respect to the age of a relict soil, except that they do give minimum ages. They are important, however, in dating buried A horizons, to obtain a limiting date on the overlying deposits, because the MRT prior to burial will be a built-in error. The error will vary with the soil and is difficult to estimate. In any sampling of buried soils, the uppermost part of the A horizon should be sampled because the MRT usually decrease upward (Fig. 5–9). Ruhe[42] discusses this as well as possible chemical techniques to identify contamination.

B-horizon radiocarbon dates on organic carbon also are contaminated with younger carbon translocated from above. In many soils the B-horizon dates are greater than the A-horizon dates (Fig. 5–9). However, with Spodosols there is essentially no relationship of radiocarbon age with depth, because the organic compounds are so mobile.[45]

Radiocarbon dates on inorganic carbon of $CaCO_3$-enriched horizons have been obtained in an effort to date soils.[5,53] The problem here is that

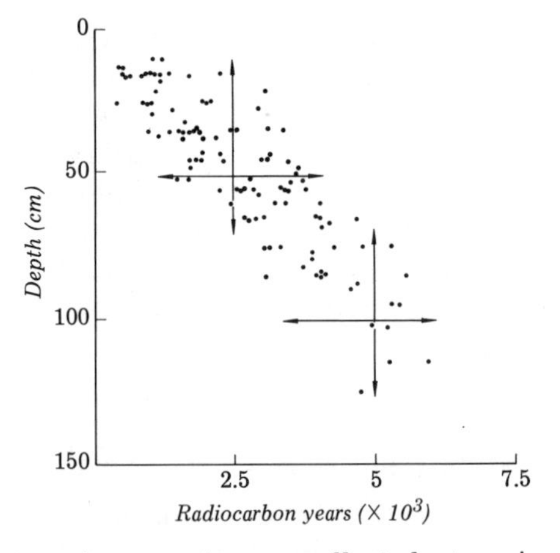

Fig. 5–9 Radiocarbon dates on humus collected at various depths from 24 Mollisols. (Taken from Scharpenseel,[45] Fig. 2; Published by Israel Universities Press, Jerusalem.)

$CaCO_3$ is readily soluble, and solution and re-precipitation can take place; every time this happens new carbon is added to the system because of the CO_2 in the soil air. Moreover, the carbon in airfall carbonate may be of any age. Gile and others[25] have studied these problems in considerable detail; dates on one of the soils they have studied are given in Fig. 5–10. Although the $CaCO_3$ is mainly of airfall origin and enters the soil by solution and re-precipitation, paired dates on initial pedogenic $CaCO_3$ in young soils and on charcoal in the deposits from which the soils have formed indicate an initial age at time of deposition of less than 3000 years. The dates on the Btca horizon are consistent, because carbonate first coats pebble bottoms and later begins to accumulate in the fine-grained matrix. The soft upper laminae at the top of the K horizon probably undergoes occasional solution and re-precipitation as water is held at that impermeable interface, and thus it is younger than the B horizon pebble coatings. Hard laminae in the upper K horizon is older, probably because percolating waters seldom have a chance to dissolve the more indurated carbonate. One important reversal in the dates is apparent here: organic carbon sealed in the hard laminae is younger than that of the surrounding carbonate sealant. This reversal brought forth the warning by Ruhe[41] to be very cautious in the use of inorganic carbonate dates. One explanation of this discrepancy might be that roots in cracks in the K horizon introduced the material later; however, this problem requires further study. At greater depth in the K horizon, dates on pebble coatings are consistent with other dates, but that on the whole soil is much younger, again

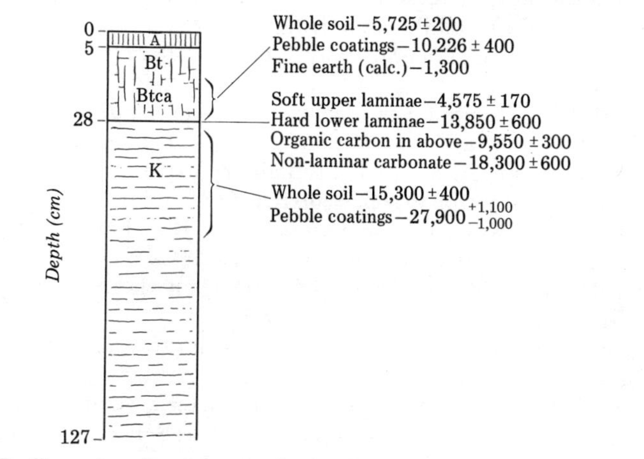

Fig. 5–10 Radiocarbon dates on a soil near Las Cruces, New Mexico. This is the Terino soil; additional laboratory data are presented in Fig. 1–2. Radiocarbon age data, from Gile and others[25] (p. A14), are on inorganic carbon of the $CaCO_3$ fraction, with the exception of one date on organic carbon.

perhaps the result of solution and re-precipitation. Stratigraphic work seems to suggest that these dates are not all worthless. Combined stratigraphic data and ^{14}C dates suggest that deposition ceased and soil formation began about 30,000 years ago. This lengthy discussion is mainly to caution workers on the use of inorganic carbon dates for carbonate, to demonstrate their potential usefulness, and to emphasize the importance of extreme care in obtaining representative samples.

REFERENCES

1. Alexander, L. T., and Cady, J. G., 1962, Genesis and hardening of laterite in soils: U.S. Dept. Agri. Tech. Bull. 1282, 90 p.
2. Arkley, R. J., 1963, Calculation of carbonate and water movement in soil from climatic data: Soil Sci., v. 96, p. 239–248.
3. Barshad, I., 1964, Chemistry of soil development, p. 1–70 *in* F. E. Bear, ed., Chemistry of the soil:Reinhold Publ. Corp., New York, 515 p.
4. Black, C. A., 1957, Soil-plant relationships: John Wiley and Sons, New York, 332 p.
5. Bowler, J. M., and Polach, H. A., 1971, Radiocarbon analyses of soil carbonates: An evaluation from paleosols in southeastern Australia, p. 97–108 *in* D. H. Yaalon, ed., Paleopedology: Israel Univ. Press, Jerusalem, 350 p.
6. Bretz, J. H., and Horberg, L., 1949, Caliche in southeastern New Mexico: Jour. Geol., v. 57, p. 491–511.
7. Brewer, R., 1955, Mineralogical examination of a yellow podzolic soil formed on granodiorite: Commonwealth Scientific and Industrial Res. Organ. (Australia), Soil Publ. no. 5, 28 p.
8. ____ 1964, Fabric and mineral analysis of soils: John Wiley and Sons, New York, 470 p.
9. ____ 1968, Clay illuviation as a factor in particle-size differentiation in soil profiles: 1968 Internat. Cong. Soil Sci. Trans., v. 4, p. 489–499.
10. ____ and Haldane, A. D., 1957, Preliminary experiments in the development of clay orientation in soils: Soil Sci., v. 84, p. 301–309.
11. Broecker, W. S., Kulp, J. L., and Tucek, C. S., 1956, Lamont natural radiocarbon measurements III: Science, v. 124, p. 154–165.
12. Brown, C. H., 1956, The origin of caliche on the northeastern Llano Estacado, Texas: Jour. Geol., v. 64, p. 1–15.
13. Buckman, H. O., and Brady, N. C., 1969, The nature and properties of soils: The Macmillan Co., Toronto, 653 p.
14. Buol, S. W., and Hole, F. D., 1959, Some characteristics of clay skins on peds in the B horizon of a Gray-Brown Podzolic soil: Soil Sci. Soc. Amer. Proc., v. 23, p. 239–241.
15. ____ and ____ 1961, Clay skin genesis in Wisconsin soils: Soil Sci. Soc. Amer. Proc., v. 25, p. 377–379.

16. Buringh, P., 1968, Introduction to the study of soils in tropical and subtropical regions: Centre for Agricultural Publishing and Documentation, Wageningen, 118 p.
17. Campbell, C. A., Paul, E. A., Rennie, D. A. and McCallum, K. J., 1967a, Factors affecting the accuracy of the carbon-dating method in soil humus studies: Soil Sci., v. 104, p. 81–85.
18. ——, ——, ——, and —— 1967b. Applicability of the carbon-dating method of analysis to soil humus studies: Soil Sci., v. 104, p. 217–224.
19. Claridge, G. G. C., and Campbell, I. B., 1968, Some features of Antarctic soils and their relation to other desert soils: 9th Internat. Cong. Soil Sci. Trans., v. 4, p. 541–549.
20. Flach, K. W., Nettleton, W. D., Gile, L. H., and Cady, J. G., 1969, Pedocementation: Induration by silica, carbonates, and sesquioxides in the Quaternary: Soil Sci., v. 107, p. 442–453.
21. Gardner, L. R., 1972. Origin of the Mormon Mesa caliche: Geol. Soc. Amer. Bull., v. 83, p. 143–156.
22. Gile, L. H., Peterson, F. F., and Grossman, R. B., 1965, The K horizon: A master soil horizon of carbonate accumulation: Soil Sci., v. 99, p. 74–82.
23. ——, ——, and —— 1966, Morphological and genetic sequences of carbonate accumulation in desert soils: Soil Sci., v. 101, p. 347–360.
24. Gile, L. H., and Grossman, R. B., 1968, Morphology of the argillic horizon in desert soils of southern New Mexico: Soil Sci. v. 106, p. 6–15.
25. ——, Hawley, J. W., and Grossman, R. B., 1970, Distribution and genesis of soils and geomorphic surfaces in a desert region of southern New Mexico: Soil Sci. Soc. Amer. Guidebook, soil-geomorphology field conferences, Aug. 21–22, 29–30, 1970, 156 p.
26. Hallsworth, E. G., 1963, An examination of some factors affecting the movement of clay in an artificial soil: Jour. Soil Sci., v. 14, p. 360–371.
27. Jenny, H., 1941, Calcium in the soil: III. Pedologic relations: Soil Sci. Soc. Amer. Proc., v. 6, p. 27–35.
28. Kononova, M. M., 1961, Soil organic matter: Pergamon Press, New York, 450 p.
29. Krauskopf, K. B., 1967, Introduction to geochemistry: McGraw-Hill, New York, 721 p.
30. Maignien, R., 1966, Review of research on laterites: UNESCO, Natural resources research, no. 4, 148 p.
31. Malde, H. E., 1955, Surficial geology of the Louisville Quadrangle, Coloardo: U.S. Geol. Surv. Bull. 996-E, p. 217–259.
32. McKeague, J. A., and St. Arnaud, R. J., 1969, Pedotranslocation: Eluviation-illuviation in soils during the Quaternary: Soil Sci., v. 107, p. 428–434.
33. Mohr, E. C. J., and van Baren, F. A., 1954, Tropical soils: N. V. Uitgeverij W. van Hoeve, The Hague, Holland, 498 p.
34. Nettleton, W. D., Flach, K. W., and Borst, G., 1968, A toposequence of soils in tonalite grus in the southern California Peninsular Range: U.S. Dept. Agri., Soil Surv. Investigation Rept. no. 21, 41 p.

35. Nettleton, W. D., Flach, K. W., and Brasher, B. R., 1969, Argillic horizons without clay skins: Soil Sci. Soc. Amer. Proc., v. 33, p. 121–125.
36. Oertal, A. C., 1968, Some observations incompatible with clay illuviation: 1968 Internat. Cong. Soil Sci. Trans., v. 4, p. 481–488.
37. Olphen, H. van, 1963, An introduction to clay colloid chemistry: John Wiley and Sons, New York, 301 p.
38. Paul, E. A., 1969, Characterization and turnover rate of soil humic constituents, p. 63–76 *in* S. Pawluk, ed., Pedology and Quaternary research: Univ. of Alberta Printing Dept., Edmonton, Alberta, 218 p.
39. Ponomareva, V. V., 1969, Theory of podzolization: Israel Program for Scientific Translations, Jerusalem, 309 p.
40. Reeves, C. C., Jr., 1970, Origin, classification, and geologic history of caliche on the southern High Plains, Texas and eastern New Mexico: Jour. Geol., v. 78, p. 352–362.
41. Ruhe, R. V., 1967, Geomorphic surfaces and surficial deposits in southern New Mexico: New Mex. Bur. Mines and Mineral Resources, Memoir 18, 66 p.
42. ——— 1969, Quaternary landscapes in Iowa: Iowa State Univ. Press, Ames, 255 p.
43. Schnitzer, M., 1969, Reactions between fulvic acid, a soil humic compound and inorganic soil constituents: Soil Sci. Soc. Amer. Proc., v. 33, p. 75–81.
44. Schuylenborgh, J. van, 1965, The formation of sesquioxides in soils, p. 113–125 *in* E. G. Hallsworth and D. V. Crawford, eds., Experimental pedology: Butterworth and Co., London, 414 p.
45. Sharpenseel, H. W., 1971, Radiocarbon dating of soils—problems, troubles, hopes, p. 77–88 *in* D. H. Yaalon, ed., Paleopedology: Israel Univ. Press, Jerusalem, 350 p.
46. Simonson, R. W., 1959, Outline of a generalized theory of soil genesis: Soil Sci. Soc. Amer. Proc., v. 23, p. 152–156.
47. Sivarajasingham, S., Alexander, L. T., Cady, J. G., and Cline, M. G., 1962, Laterite: Advances in Agronomy, v. 14, p. 1–60.
48. Soil Survey Staff, 1960, Soil classification, a comprehensive system (7th approximation): U.S. Dept. Agri., Soil Cons. Service, 265 p.
49. ——— 1967, Supplement to soil classification system (7th approximation): U.S. Dept. Agri., Soil Cons. Service, 207 p.
50. Stobbe, P. C., and Wright, J. R., 1959, Modern concepts of the genesis of podzols: Soil Sci. Soc. Amer. Proc., v. 23, p. 161–164.
51. Tedrow, J. C. F., 1966, Polar desert soils: Soil Sci. Soc. Amer. Proc., v. 30, p. 381–387.
52. Ugolini, F. C., 1970, Antarctic soils and their ecology, p. 673–692 *in* Holdgate, M. W., ed., Antarctic ecology, v. 2: Academic Press, New York.
53. Williams, G. E., and Polach, H. A., 1971, Radiocarbon dating of arid-zone calcareous paleosols: Geol. Soc. Amer. Bull., v. 82, p. 3069–3086.
54. Wright, J. R., and Schnitzer, M., 1963, Metallo-organic interactions associated with podzolization: Soil Sci. Soc. Amer. Proc., v. 27, p. 171–176.
55. Yaalon, D. H., 1963, On the origin and accumulation of salts in groundwater and in soils of Israel: Bull. Res. Council of Israel, v. 11G, p. 105–131.
56. ———, ed., 1971, Paleopedology: Israel Univ. Press, Jerusalem, 350 p.

CHAPTER 6

Factors of soil formation

Since the early work of Dokuchaev in Russia and Hilgard in the United States, pedologists have been trying to describe the main factors that define the soil system and to determine mathematically the relationship between soil properties and these factors. Jenny[6,8,9] has made many important contributions to this facet of pedology; he has also traced its historical development. Major[11] applied the same principles to plant ecology, and Crocker[3] critically analyzed Jenny's factors, especially the biotic factor, and reviewed some of the Russian thinking on the factorial approach.

Factors and the fundamental equation

The factors that define the state of the soil system are commonly called state factors. The five basic factors recognized are climate, organisms, topography, parent material, and time. Other factors may be important locally, but these five adequately define the state of most soils. The factors theoretically are independent variables, in that field sites can be found in which the factors vary independently of each other. Although the factors can be dependent variables in some field sites, their only real value in a rigorous quantitative factorial treatment is as independent variables.

A clear distinction should be made between factors and processes. In previous chapters, we have been concentrating on the processes that are operative in soils. These processes form the soil. The factors, in contrast, define the state of the soil system. If one knew precisely the combination of factors that describe a soil system, the soil properties would be precisely known. A change in any factor would result in a change in the soil. However, we are not yet to the point in our knowledge of the factors where they can be defined this precisely, and we may never be. Despite this, some valid qualitative predictions can be made.

With the recognition of the factors, equations were formulated to establish the dependence of certain soil properties on the factors. The latest of these, the fundamental equation of Jenny,[6] is

$$S \text{ or } s = f(cl, o, r, p, t, \ldots,)$$

where S denotes the soil, s any soil property, cl the climatic factor, o the biotic factor, r the topographic factor, p the parent material, t the time factor, and the dots after t represent unspecified factors, such as airfall salts, that might be important locally. In this equation, S and s are the dependent variables and the factors are the independent variables. More elaborate forms of this equation have been proposed, but none have been solved, for the reason that, if all factors are allowed to vary, it would not be possible to sort out the effect of each factor on the soil property studied. Moreover, if an equation cannot be solved we have not gained much by merely listing the factors.

Jenny overcame this dilemma of solving the equation by solving the equation for one factor at a time. To do this, one factor is allowed to vary while the others are held constant. He established the following functions

$$s = f(\underline{cl}, o, r, p, t, \ldots,) \text{ climofunction}$$
$$s = f(\underline{o}, cl, r, p, t, \ldots,) \text{ biofunction}$$
$$s = f(\underline{r}, cl, o, p, t, \ldots,) \text{ topofunction}$$
$$s = f(\underline{p}, cl, o, r, t, \ldots,) \text{ lithofunction}$$
$$s = f(\underline{t}, cl, o, r, p, \ldots,) \text{ chronofunction}$$

To solve each function the first factor listed (underlined) is allowed to vary while the others remain constant; one therefore determines the dependency of one (or more) soil property on a single factor by appropriate statistical methods. This can be extended to include more factors, and eventually it might be possible to rank them on their relative importance to that soil property or properties.

There are two ways in which a factor can be considered constant: (a) if the range in the state factor is quite small and (b) if variation in the state factor is large, yet has a negligible effect on the soil property. Figure 6–1 illustrates this latter case. The functional relationship between a soil property and factor $\underline{a}$ is to be determined. In the same sampling area, factor $\underline{b}$ also varies, so its relationship to the soil property also must be established. The plot indicates a close dependence of the property on both factors between 0 and x on the horizontal axis, and a simple functional relationship cannot be established.

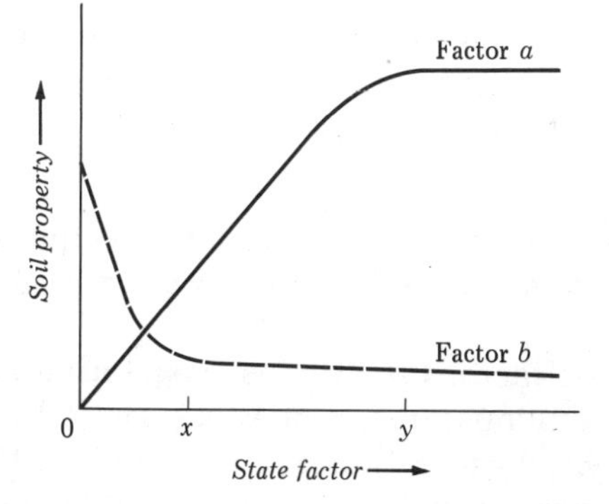

Fig. 6–1 Hypothetical variation in one soil property with variation in two state factors. The vertical scale is not necessarily the same for both plots. On the horizontal axis, x and y are arbitrary values for the state factors.

Between x and y, however, the slope of ds/dF_b approaches zero, and thus variation in factor $\underline{b}$ has little effect on the soil property. It can be considered a constant. Variation in s, therefore, between x and y is ascribed to variation in factor $\underline{a}$ in this simplified model. Beyond y, both factors can be considered constants, because the slopes of the functions are close to zero. To establish the dependence of s on factor $\underline{a}$ between 0 and x, sites would have to be chosen in which variation in factor $\underline{b}$ is small.

The relative importance of the factors varies from soil to soil. In the early days of soil science, however, rock type was generally considered to be most important. Following the early Russian work, especially that of Dokuchaev, the effect of climate was considered to be most important, and many soil-classification schemes have been based on climate and processes thought to be related to climate. Although climate may be the most important factor in the gross, world-wide distribution of soils, the other four factors are equally important in describing soil variation in a landscape.[5,7]

One problem in solving the functions set down by Jenny is to find appropriate field sites at which the factors operate as independent variables. Crocker[3] attributes a part of Jenny's success in solving the functional equations to the definitions given the state factors, definitions that assure that they could be independent of each other. The definition of the climatic factor will illustrate the problem. One can speak of the regional climate or the climate at the ground surface immediately above the soil. The two climates are measurably different. The regional climate is measured above the canopy of vegetation, and so it is less dependent on any ecosystem property. The climate at the ground surface, however, is dependent on the vegetation at the site or on the orientation of slope with respect to position of the sun. In this example, the regional climate is the independent variable, whereas the soil climate is a dependent variable. Thus, in solving for the functions the regional climate must be used. The definitions of Jenny are as follows.

Climate (*cl*): The regional climate; because precipitation and temperature are not interdependent, they can be treated as separate functions.

Organism or biotic factor (*o*): Because vegetation is most important, this factor is the summation of plant disseminules reaching the soil site or the potential vegetation; it is approximated by a list of species growing in the surrounding region that could gain access to the site under appropriate conditions.

Topographic (*r*): Included are the initial shape and slope of the landscape related to the soil, the direction the slopes faces, and the effects of a high water table, the latter being commonly related to the topography.

Parent material or initial state of the system (*p*): Included are materials, both weathered and unweathered, from which the soil formed. Parent material could also be a soil in the case where one wishes to study the effect of climatic change on a pre-existing soil.

Time (*t*): Elapsed time since deposition of material, the exposure of the material at the surface or formation of the slope to which the soil relates; if a study is being made of the effect of climatic change on a pre-existing soil, the time since the change.

These definitions raise several problems, and these have been discussed at length by Jenny[6,8] and Crocker.[3] One problem is that the biotic and parent-material factors cannot be quantified; for the most part they can only be described in qualitative terms. An example of what can be done, however, is to determine the function for each of several rock or vegetation types and then compare one function with the other. Some workers have criticized Jenny for including time as one of the factors, because by itself time does nothing to a soil. Its importance, however, lies in the fact that most soil-forming processes are so slow that their effect on the soil is markedly time-dependent. The one definition that probably has been most controversial, however, is that of the biotic factor, because it is defined and used in two different ways. When deriving functions for the other factors, the biotic factor is taken as the potential vegetation or species pool. The actual vegetation at the site is a function of the same set of factors as is the soil[8,11]; both vegetation and soil develop concurrently and can and do influence each other. Therefore vegetation cannot be taken as an independent variable. Biofunctions can be derived, however, for areas in which vegetation is the variable and all other factors are constant and thus do not influence the vegetation. Instances of this are quite rare, but a commonly studied example is the prairie-forest transition.

Arguments for and against the use of factors

Several criticisms have been made of the use of factors in the study of soils, some of which are discussed by Crocker[3] and Bunting.[2] One is that the general equation has never been solved. By this I mean that we cannot apply quantitative data on the factors to predict the resulting soil or soil properties adequately. This is true now, and it may never be otherwise. Moreover, many soils have formed under conditions in which several factors have varied, and functions for these soils may never be derived. Individual functions for other soils have been derived, however, and they are very useful in pedology.

Another problem is that many soils are polygenetic, that is, they have formed under more than one set of factors, and thus present-day factors should not be used to define the state of these particular soils. Monogenetic soils (soils that have formed under one set of factors) are certainly preferred; however, these soils are not abundant and generally are restricted to those formed in postglacial time. There are two ways to surmount this problem. One would be to study only those soil properties that form rapidly and reach a steady state in a rather short period of time, that is less than the duration of the postglacial time span. This would limit the kinds of properties that could be studied by the use of factors. Many argillic B horizons, for example, take longer than that span of time to form. The other solution is to allow for climatic change and assume that at all sampled sites in a particular region the differences between the climates of the past and the present climate were relatively constant; that is, all sites experienced a similar variation in climate.[4] This seems reasonable as long as soils of similar age are being studied, because all these soils may have gone through similar climatic cycles. And, as pointed out by Jenny,[6] in solving for climofunctions the main interest is in gradients rather than in absolute values of precipitation and temperature. Past climatic gradients may have been similar to present ones in some, but probably not all regions. It might be difficult to solve some chronofunctions, however, because older soils probably have been subjected to more changes in climate than have younger soils. Perhaps these problems are not insurmountable when one considers the problems in locating and collecting truly representative samples in the field. Furthermore, as more data become available on past climates and their duration, perhaps these can be taken into account in quantitative studies.

Another problem cited in the use of factors is that one learns a lot about the factors, but little about the soil.[2] However, Barshad[1] feels that this approach is important in studying soil-forming processes, and the rates at which they proceed, and that it is valuable in predicting the soil properties that might be encountered at a particular site. As an example, we can consider the origin of a clay-enriched horizon near the surface. It could be an argillic B horizon or a depositional layer. A knowledge of the variation in clay-chronofunctions with other factors, and the factors at the locality studied, would help set upper and lower limits on the probable amount of pedogenic clay. If further study indicates that more clay is pedogenic than at first was thought probable, other factors, such as time, would have to be reassessed. Another example might be the determination of the origin of an E horizon in a soil. It could result from podzolization; if it did, it usually would be indicated by a combination of factors. Or, the E could be a carry-over from a past vegetation or climatic factor. It might be found, however, that the E horizon is not related to podzolization but rather is due to either a layered parent material or a strongly differentiated soil with a perched water table that produced the E horizon by removal of Fe^{2+} in laterally draining soil water. Although more examples could be cited, the point is made. The use of factors, along with geologic data and data on soil processes, provides one with the tools necessary to evaluate the origin of the soil in the field partly because it can keep the mind open to alternate hypotheses.

Although the derived functions are few, and although one may question how quantitative the use of factors is, the approach is basic to geomorphological research. Derived functions can be used in a qualitative sense, and they are commonly used to indicate the trends that one can expect throughout a region. Such qualitative expressions of the variables give rise to what Jenny[8] would call sequences rather than functions. One can study chronosequences or climosequences, and these are not without value in research. If the trends are not those that were predicted, perhaps the factors for that site will have to be redefined.

Steady state

Soils often are described as being mature or in equilibrium with their enviroment. Lavkulich[10] argues that equilibrium is not a good term in the sense that reactions at dynamic equilibrium go in both directions with no apparent change in the system. This condition is probably uncommon in the soil, because many reactions, such as primary mineral → clay, are not reversible in the soil environment.

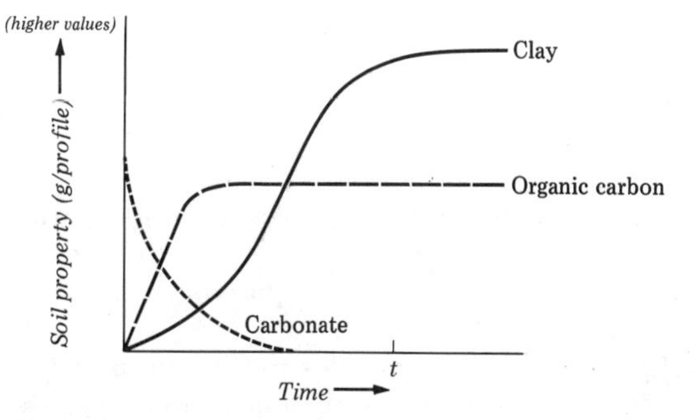

Fig. 6–2 Hypothetical variation in several soil properties with time. The parent material contained carbonate which has been leached from the soil. Organic carbon reaches the steady-state condition before clay, and the soil profile might be considered at the steady state at time *t*.

He believes steady state is a better term to describe soil conditions. At the steady state, energy is being applied to the system, and reactions are going on, but properties are either not changing or their rate of change is too slow to be measured. This is similar to the concept of a soil-forming factor being considered constant even though it varies. That is, for a change in the factor, the soil property does not change.

The time factor is a familiar example of soil steady state. If one plots a soil property as a function of time, the curve is usually steep during the initial stages of soil development (Fig. 6–2). After some time the soil property shows little change, even though reactions are still going on. We can say that the soil property has reached a steady state with the environmental factors. If we consider a soil at one locality, each soil property commonly will require a different amount of time to attain the steady-state condition (Fig. 6–2). The soil profile could be said to be in a steady state when its diagnostic properties are each in a steady state. This seems better than labeling a soil mature, because that term is not well defined.

REFERENCES

1. Barshad, I., 1964, Chemistry of soil development, p. 1–70 *in* F. E. Bear, ed., Chemistry of the soil: Reinhold Publ. Corp., New York, 515 p.
2. Bunting, B. T., 1965, The geography of soil: Aldine Publ. Co., Chicago, 213 p.
3. Crocker, R. L., 1952, Soil genesis and the pedogenic factors: Quat. Rev. of Biology, v. 27, p. 139–168.
4. Harradine, F., and Jenny, H., 1958, Influence of parent material and climate on texture and nitrogen and carbon contents of virgin California soils: 1: Texture and nitrogen contents of soils: Soil Sci., v. 85, p. 235–243.

5. Harris, S. A., 1968, Comments on the validity of the law of soil zonality: 9th Internat. Cong. Soil Sci. Trans., v. 4, p. 585–593.
6. Jenny, H., 1941, Factors of soil formation: McGraw-Hill, New York, 281 p.
7. ——— 1946, Arrangement of soil series and types according to functions of soil-forming factors: Soil Sci., v. 61, p. 375–391.
8. ——— 1958, Role of the plant factor in the pedogenic functions: Ecology, v. 39, p. 5–16.
9. ——— 1961, Derivation of state factor equations of soils and ecosystems: Soil Sci. Soc. Amer. Proc., v. 25, p. 385–388.
10. Lavkulich, L. M., 1969, Soil dynamics in the interpretation of paleosols, p. 25–37 *in* S. Pawluk, ed., Pedology and Quaternary research: Univ. of Alberta Printing Dept., Edmonton, Alberta, 218 p.
11. Major, J., 1951, A functional, factorial approach to plant ecology: Ecology, v. 32, p. 392–412.

CHAPTER 7

Influence of parent material on weathering and soil formation

Parent material influences many soil properties to varying degrees. Its influence is greatest in drier regions and in the initial stages of soil development. In wetter regions, and with time, other factors may overshadow the influence of parent material. Here we will deal with the properties of the rocks and minerals in crystalline, consolidated, and unconsolidated deposits and their effect on weathering and soils. We also will explore briefly the genesis of soils in areas characterized by a slow rate of sedimentation.

Mineral stability

Minerals vary in their resistance to weathering; some weather quite rapidly (in 10^3 years), whereas others weather so slowly (10^5 to 10^6 + years) that they persist through several sedimentary cycles. Minerals can be ranked by their resistance to weathering in several ways. One way is to study the soil or weathered material and compare the minerals there with minerals in the unweathered material on the basis of their respective depletion, evidence for alteration to clay, or etching of individual grains (Fig. 7–1). Goldich[15] established the mineral stability series for the common rock-forming minerals in this way. His series is given in Table 7–1, modified somewhat by the work of Hay.[17] This latter work is added because it provides a cross-link between the weathering series of the ferromagnesian and non-ferromagnesian minerals, based on plagioclase composition. Although Hay did not find a difference between the weathering of hypersthene and augite, other stability

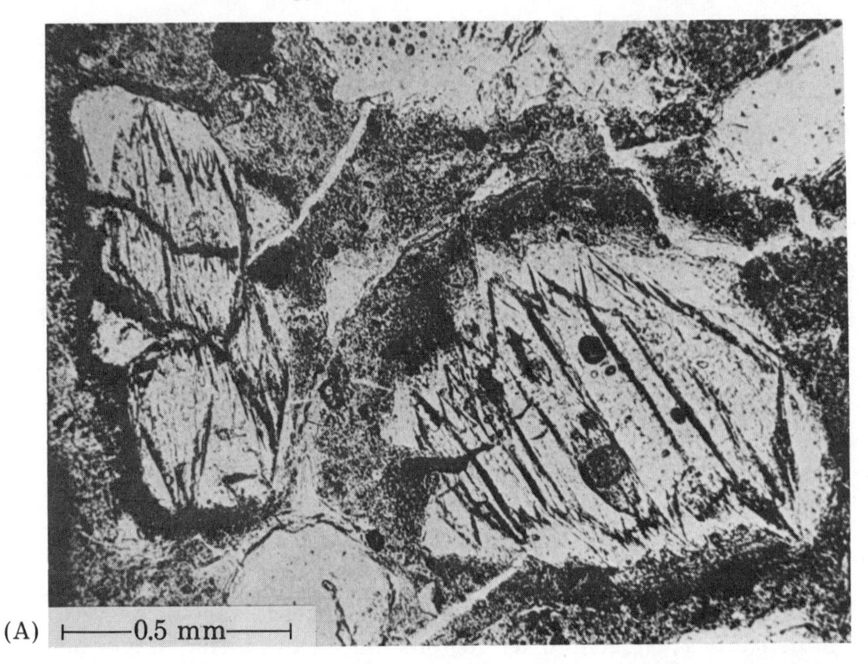

Fig. 7–1 Photomicrographs of thin sections showing mineral weathering. (A) Two etched pyroxene grains in weathered ash. (Taken from Hay,[17] pl. 2, D, Jour. of Geology, © 1959, The University of Chicago.) *Left.* Hypersthene occupies a cavity that retains the original shape of the crystal. *Right.* Augite is surrounded by halloysite subsequently deposited in the cavity formed during grain weathering. Such sharp cockscomb terminations are characteristic of

studies have ranked augite as the more stable of the two, and that ranking is preserved in the table. Hay also found that iron-rich olivine grains weather more rapidly than magnesium-rich grains, and that the latter compare favorably with augite and hypersthene in resistance to weathering. This general sequence is in accord with many data, although the order may be changed somewhat under certain environmental conditions.

The stability of minerals also may be assessed by their occurrence in sedimentary rocks of varying ages; older rocks should have relatively greater amounts of the more resistant minerals. Pettijohn[29] has drawn up a list for the ferromagnesian and accessory minerals, and, although it includes more minerals than does Table 7–1, the ranking of the common rock-forming minerals is similar. Other persistence schemes are compared with Pettijohn's by Brewer,[13] and these data indicate that rankings of mineral stability differ. Most studies, however, indicate that zircon and tourmaline compare with quartz in their high resistance to weathering, and therefore any of these three minerals can be

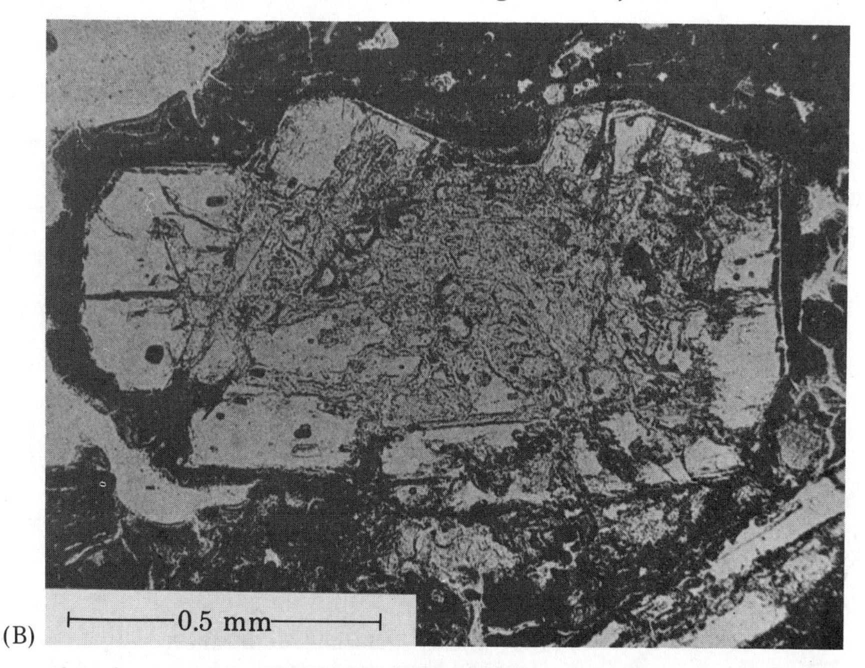

(B)

weathered pyroxene and hornblende grains in many areas. Lack of clay formation in place here might be related to insufficient aluminum in the parent grain. (B) Halloysite replacing interior of a zoned plagioclase. (Taken from Hay,[17] pl. 3, A, Jour. of Geology, © 1959, The University of Chicago.) The core of this grain probably was more calcic than the rim, thus the selective replacement.

used as the stable component in mineral-depletion weathering studies. Raeside[30] points out, however, that some seemingly stable minerals may not always be stable for various mineralogical reasons.

Several factors account for the stability of the common rock-forming minerals. Some are related to the mineral itself, and some are related to the weathering environment. Only those related to the mineral will be considered here.

Mineral structure plays a major role in the resistance of a mineral to weathering. From the data in Table 7–1, it is noted that for some minerals the mineral-stability ranking parallels a ranking based on mineral structures: that is, the progressive increase in sharing of oxygens between adjacent silica tetrahedra correlates with increased resistance to weathering. This is especially true for the sequence olivine-pyroxene-hornblende-biotite-quartz. The energies of formation of the cation-oxygen bond seem to explain the above sequence. Keller[22] calculated the energies, and the data indicate that Si—O is the strongest bond (> 3000 kg cal), Al—O the next strongest bond (< 2000 kg cal),

Table 7–1
Ranking of stability of common rock-forming minerals, their structural classification and bonding energies*

	FERROMAGNESIAN MINERALS		NON-FERROMAGNESIAN MINERALS
↓ *Increased stability*			Ca^{2+}–plagioclase[4] (32)
			An_{80-100}
	Olivine[1] (29)	Fe-rich	An_{65-80}
		Mg-rich	
	Hypersthene[2a]		
	Augite[2a] (31)		An_{50-65}
	Hornblende[2b] (32)		
	Biotite[3] (30)		Na^{+}–plagioclase[4] (34)
			K^{+}–feldspar[4] (34)
			Muscovite[3]
			Quartz[4] (37)

Number of oxygens shared between adjacent silica tetrahedra		ALUMINOSILICATE MINERAL STRUCTURES
↓ *Increased sharing*	None	1. Nesosilicate (independent tetrahedrons)
	2	2a. Inosilicate (single chain)
	2 and 3	2b. Inosilicate (double chain)
	3	3. Phyllosilicate (sheet structure)
	4 (all)	4. Tectosilicate (interlinked network of tetrahedrons)

*Superscripts with mineral name are keyed to the numbered aluminosilicate structures. Numbers in parentheses are bonding energies (× 1000 kg cal) for 24 oxygens in each mineral structure, calculated by Keller[22] (1954), and rounded to the nearest 1000. (Use granted by permission of the Mineralogical Society of America.)
(Taken from Goldich,[15] Table 18, Jour. of Geology, published by The Univ. of Chicago and Hay,[17] Fig. 8, Jour. of Geology. © 1959, The Univ. of Chicago.)

and the bond between the common base ions and oxygen is the weakest (< 1000 kg cal). Therefore, mineral resistance to weathering should be reflected in the structure, composition, and total energy of all the bonds in the mineral. Keller[22] calculated the energies of formation of all the bonds for a variety of minerals, based on a mineral unit with 24 oxygens (Table 7–1). The relationship between energies of the total bonds and the mineral-stability ranking is fairly good both for the non-ferromagnesian minerals and the ferromagnesian minerals, and the only mineral out of sequence is biotite. The correlation between the ferromagnesian and non-ferromagnesian minerals, however, does not correspond with that based on field data. From the bonding energy data, one could predict that volcanic glass should be highly susceptible to weathering, and that appears to be the case.

Mineral breakdown during weathering probably is initiated at the site of the weakest bond, and the more sites that are exposed to weathering solutions, the more rapid should be the weathering.[3] Olivine weathers rapidly because the separate tetrahedra are linked together only by the fairly weak cation-oxygen bonds. The other minerals in the stability series contain tetrahedra linked together by the stronger Si—O and Al—O bonds and so are more resistant to weathering. In these minerals weathering is most effective if the weathering solutions have access to sites of the weak cation-oxygen bonds. Such sites are those that link the chains of the inosilicates together, and the sheets of the phyllosilicates together, or sites that offset the charge deficiency brought about by the substitution of Al^{3+} for Si^{4+} in the tectosilicates.

Other factors seem to explain the relative resistance of minerals classified in the same structural groups.[3] The plagioclases, for example, show a ranking from the more stable sodium-plagioclases to the least stable calcium-plagioclases. This sequence is due in large part to the substitution of Al^{3+} for Si^{4+} in the tetrahedral position as the plagioclase becomes more calcic; thus, stronger Si—O bonds are replaced by weaker Al—O bonds, and this is reflected in a decrease in the total bonding energies. Biotite and muscovite, although both phyllosilicates, differ in their weathering. Biotite is more susceptible to weathering, probably because the Fe^{2+} is readily oxidized to Fe^{3+}. This change in valency disrupts the electrical neutrality of the mineral, which in turn causes other cations to leave and thus weaken the crystal lattice. The tightness of packing of the oxygens might also be a factor with some minerals. In both of two mineral pairs, sodium-plagioclase and orthoclase, and olivine and zircon, the latter minerals have the tighter packing and are more resistant to weathering. This is especially important in the olivine-zircon pair, because it includes one of the least and one of the most resistant minerals.

Susceptibility of rocks to chemical weathering

Rocks weather and erode at different rates, as can be seen in the variations in topographic relief that accompany variations in rock type. For rocks from widely spaced localities, however, factors other than rock type might influence the weathering variations observed. To be able to compare rocks of differing lithology under similar conditions of weathering, it might be best to study a sedimentary deposit, such as bouldery till or outwash, which includes a variety of rocks types. Rocks from the same depth below the surface should have weathered under as similar conditions of weathering as one could hope to find.

Goldich's stability series for minerals can be used to predict igneous rock stability in the weathering environment. Rocks with a high content of more weatherable minerals should weather more rapidly than rocks with a high content of minerals resistant to weathering. To make a valid comparison, however, the rocks should be similar in crystallinity and grain size. For igneous rocks, therefore, resistance to weathering should increase in the order gabbro $\longrightarrow$ granite or basalt $\longrightarrow$ rhyolite. Clay production should follow these trends, and it commonly does.[2]

Grain size has an effect on the rate of weathering, for it is observed that coarser-grained igneous rocks commonly weather more rapidly than finer-grained rocks.[39] This is readily seen in many tills in the Cordilleran Region. In the Sierra Nevada, for example, till of probable pre-Wisconsin age has the following variation in weathered clasts: coarse-grained granitic stones are weathered to grus, coarse-grained porphyritic andesitic pebbles and cobbles have thick weathering rinds or are weathered to the core, and fine-grained volcanic rocks of intermediate to basic composition have weathering rinds up to 4 mm thick.[7] Moreover, it is a common observation in large boulders that mafic inclusions, generally of finer grain size than the enclosing granitic rock, stand in relief above the latter due to weathering.[11] In the Colorado Rocky Mountains, granodiorite clasts on the surface of rock glaciers on Mt. Sopris have oxidation rinds up to 45 mm thick, which probably formed in late-Wisconsin and postglacial time,[9] whereas basalt clasts in early-Wisconsin till on Grand Mesa,[43] in what might be a wetter climate, have rinds about 2 mm thick.

These studies suggest that, other factors being equal, rocks with high amounts of glass weather more rapidly than rocks with low glass contents, and rocks of finer grain size weather more slowly than rocks of coarser grain size. One reason for the latter may be that intergranular surface area increases with decrease in grain size; hence more energy probably would be required to disintegrate the finer-grained rock.

The weathering process that goes on in rocks containing biotite differs somewhat from that in biotite-free rocks. The biotite-bearing rocks commonly weather more rapidly, as seen in many outcrops of igneous and metamorphic rocks. The reason for this behavior was presented earlier (Ch. 3). Biotite, in addition to being the first mineral to weather in granitic rocks, forms alteration products that can occupy a greater volume than did the original biotite[26]; the result is mineral expansion, with numerous localized points of stress within the rock that eventually shatter the rock and form grus. This mechanism probably explains in part why the zone between fresh and weathered granitic rock usually is gradational over about 0.5 m or more. Basalt is an extreme case of a rock lacking biotite. Weathering proceeds inward, grain by grain, and the boundary between fresh and chemically weathered rock can be quite sharp (< 1 mm).[14] The point to be made is that the chemical alteration necessary to disaggregate a biotite-bearing granitic rock is not comparable to that necessary to weather basalt to a similar depth. Much of granitic rock weathering is mechanical shattering induced by slight chemical weathering, whereas basalt rock weathering is mainly chemical and, as noted, proceeds slowly inward. This, along with the variation in susceptibility of minerals to weathering, explains the common observation that soils formed from granitic rock are higher in sand and lower in clay than are soils formed from basalt.

The weathering of sedimentary rock differs from that of igneous rock, but there is no fast rule to determine which rocks are most susceptible to weathering. In some areas the evidence favors more rapid weathering of sedimentary rocks,[18] but this can be reversed in other areas with other rock types. Sedimentary rocks can have any combination of nonclay and clay minerals, and the weathering of these rocks probably would proceed as predicted by the Goldich stability series or the Pettijohn persistence series. Rocks with a considerable clay-size component may break down more rapidly than rocks low in clays because clays by themselves do not bind minerals tightly together and they might expand and contract with variation in moisture content. The cementing agent for sedimentary rocks, however, has a marked effect on the weathering of the rock (Fig. 7–2). In pre-Wisconsin stream terrace deposits in the Colorado Piedmont area, for example, some clasts of silica-cemented arkoses in soils are intact, whereas adjacent granitic clasts of seemingly similar mineralogy are weathered to grus.[19,24,36] Either the silica cement keeps weathering solutions from reaching the biotites, or biotites were weathered prior to deposition, or the cement is strong enough to withstand stresses set up by weathering biotites. In the same soils, clasts of silica-cemented quartz sandstones also are intact, but clasts of shale are weathered.

Limestones present a special case in the weathering of sedimentary rock. The rock weathers by solution of the readily dissolved carbonate minerals, and the rate of weathering is determined by carbonate equilibria. The soil formed and its characteristic properties commonly indicate derivation as the insoluble residue that accumulates at the surface upon removal of the carbonate fraction.[5,28,40]

Till deposits in Illinois contain a wide variety of rock types and thereby provide field sites for relative weathering rates. The data of Willman and others[42] suggest the following rock-stability ranking, from more stable to less stable

Fig. 7–2 Variation in boulder weathering as a function of rock type. (A) Eagle River terrace gravel, western Colorado. Some granitic stones have weathered to grus (g), whereas other granitic stones (light tone) and sedimentary stones (dark tones) are unweathered. (B) Pre-Wisconsin stream gravel near Boulder, Colorado. Coarse-grained granitic stones (g) are weathered, whereas granitic stones that are finer grained, metamorphic stones, and sandstones display little weathering. (Photo by Alayne Street.)

Quartzite, chert > granite, basalt > sandstone, siltstone > dolomite, limestone

Although there are local variations in the ranking due to variations in specific lithologies, the overall trend seems reasonable.

Influence of parent material on the formation of clay minerals

The parent material, whether mineral or rock, exerts some control on the clay minerals that form, because weathering releases constituents essential to the formation of the various clays. If the required constituents for a particular clay are not present in the parent material, it is obvious that the mineral cannot form. The specific ions, their concentration, and the molar $SiO_2:Al_2O_3$ ratio are all important aspects of the parent material. However, once weathering release has occurred, the micro- and macro-environments within the soil determine whether certain ions or other constituents are selectively removed or remain behind, and this then determines which clay mineral forms. In order to keep factors constant, samples should be collected from a certain depth within the soil, and comparisons should be made between the weathering products and the parent minerals or rocks. Most results substantiate the conditions that were discussed earlier as necessary for the formation of the clay minerals.

Even the clay mineral that forms may not be a stable end product, because it too can subsequently change to other products with changing conditions within the soil. There is a large literature on clay-mineral formation relative to the parent mineral or rock, and almost any product is possible, probably because of the large variety of environments that are possible. These will not be reviewed here because many of the relationships could involve parent material masked by other environmental factors. A review of some of this work is given in Loughnan.[23]

Rocks or minerals of varying composition weathering in the same environment can produce different clay minerals. Barnhisel and Rich,[1] for example, sampled boulders of different lithologies within the same weathered zone and analyzed the clays that formed. Granites and gneisses, rocks low in bases, produced kaolinite predominately, whereas gabbro, a rock high in bases, produced mostly montmorillonite. These same extremes in clay-mineral formation can be seen at the mineral level. In Hawaii, Bates[6] reports plagioclase altering to halloysite, whereas adjacent olivine grains alter to montmorillonite. As a generalization the parent mineral will greatly influence the mineralogy of the weathering by-product in those soils, or places within soils, characterized by low leaching, and the influence of the parent rock or mineral will gradually diminish with greater leaching.

Parent-material influences can also be studied by sampling over large regions. Barshad[4] has reported on such a study in California. Samples of the uppermost 15 cm of the soil were collected from a large part of the state. This assured representation of a large range in climate and in parent materials. Kaolinite, halloysite, montmorillonite, illite, vermiculite, and gibbsite are the major minerals present. In his data analysis, Barshad found that the main variation in clay minerals is a function of precipitation (Fig. 11–12) but that parent material affected some clay-precipitation relationships. The parent materials form two main groups: mafic and felsic igneous rocks. The influence of parent material was recorded in two ways. Illite is associated with felsic rocks only, probably because of their higher mica content as well as the availability of potassium. Montmorillonite, although abundant at low precipitation with both rock types, persists as a prominent clay mineral at higher precipitation in the mafic parent materials than in the felsic parent materials, probably because of the greater content of bases in the mafic rocks.

The eastern side of the Sierra Nevada offers another sampling region suitable for studying parent-material effects on clay mineralogy because tills and stream gravels with a variety of rock types are abundant.[8,10] Again, the major effects probably are climatic (Ch. 11), although parent-material influences are recognized in the drier areas. In soils formed from a parent material that is partly andesitic, montmorillonite can be present at a mean annual precipitation of 48 cm. In contrast, soils formed from materials of predominantly granitic lithology, at even lower rainfalls, contain illite and kaolinite, and montmorillonite only at depth. Again, the parent material helps determine the kind of clay that forms through availability and kind of bases.

Influence of original texture of unconsolidated sediments on soil formation

Much of the research undertaken by geomorphologists is with unconsolidated deposits of Quaternary age; hence, a brief treatment of the effect of textural variation of this material on soil formation seems in order. Although we shall try to keep lithology and mineralogy of parent material constant, this is not always possible in the field. When all other factors are equal, however, the texture of the parent material has a great influence on the course of soil formation.

Texture influences the rate and depth of leaching, and this is related to many soil properties. Textural influence can be such that soils that generally occur only in different climatic regions occur side by side, and yet each is stable for the prevailing site conditions. Figure

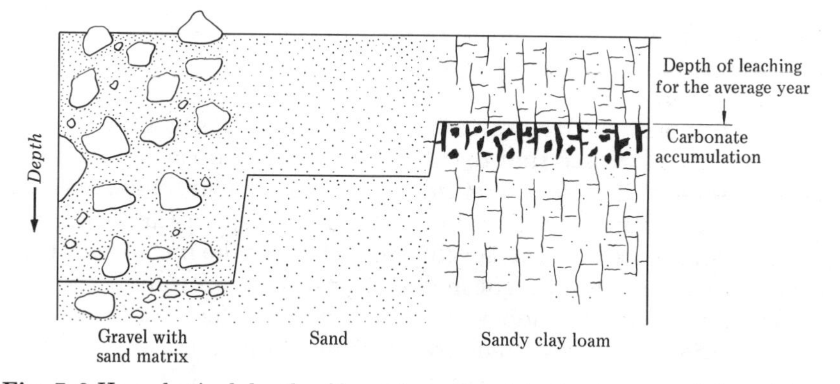

Fig. 7–3 Hypothetical depth of leaching related to the texture of the original parent material.

7–3 depicts some possible field conditions. The depth of leaching is governed by the texture, being greater in gravel, less in non-gravelly sand, and least in finer-textured material. For the same duration of soil formation, an argillic B horizon could form in the finer-textured material but not in the sandy material, because it could only be a function of translocation of original clay-size material. If sufficient time has passed for clay formation to be noticeable, it probably would proceed more rapidly in the finer-textured material, because weathering rate increases with greater surface area per unit volume due to the availability of more water for weathering over a longer period of time each year. Clay formation by precipitation from solutions is also enhanced by increased surface area, because surfaces promote retention of clay-forming constituents. In contrast, constituents released by weathering in the more gravelly materials may be leached from the soil before they have the opportunity to react to form clays. This observation may help explain the presence of well-expressed argillic B horizons in late-Wisconsin tills and loesses in the midcontinent and the general lack or poor expression of such horizons in gravelly tills and outwashes of the same age in the Sierra Nevada and other parts of the Cordilleran Region (Fig. 8–10). In addition, there might be more release by weathering of Ca^{2+} in the finer-textured material, which, in combination with less leaching, might lead to formation of a Cca horizon. Greater leaching in the coarser-grained materials, especially during the wetter years, might tend to keep the Ca^{2+} content too low to form a Cca horizon. The clay minerals that form also follow these trends. Those in the finer-textured material might have a higher ratio of 2:1 layer clay to 1:1 layer clay than those in the more permeable material. These relationships can be seen in soil samples spaced closely together; in Hawaii, kaolinite can occur in freely drained parts of the soil, whereas montmorillonite can form in the same soil in local areas of slightly restricted drainage beneath stones.[37] Finally, organic-matter contents should be higher with the finer-textured material, a relationship that has been demonstrated in California[16] and other areas (Ch. 9).

It should be stressed that the above statements are generalizations and that some of the effects mentioned probably are not the same from one climatic region to another. Furthermore, any quantitative classification scheme of soil-profile development should take into account the original texture of the parent material, just as stratigraphic correlations based on soil development should, because finer-textured soils commonly develop profile characteristics more rapidly.

Test for uniformity of parent material

In any soil study one should establish, beyond reasonable doubt, the identity of the parent material from which the soil formed. In some places this is no easy task, especially if the soil is quite weathered. If parent material cannot be identified in the field, several mineralogical tests for parent-material uniformity can be made.

In the field, one carefully examines the least weathered material, the C or R horizon, to determine if the soil could have formed from that material. For example, if the C or R horizon is granitic bedrock or colluvium the soil should contain only that material; rounded gravel should be absent unless it can be shown that it resulted from weathering. Stone lines have been used to recognize unconformities in lateritic and other materials—unconformities that might otherwise be overlooked.[33] A textural B horizon, if present, should have a position and relationship to both overlying and underlying horizons that is pedogenic in character (see criteria for buried soils, Ch. 1). In stream gravel, uniformity would be suggested by a more or less constant gravel content to the surface. Actually, most soil forms from material of a size smaller than gravel, and so it is especially important to determine uniformity within those size fractions. Field and laboratory determination of textural class may indicate eolian or overbank deposits overlying the assumed parent material, with the soil extending through all of the deposits. Even high-altitude regions are not beyond eolian influences. Loess forms a thin surface layer in parts of the Alps[12]; if the material is not recognized as loess, one might erroneously ascribe the textural variation to weathering. Loess in the Colorado Rocky Mountains on rock-glacier deposits is recognized by its fine texture and the absence of large quartz and feldspar grains that impart a grittiness to the field texture. Soil beneath the loess is course grained and gritty; it is derived from the alteration of the matrix of the rock-glacier deposit (Fig. 7–4).

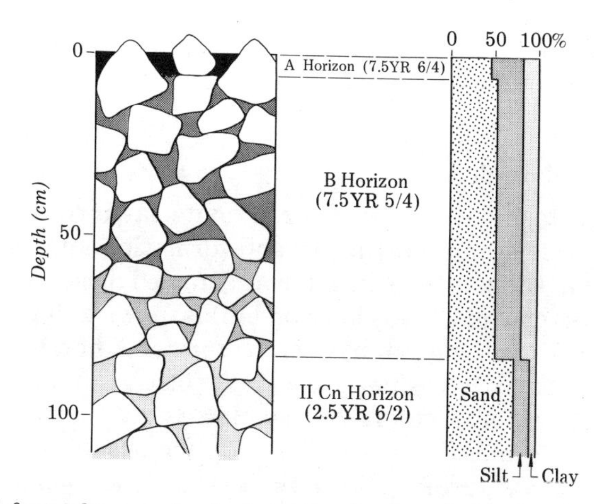

Fig. 7–4 Soil formed on rock-glacier deposit of probable late-Wisconsin age, Mt. Sopris, western Colorado (unpubl. data of the writer). Field and laboratory analysis of particle-size distribution indicates that the upper 83 cm is loess and that it overlies material derived from the rock glacier, because the large sand grains in the lower material are made up of both quartz and feldspar which could not have broken down to the sizes found in the upper layer in the time available.

Microscopic examination of the soil and parent material might also indicate influx of foreign material either by eolian or subaerial processes. Contamination is recognized by a nonclay mineralogy of the soil that does not match that of the parent material or that cannot be explained by weathering of the original grains of the parent material. In places, this kind of contamination has been shown to account for as much as one-half of the soil material.[25] More subtle eolian influxes can be determined by oxygen-isotope data on quartz in the soil and in the assumed parent material.[41]

Once the more obvious ways to ascertain uniformity have shown that the material is probably uniform, a final mineralogical check can be made. Within a specific sand or silt fraction, minerals that are relatively resistant to weathering should show a relative increase in abundance from the parent material to the surface, whereas more weatherable minerals should show a relative decrease in the same direction. According to Brewer,[13] plots such as these should change gradually and uniformly with depth; sharp inflections or reversals in trends might indicate lack of uniformity. Ratios of resistant mineral to nonresistant mineral also can be used to indicate the same trends (Fig. 8–4). Another technique is to determine the ratio of two resistant minerals; if the ratio remains nearly constant with depth and matches that of the parent material, uniformity is strongly suggested.[3] This may not be true in every case, however, because sedimentary layers of slightly different textures could have similar ratios of resistant mineral to resistant mineral in the same size fraction.

Cumulative soil profiles

Some soils receive influxes of parent material at the same time that soil formation is going on; that is, soil formation and deposition are concomitant at the same site. Nikiforoff[27] named these soils cumulative (Fig. 7–5). In such soils, the A horizon builds up with the accumulating parent material, and the material in the former A horizon can eventually become the B horizon. In contrast, other soils gradually lose material through surface erosion so that the A horizon eventually forms from the former B horizon (Fig. 7–5). These soils were called non-cumulative by Nikiforoff.[27] Both kinds of soils are common in many landscapes. Some examples of cumulative profiles will be given here, and others are given in Chapter 9, where lateral soil variations across different landscapes are discussed.

Because cumulative soils have parent material continuously added to their surfaces, their features are partly sedimentologic and partly pedogenic. In a soil study, therefore, it is important that sedimentologic features are not ascribed to pedogenesis.

Some topographic positions are favorably situated for cumulative profile formation. They are especially common in colluvial and fan deposits at the base of hillslopes,[31] and along river floodplains.[21] Uplands receiving increments of loess during soil formation also can have cumulative profiles.

Cumulative profiles are sometimes recognized by properties that are not consistent with those of the soils in the surrounding region. A common property is an overthickened A horizon due to deposition of material rich in organic matter, to organic matter at the site being continually buried, or to a combination of both processes. At any rate, the gains in organic matter may not be balanced by decompositional losses, and these gains, combined with deposition at the surface, produce thick A horizons. Cumulative profiles also may contain more clay to a greater depth than adjacent non-cumulative soils from which they derive their surface increments. This occurs because clay content in the cumulative soil is a function of clay formation at that site in addition to the clay delivered from upslope sites by erosion.

Some loess areas in the midcontinent are characterized by cumulative profiles. Smith,[38] in studying loess in Illinois, suggested an influence of parent material on the soil pattern. Many properties of the parent material loess are related to distance from the source of the loess, which is usually a river floodplain of glacial age. The loess is

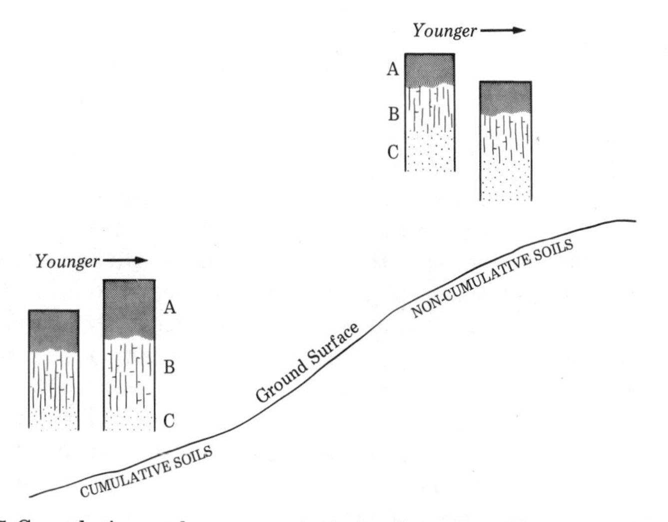

Fig. 7–5 Cumulative and non-cumulative soil profiles. If removal of material at the non-cumulative site surface is matched by soil-profile development, the profile characteristics do not vary with time. In contrast, the horizons in the cumulative profile commonly are thickened due to processes at the site and influx from upslope of organic matter, clay-size material, or other material. Because cumulative profiles are at least in part depositional, subtle sedimentary layering might be discernible.

commonly thicker and coarser grained near the source and becomes thinner and finer grained downwind. The soils developed on the loess show a downwind increase in clay content, accompanied by the gradual loss of $CaCO_3$. Smith reasoned that loess deposition and soil formation were intimately involved in producing the soil pattern. Close to the source rivers, the loess was deposited too rapidly for soil formation to keep up, and mostly unweathered calcareous loess was deposited. At localities farther from the rivers, however, soil formation could keep up or even exceed the rate of deposition, so that carbonates, if originally present, would have been leached as rapidly as the material was being deposited. Moreover, weathering could have gone on for longer periods of time on loess materials away from the river where the supposed depositional rate was less. Thus cumulative soil profiles could be found at some distance from the rivers but not close to the rivers. The effect of parent material, therefore, is twofold. Downwind decrease in particle size results in finer-textured parent material and soils, and downwind decrease in rate of deposition results in cumulative soils in which carbonate leaching and other soil-forming processes can go on concomitant with deposition; thus, soil formation of longer duration is suggested for the downwind localities. Hutton[20] explains the loess-soils distribution in Iowa in a somewhat similar manner. In Alaska,[32] variation in Spodosol soil development with distance from the loess source is attributed to variations in the rate of loess deposition.

Recent detailed work by Ruhe[34,35] demonstrates the complexity of the loess landscape and suggests that the above discussion might be an oversimplification. In particular, he demonstrated that, in northeastern Kansas, the decrease in parent-material particle size away from the loess source is not the sole cause of regional variation in B-horizon clay contents. He showed that the age of the base of the loess in Iowa is younger away from the source area, and therefore it no longer can be assumed that the more distant soils in a loess landscape have weathered longer. Moreover, some of the variation in Iowan loess soils with distance from the source river is due to greater moisture in the thinner loess sheets because of a perched water table on the underlying buried soils. The explanation of the soil landscape as a result of this study involves many factors, both geologic and pedologic, and parent material is only one of many factors.

Alluvial fans also offer an opportunity to study the genesis of cumulative soils. One fan studied in southern California will be described here (Yerkes and Wentworth,[44] and unpubl. work of the writer). The fan is at the mouth of a very small watershed underlain with sedimentary rocks, some of which are calcareous. The cumulative origin of the soil profile on the fan is suggested by the distribution of organic carbon with depth in the profile (Fig. 7–6). Organic carbon in the cumulative soil is higher in content, and the depth to a level that is less than 1 per cent organic carbon is greater than it is in a nearby stable, non-cumulative soil profile. Radiocarbon dates suggest that deposition commenced about 10,000 years ago, and this agrees with stratigraphic and radiocarbon data from nearby areas. Textural variation with depth is rather uniform, implying that deposition has been rapid and continuous enough for sedimentation to mask any B-horizon development.

Carbonate in the fan deposit does not help identify it as a cumulative profile, however (Fig. 7–6). The soil carbonate originates from the weathering of rock fragments and calcareous sediments transported from the drainage basin. The radiocarbon date on the carbonate, although subject to error, is a minimum for sediment in that part of the profile, and it is consistent with the other dates because carbonate can undergo solution and re-precipitation (Ch. 5). Carbonate translocation may have gone on during sediment deposition, or the bulk of it may have been translocated at a later time. If carbonate translocation went on during deposition, thick fan deposits should have over-thickened Cca horizons that extend to the base of the deposit. A deep artificial exposure in colluvium adjacent to the fan in similar parent material

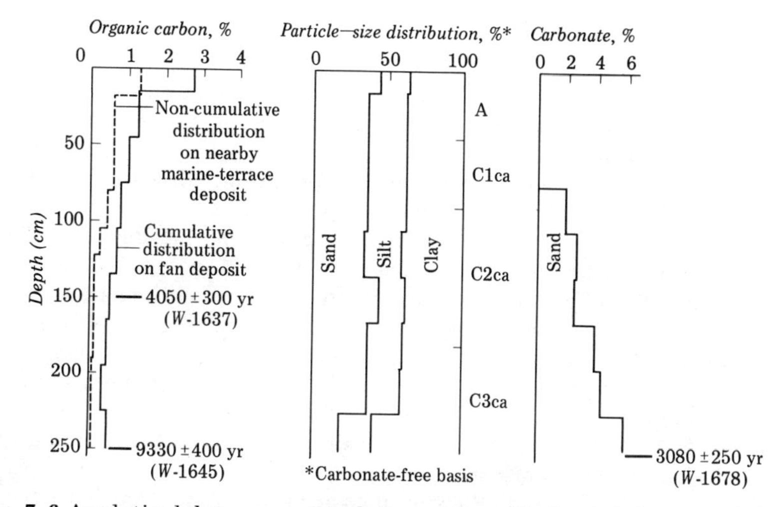

Fig. 7–6 Analytical data on a cumulative soil profile formed from an alluvial fan deposit. The fan deposit overlies a coastal marine terrace west of Santa Monica at the mouth of Coral Canyon. Radiocarbon dates are not from this profile but from a nearby profile with similar morphology; horizontal lines indicate depth from which dated samples were taken. Two samples are on organic carbon, and one is on carbonate. (Soil data from unpublished work of the writer. Radiocarbon dates by M. Rubin, U.S. Geological Survey.)

disclosed that the Cca horizon does not extend throughout the deposit, but has a base, below which is colluvium with a relatively low carbonate content. This can be taken to mean that carbonate did not accumulate with the buildup of sediment on the surface or that its absence at depth is explained by other factors, which may be paleoclimatic or lithologic. That some other factor is involved is supported by rough calculations on rates of sedimentation and carbonate translocation that suggest that translocation can keep pace with sedimentation under the present climatic conditions.

These data demonstrate the intimate interrelationship between geology and pedology in fan and colluvial deposits that form at the bases of slopes or at the mouths of stream valleys. These and loess landscapes are some of the more difficult areas to work in because of the dynamic nature of the geomorphic and pedologic systems.

REFERENCES

1. Barnhisel, R. I., and Rich, C. I., 1967, Clay mineral formation in different rock types of a weathering boulder conglomerate: Soil Sci. Soc. Amer. Proc., v. 31, p. 627–631.
2. Barshad, I., 1958, Factors affecting clay formation: 6th Natl. Conf. Clays and Clay Minerals Proc., p. 110–132.
3. ____ 1964, Chemistry of soil development, p. 1–70 *in* F. E. Bear, ed., Chemistry of the soil: Reinhold Publ. Corp., New York, 515 p.
4. ____ 1966, The effect of a variation in precipitation on the nature of clay mineral formation in soils from acid and basic igneous rocks: 1966 Internat. Clay Conf. (Jerusalem), Proc. v. 1, p. 167–173.
5. ____, Halevy, E., Gold, H. A., and Hagin, J., 1956, Clay minerals in some limestone soils from Israel: Soil Sci., v. 81, p. 423–437.
6. Bates, T. F., 1962, Halloysite and gibbsite formation in Hawaii: 9th Natl. Conf. Clays and Clay Minerals Proc., p. 307–314.
7. Birkeland, P. W., 1964, Pleistocene glaciation of the northern Sierra Nevada, north of Lake Tahoe, California: Jour. Geol. v. 72, p. 810–825.
8. ____ 1969, Quaternary paleoclimatic implications of soil clay mineral distribution in a Sierra Nevada–Great Basin transect: Jour. Geol., v. 77, p. 289–302.
9. ____ 1972, Late Quaternary rock glacier stratigraphy on Mt. Sopris, Colorado, and its bearing on Pinedale-neoglacial deposit differentiation in the Rocky Mountains: Geol. Soc. Amer., Absts. with programs, v. 4, no. 6, p. 367.
10. ____ and Janda, R. J., 1971, Clay mineralogy of soils developed from Quaternary deposits of the eastern Sierra Nevada, California: Geol. Soc. Amer. Bull., v. 82, p. 2495–2514.
11. Blackwelder, E. B., 1931, Pleistocene glaciation in the Sierra Nevada and Basin Ranges: Geol. Soc. Amer., Bull., v. 42, p. 865–922.
12. Bouma, J., Hoeks, J., van der Plas, L., and Scherrenburg, B. van, 1969, Genesis and morphology of some alpine podzol profiles: Jour. Soil Sci., v. 20, p. 384–398.
13. Brewer, R., 1964, Fabric and mineral analysis of soils: John Wiley and Sons, New York, 470 p.
14. Cady, J. G., 1960, Mineral occurrence in relation to soil profile differentiation: 7th Internat. Cong. Soil Sci. Trans., v. 4, p. 418–424.
15. Goldich, S. S., 1938, A Study in rock-weathering: Jour. Geol., v. 46, p. 17–58.
16. Harradine, F., and Jenny, H., 1958, Influence of parent material and climate on texture and nitrogen and carbon contents of virgin California soils. I. Texture and nitrogen contents of soils: Soil Sci., v. 85, p. 235–243.
17. Hay, R. L., 1959, Origin and weathering of late Pleistocene ash deposits on St. Vincent, B.W.I.: Jour. Geol., v. 67, p. 65–87.
18. Hembree, C. H., and Rainwater, F. H., 1961, Chemical degradation of opposite flanks of the Wind River Range, Wyoming: U.S. Geol. Surv. Water Supply Pap. 1535-E, 9 p.

19. Hunt, C. B., and Sokoloff, V. P., 1952, Pre-Wisconsin soil in the Rocky Mountain region, a progress report: U.S. Geol. Surv. Prof. Pap. 221-G, p. 109–123.
20. Hutton, C. E., 1951, Studies of the chemical and physical characteristics of a chrono-litho-sequence of loess-derived prairie soils of southwestern Iowa: Soil Sci. Soc. Amer. Proc., v. 15, p. 318–324.
21. Jenny, H., 1962, Model of a rising nitrogen profile in Nile Valley alluvium, and its agronomic and pedogenic implications: Soil Sci. Soc. Amer. Proc., v. 26, p. 588–591.
22. Keller, W. D., 1954, Bonding energies of some silicate minerals: Amer. Mineralogist, v. 39, p. 783–793.
23. Loughnan, F. C., 1969, Chemical weathering of the silicate minerals: American Elsevier Publ. Co., Inc., New York, 154 p.
24. Malde, H. E., 1955, Surficial geology of the Louisville Quadrangle, Colorado: U.S. Geol. Surv. Bull. 996-E, p. 217 259.
25. Marchand, D. E., 1970, Soil contamination in the White Mountains, eastern California: Geol. Soc. Amer. Bull., v. 81, p. 2497–2506.
26. Nettleton, W. D., Flach, K. W., and Nelson, R. E., 1970, Pedogenic weathering of tonalite in southern California: Geoderma, v. 4, p. 387–402.
27. Nikiforoff, C. C., 1949, Weathering and soil evolution: Soil Sci., v. 67, p. 219–223.
28. Norrish, K., and Rogers, L. E. R., 1956, The mineralogy of some terra rosa and rendzinas in South Australia: Jour. Soil Sci., v. 7, p. 294–301.
29. Pettijohn, F. J., 1941, Persistence of heavy minerals and geologic age: Jour. Geol., v. 49, p. 610–625.
30. Raeside, J. D., 1959, Stability of index minerals in soils with particular reference to quartz, zircon, and garnet: Jour. Sed. Petrol., v. 29, p. 493–502.
31. Riecken, F. F., and Poetsch, E., 1960, Genesis and classification considerations of some prairie-formed soil profiles from local alluvium in Adair County, Iowa: Iowa Acad. Sci., v. 67, p. 268–276.
32. Rieger, S., and Juve, R. L., 1961, Soil development in Recent loess in the Matanuska Valley, Alaska: Soil Sci. Soc. Amer. Proc., v. 25, p. 243–248.
33. Ruhe, R. V., 1959, Stone lines in soils: Soil Sci., v. 87, p. 223–231.
34. ——— 1969a, Quaternary landscapes in Iowa: Iowa State Univ. Press, Ames, 255 p.
35. ——— 1969b, Application of pedology to Quaternary Research, p. 1–23 *in* S. Pawluk, ed., Pedology and Quaternary Research: Univ. of Alberta Printing Dept., Edmonton, Alberta, 218 p.
36. Scott, G. R., 1963, Quaternary geology and geomorphic history of the Kassler Quadrangle, Colorado: U.S. Geol. Surv. Prof. Pap. 421-A, 70 p.
37. Sherman, G. D., and Uehara, G., 1956, The weathering of olivine basalt in Hawaii and its pedogenic significance: Soil Sci. Soc. Amer. Proc., v. 20, p. 337–340.
38. Smith, G. D., 1942, Illinois loess—variations in its properties and distribution: A pedologic interpretation: Univ. Illinois Agri. Exp. Sta. Bull. 490, p. 137–184.

39. Smith, W. W., 1962, Weathering of some Scottish basic igneous rocks with reference to soil formation: Jour. Soil Sci., v. 13, p. 202–215.
40. Stace, H. C. T., 1956, Chemical characteristics of terra rosa and rendzinas in South Australia: Jour. Soil Sci., v. 7, p. 280–293.
41. Syers, J. K., Jackson, M. L., Berkheiser, V. E., Clayton, R. N., and Rex, R. W., 1969, Eolian sediment influence on pedogenesis during the Quaternary: Soil Sci., v. 107, p. 421–427.
42. Willman, H. B., Glass, H. D., and Frye, J. C., 1966, Mineralogy of glacial tills and their weathering profile in Illinois II. Weathering profiles: Illinois State Geol. Surv. Circ. 400, 76 p.
43. Yeend, W. E., 1969, Quaternary geology of the Grand and Battlement Mesas area, Colorado: U.S. Geol. Surv. Prof. Pap. 617, 50 p.
44. Yerkes, R. F., and Wentworth, C. M., 1965, Structure, Quaternary history, and general geology of the Corral Canyon area, Los Angeles County, California: U.S. Geol. Surv. open-file report, 215 p.

CHAPTER 8

Weathering and soil development with time

Rock and mineral weathering and the development of prominent soil features are time-dependent. The time necessary to produce various weathering and soil features varies, however; those soil properties associated with organic-matter buildup develop rapidly, but those associated with the weathering of the primary minerals develop rather slowly.[72] In this chapter data are presented on the rates of development of various weathering and soil features, soil orders, and clay-mineral alteration products.

The time scale

In order to compare soil data on even a semiquantitative basis, a time scale must be adopted. The dating of the deposits and soils discussed in this chapter, and in others, will vary in accuracy. Some deposits are dated directly by radiometric methods, others are bracketed by radiometric dates, whereas ages of other deposits are known only in a relative way or by correlation with dated deposits elsewhere. Although there is no consensus of opinion on the absolute ages for the deposits I will discuss, the ages presented by Birkeland and others[10] will be used for the Cordilleran Region. Correlation of Cordilleran deposits with those of the midcontinent is speculative beyond the range of radiocarbon dating (about 40,000 years). Where time-stratigraphic names are used, the following *approximate* ages in years before the present are suggested

Holocene (0–10,000)
Late Wisconsin (10,000–30,000)
Middle Wisconsin (30,000–40,000)
Early Wisconsin (40,000–130,000)
Sangamon (130,000–250,000+)
Illinoian (pre-250,000?)

It should be noted that although these figures do not correspond with the post-Sangamon ages tentatively proposed by Flint,[22] they are of value in that they indicate order-of-magnitude ages. It is these ages that are used to plot data for the figures in this chapter. An alternative would be to use local geologic names, but this would confuse the reader not familiar with local successions.

One problem should be brought up with respect to the use of this time scale. The early Wisconsin as used here is considered to be those deposits laid down just after the last long major interglaciation, here thought to be the Sangamon. The Sangamon of the midcontinent, however, is poorly dated, as is the Illinoian. In fact, the older part of what I have called the early Wisconsin could include both the Illinoian and the Sangamon.

Rate of rock and mineral weathering

As discussed in Chapter 7, rocks and minerals weather at different rates. Here I will present some quantitative data on the rate of weathering. Some data on initial rates of weathering come from the study of tombstones and other cultural features, but the majority of the data on longer durations of weathering for which the time factor is reasonably well known come from the study of Quaternary unconsolidated deposits.

Tombstones or other man-made structures are good indicators of the rate at which weathering can proceed above the ground surface. Data are not too numerous, however; they cover a variety of rock types in a variety of climates, and they are commonly referred to in most physical geology textbooks, as well as in Jenny[37] and Ollier.[48] At any rate, rocks that weather quite rapidly, such as some limestones, can lose their tombstone inscriptions in as little as 100 years in a humid climate. With more resistant rock, such as some sandstones and igneous rocks, it may take several centuries before the tombstone shows distinct signs of weathering.[53] It would seem therefore that several centuries are sufficient for weathering to be visible on almost any rock type in a humid climate. Arid-climate weathering proceeds at much slower rates.[3]

Studies of glacial tills and outwash deposits are ideally suited to the determination of weathering of different lithologies as a function of time because all factors, except paleoclimate, can be kept constant. Because quantitative weathering studies usually are made in conjunction with stratigraphic studies, there are data from which fairly sound conclusions can be drawn. Most of these studies were pioneered by Blackwelder[12] and have been continued by other workers. Tills with deep exposures are best for these comparisons, because in places where only surface boulders are present one cannot be certain that any lithologic variations that are recognized are due solely to weathering subsequent to deposition. Some variation in the percentage of fresh boulders and in boulder frequencies at the surface, for example, could be a function of the weathered nature of the terrain over which the glaciers advanced. If the terrain had been highly weathered, only the more resistant lithologies could be picked up, transported, and deposited as boulders. If the terrain had not been weathered, boulders of all lithologies might be plentiful. Little is known of the amount of weathering of landscapes between glacial advances. In a study in Idaho, however, concentration of thorium-bearing minerals as placer deposits in outwash seems to be correlated with the duration of the interglaciation immediately prior to the glacial advance that produced the outwash.[61]

In weathering studies of glacial tills, one can compare the amounts of subaerial versus subsurface weathering with time (Fig. 8–1). Many tills in the Cordilleran Region display the weathering succession shown in Fig. 8–2, at least for granitic rock types. Weathering of late-Wisconsin tills has taken place on the exposed surfaces of all stones so that they show signs of oxidation and have a micro-relief due to weathering. Granitic stones on rock-glacier surfaces of late-Wisconsin age in western Colorado have weathering-oxidation rinds as thick as 45 mm. The same rock types in glacial till at depth, however are less weathered and retain glacial polish and striations[6]; in addition, volcanic clasts at depth have extremely thin weathering rinds or none at all. Hence, for that period of time subaerial weathering exceeds subsurface weathering. In contrast, tills of early-Wisconsin age contain granitic stones weathered at the surface and partly or wholly weathered to grus at depth. Therefore, for this period of time, weathering of granitic clasts at depth is equal to or exceeds that at the surface. Volcanic stones in soil on early-Wisconsin deposits show little weathering or have 0.4 to 0.7 mm rinds in the Sierra Nevada, but clasts in till near Lassen Peak, California, have rinds up to 2 mm thick,[18] andesitic clasts in the western Cascade Range south of Seattle have rinds 2 to 5 mm thick (D. R. Crandell, pers. commun., 1972), basalt clasts on the east slope of the Cascade Range have rinds about 1 mm thick,[52] and those in western Colorado in deposits mapped by Yeend[73] have rinds

(A) (B)

Fig. 8–1 Variation in stone weathering in the Rocky Mountains with age of till and position with respect to surface. (A) Rotated boulder on late-Wisconsin till; that part exposed to surface weathering (left) is quite weathered, whereas that part that was buried beneath the surface (right) is virtually unweathered. (B) Rotated boulder on early-Wisconsin till; weathering is extensive both on parts of the boulder that have remained at the surface (right) and on parts that have been beneath the surface (left).

2 mm thick. Tills of the youngest pre-Wisconsin glaciation in most areas generally have granitic boulders that are highly weathered at the surface and weathered to grus at depth, and dense volcanic stones like basalt in soils in the Sierra Nevada and the eastern Cascade Range have rinds about 2 mm thick.

Quantitative data on surface-boulder weathering can be provided by the relief between protrusions of mafic inclusions and aplite and pegmatite dikes above the surface of the weathering boulder. These give minimum rates on the subaerial weathering of granitic boulders. Data from the central Sierra Nevada indicate that such relief is slight on late-Wisconsin till stones, as much as 7.5 cm on early-Wisconsin stones, and 15 cm on pre-Wisconsin stones.[11]

To summarize, initial weathering of granitic clasts is characterized by rates of subaerial weathering that exceed those of subsurface weathering; with time, however, subsurface rates are greater than subaerial rates. Thus, in old tills, most granitic stones at the surface have been at the surface for a long time; they are not lag gravel from

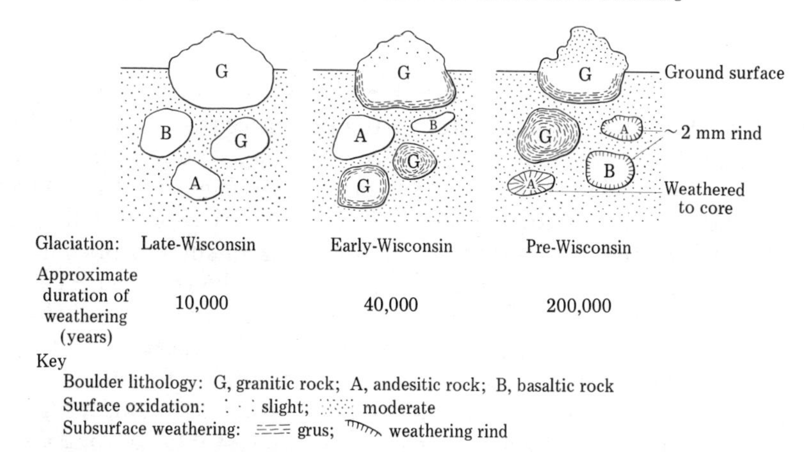

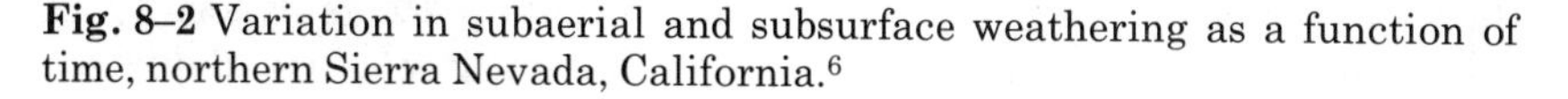

Fig. 8–2 Variation in subaerial and subsurface weathering as a function of time, northern Sierra Nevada, California.[6]

depth. This difference in weathering with position relative to the surface is basic to an understanding of the origin of topographic features in granitic terrain.[70] Furthermore, it is important to remember that in the time necessary to alter a sound granitic boulder to grus in a soil, dense volcanic rocks may show only thin weathering rinds.

The rate of subsurface weathering of granitic stones east of the Rocky Mountains is somewhat similar to that reported above. Granitic clasts in outwash in Ohio show progressively greater alteration with age, but only in possible pre-Illinoian outwash are they thoroughly decomposed.[38]

Surface-weathering studies can give an indication of the amount of granitic-rock breakdown with time. In such studies it is important that the definition of fresh and weathered boulders be readily applicable and used consistently over large areas. An adequate definition of a weathered granitic clast is that on over one-half of the exposed surface weathering penetrates the depth of the average grain diameter.[11] These surfaces are characterized by loose minerals or minerals that can be readily loosened with a light hammer blow. It should be noted that the weathered boulders may have spalled subsequent to deposition and still be considered fresh; the only requirement is the condition of the mineral grains on the present surface. Data collected from widely separated areas give comparable results in both per cent fresh boulders and frequency of boulders per unit area (Fig. 8–3). We can conclude that 10^5 to 10^6 years are necessary for most stones to obtain weathered surfaces, and over 10^5 years are necessary to decompose all surface granitic stones. In many places, however, surface granitic stones persist for much longer periods of time.

There are few data on surface weathering of non-granitic clasts, but several workers have published data on the ratios of granitic to non-granitic lithologies of surface stones. In most cases the granitic stones are the more readily decomposed. For example, in one area in the eastern Sierra Nevada the granitic to metamorphic clast ratio is 80 to 20 for late-Wisconsin till and 30 to 70 for early-Wisconsin till.[63] In northeastern Oregon, the surfaces of early-Wisconsin moraines have 70 to 90 per cent basalt and metamorphic clasts, the remainder being granodiorite; pre-Wisconsin moraines, in contrast, contain everything but granodiorite.[17] Thus, resistant non-granitic rocks persist at the surface long after the granitic rocks have been reduced to grus.

The rate of mineral weathering can be determined in two ways. In one, the uniformity of the parent material can be established, and then the ratio of resistant to nonresistant minerals for dated surfaces gives the rates at which the more weatherable minerals are depleted. Ruhe[55,57] has done this for an area in Iowa in which erosion surfaces of varying age are cut on Kansan Till (Fig. 8–4). Soils that formed over a longer period of time have higher ratios of resistant to nonresistant minerals. The duration of weathering is difficult to estimate because the landscape has undergone burial by loess, and, in some places, it has been stripped by erosion. Nevertheless, Ruhe[55] estimates that the soil

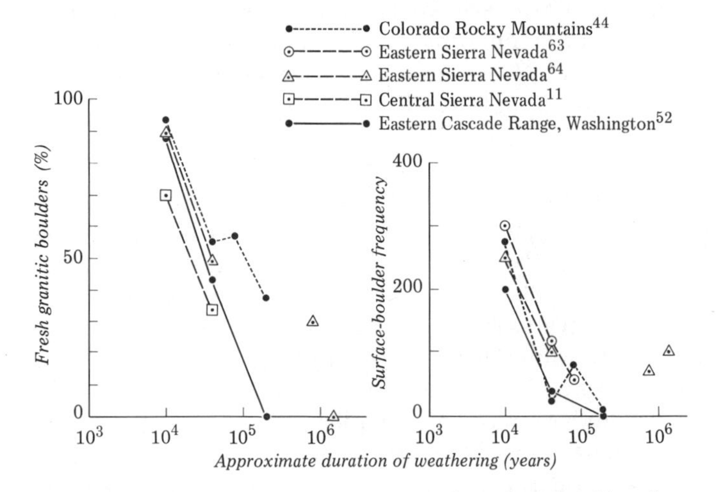

Fig. 8–3 Per cent fresh granitic boulders and surface-boulder frequencies (number in 186 m^2) for several areas in the Cordilleran Region. See text for the definition of a fresh granitic boulder. In both diagrams, the two older eastern Sierra Nevada data points are not from the same drainage basin as the other eastern Sierra Nevada data; hence, they are not connected by dashed lines.

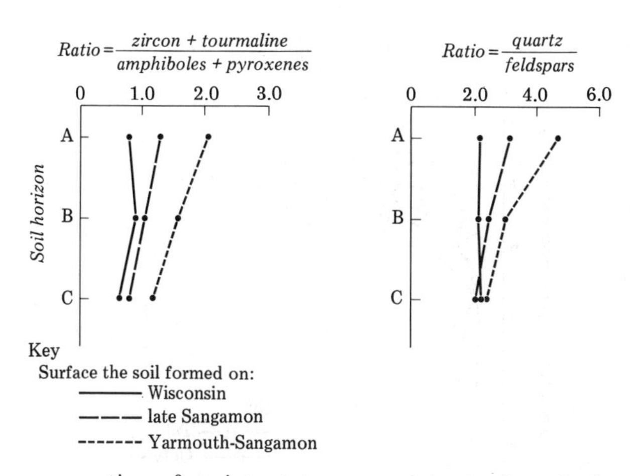

Fig. 8–4 Average ratios of resistant to nonresistant minerals for soils developed on surfaces of different age from Kansan Till in Iowa. (Taken from Ruhe,[55] Figs. 5 and 6, © 1956, The Williams & Wilkins Co., Baltimore, and Ruhe,[60] Table 3.1, reprinted by permission from *Quaternary Landscapes in Iowa*, © 1969 by the Iowa State University Press, Ames.)

related to the Wisconsin surface may have weathered 6800 years, the soil related to the late-Sangamon surface no less than 13,000 years and probably much longer, and the soil related to the Yarmouth-Sangamon surface 10^5 years or more. The change with time is considerable, and 10^4 years or more are necessary to record detectable variations in the ratios. The ratios also show that the surface horizons are the most strongly weathered and that weathering extends into the C horizons of the older soils. Brophy[16] analyzed Sangamon soils in Illinois in a similar way and described the influence of parent material texture on weathering in addition to that of age (Fig. 8–5). Outwash has lost 90 per cent of the original hornblende compared to a 60 per cent loss in till. The reason for the differences with texture is that outwash deposits are more permeable than tills, and therefore depletion in hornblende can go on at a more rapid rate than it can in till. Tills in Indiana also show some mineral depletion with time.[5] Heavy minerals in soils on Wisconsin Tills display little alteration, whereas those on Illinoian and Kansan Tills display a lot; these two older tills cannot be differentiated by their weathering alone, however.

Another way to study rates of mineral weathering is to examine individual mineral grains for signs of weathering, such as cockscomb terminations (Fig. 8–6). This method may be more sensitive than the ratio method in showing change, at least in the earlier stages of weathering, because minerals can show signs of weathering and yet not be depleted. Data on individual grains are not always reported, so little

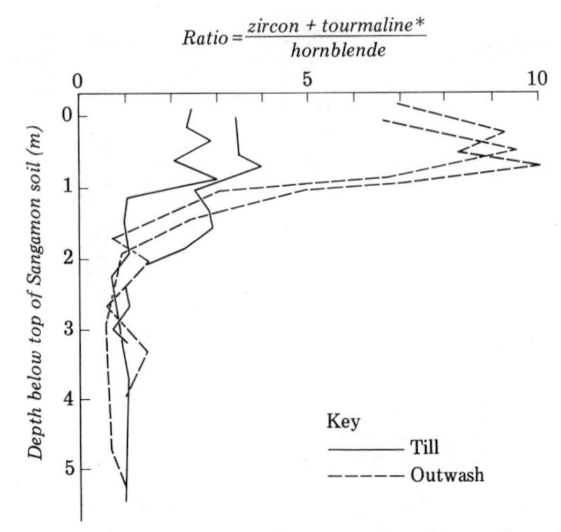

*Ratio is for indicated depth relative to that for lowest 3 samples

Fig. 8–5 Relationship of resistant to nonresistant mineral ratios with depth and with texture of the parent material. (Taken from Brophy,[16] Fig. 11.) The ratio is for the indicated horizon relative to that for the three lowest samples. Clay contents in the assumed parent materials are 14 to 25 per cent for the two tills and 3 to 10 per cent for the two outwash deposits. If the data of Ruhe for the oldest soils (Fig. 8–4) are recalculated on the same basis as the ratios here, the ratios for till in both areas are comparable.

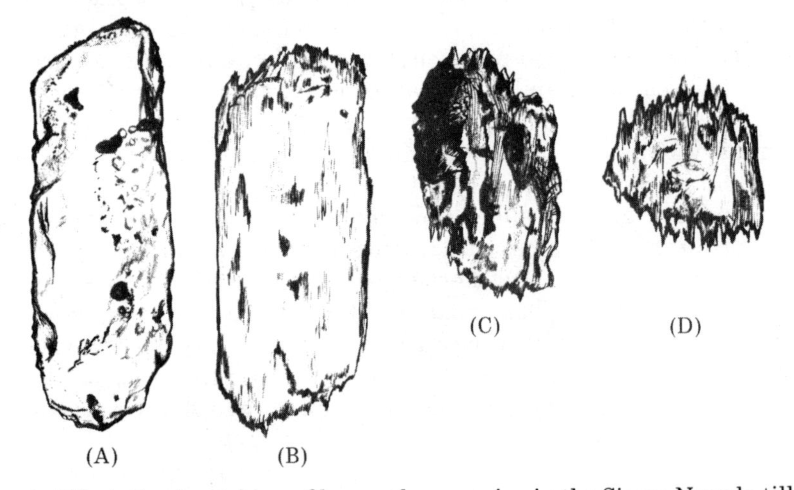

Fig. 8–6 Variation in etching of hypersthene grains in the Sierra Nevada tills described by Birkeland.[6] J. G. LaFleur (unpub. report, 1972) has studied the relationship between degree of grain etching and age of soil to find: (A) unetched or (B) slightly etched grains that are characteristic of soils formed on early- and late-Wisconsin deposits; (C) and (D) are highly etched forms that are common in soils formed from the youngest pre-Wisconsin till.

information is available. Data from several sources (Fig. 8–7), however, suggest age limits for mineral weathering. The weathering in soil on St. Vincent, British West Indies, is one of rapid alteration and depletion. Most of the minerals show some alteration in 4000 years, and calcic plagioclase is nearly gone in several tens of thousands of years. Mineral weathering in soils of central Europe seems at least comparable to that on St. Vincent; in addition, hornblende alteration proceeds at a rapid rate in Europe, probably at about the same rate as in Michigan. The data for the central and southern California coast are for a soil dated at 10^4 years and for samples collected below soils for the older deposits. Noticeable weathering of pyroxenes takes more than 10^4 years, and hornblende etching and the depletion of pyroxenes take more than 10^5 years. Hornblende may weather more rapidly in the soil environment at these California localities, but no data are available. Data on pyroxene weathering in soils of the eastern Sierra Nevada and the Wallowa Mountains show very little etching in 10^4 years, but extensive etching in 10^5 years or more. Hornblende alteration also is slow in the Colorado Piedmont, with 10^5 years or more being required

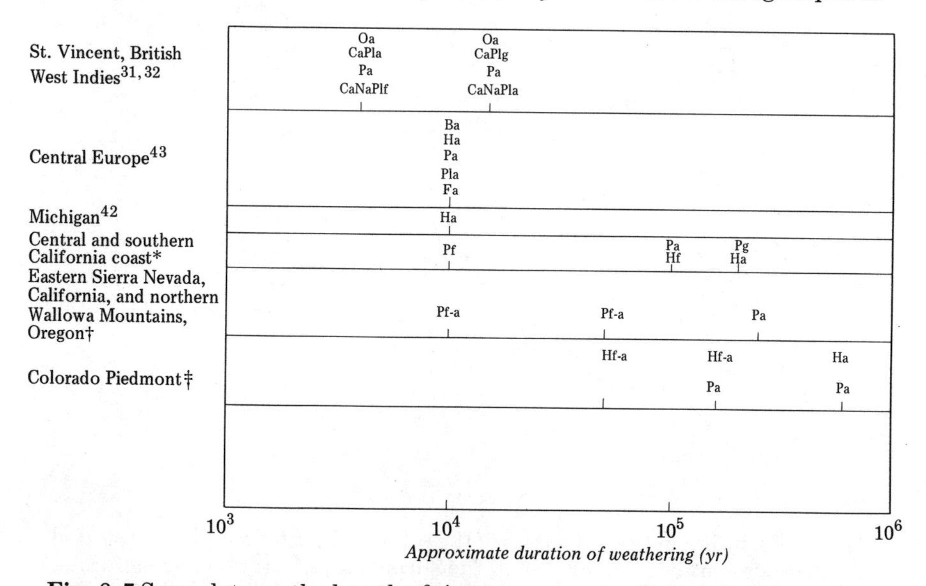

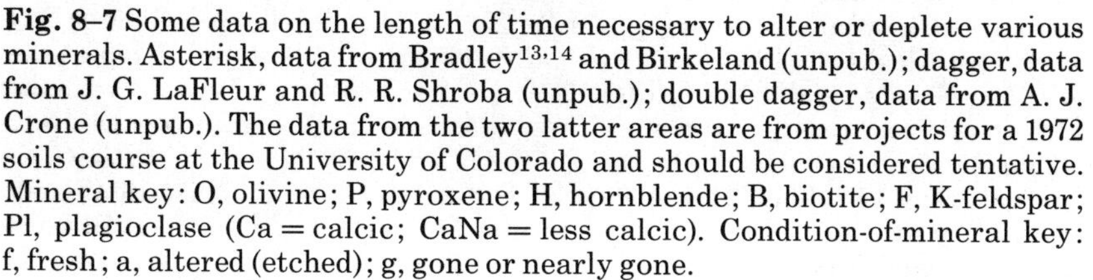

Fig. 8–7 Some data on the length of time necessary to alter or deplete various minerals. Asterisk, data from Bradley[13,14] and Birkeland (unpub.); dagger, data from J. G. LaFleur and R. R. Shroba (unpub.); double dagger, data from A. J. Crone (unpub.). The data from the two latter areas are from projects for a 1972 soils course at the University of Colorado and should be considered tentative. Mineral key: O, olivine; P, pyroxene; H, hornblende; B, biotite; F, K-feldspar; Pl, plagioclase (Ca = calcic; CaNa = less calcic). Condition-of-mineral key: f, fresh; a, altered (etched); g, gone or nearly gone.

for noticeable alteration. This is supported by work in Rocky Mountain National Park, Colorado, in which the hornblendes in soils formed on both early- and late-Wisconsin moraines are unetched (R. R. Shroba, unpub. report, 1972). Hay[32] makes the point that in comparing the rate of weathering of two profiles by mineral alteration the depth through which weathering is noticeable should be considered. That is, intensive weathering to a shallow depth may be equal in the total amount of material altered by less intense weathering to greater depth.

Soil morphology and time

Many of the prominent properties of soil profiles require a fairly

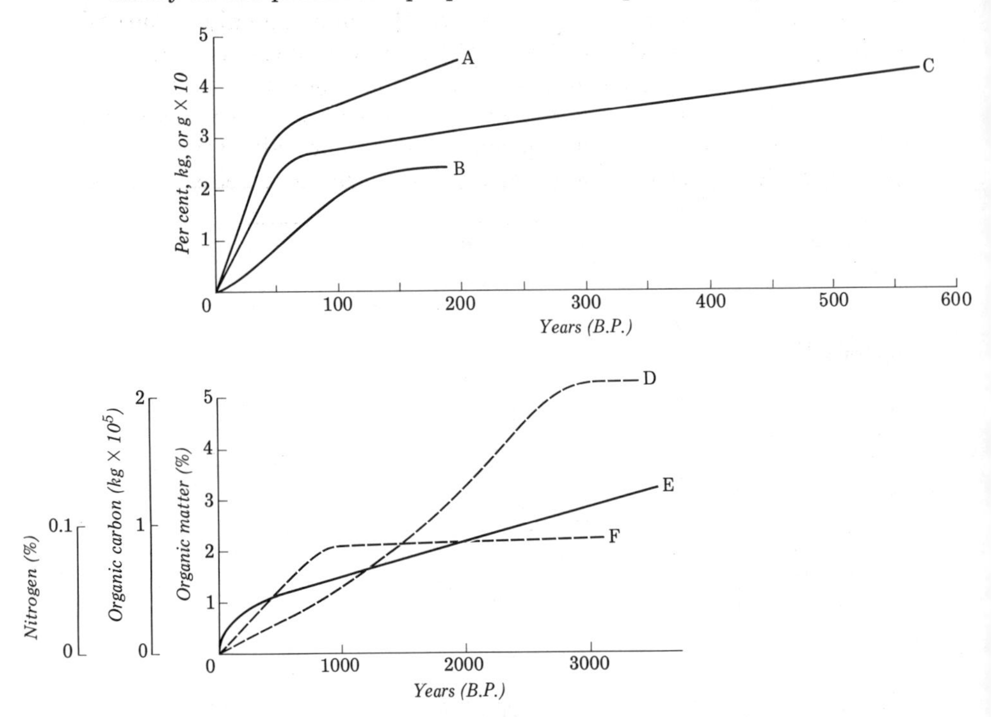

Fig. 8–8 Variation in the properties of the A horizon with time. Key for plotted data: (A) per cent organic matter in the surface layer, Størglaciär moraines, Sweden[65]; (B) organic carbon (kg/m^2) for the top 46 cm of the soil, Glacier Bay moraines, Alaska[19]; (C) organic carbon (g/46 cm^2) for the top 91.5 cm of the soil, Mt. Shasta mudflows, California[21]; (D) per cent organic matter in the surface horizon, Front Range moraines, Colorado[40]; (E) organic carbon (kg/hectare) for the top meter of the soil, New Zealand sand dunes[66] (note that this curve increases to about 2.1 at 10,000 years; thus the rate of increase diminishes beyond the last plotted point); (F) nitrogen in the top 10 cm of the soil, Lake Michigan sand dunes.[49]

long time to form. The time it takes to acquire particular soil features varies with the feature and is of considerable interest to pedologists who work on the soil-forming processes and to geomorphologists who might use the data to date deposits or surfaces.

A-horizon organic matter and associated properties

Organic matter probably reaches a steady state more rapidly than any other property of the soil. Data are given in Fig. 8–8 for widely spaced localities, different parent materials, different properties of the A horizon, and different sampling depths. Nevertheless, the data suggest that the time to achieve steady state may range from as little as 200 to over 3000 years. Data from the Willamette Valley, Oregon,[51] and from Iowa[50] indicate that the steady state is reached within 550 years and in 1000 years or less, respectively. The data also suggest that the A horizon can reach a new steady state quite rapidly if conditions change. Hence, it is one soil property that is amenable to the functional use of factors because it can be quantified, and polygenesis is not an insolvable problem. The A horizon properties of most soils are probably in a steady state with prevailing conditions and therefore are of limited use in stratigraphic studies except for very young deposits or surfaces.

Several other soil properties develop as rapidly as the trends of the organic-matter constituents of the A horizon, and they probably respond to these trends. These changes in properties are reported in many of the articles referenced in Fig. 8–8, as well as in Crocker and Dickson.[20] Soil pH commonly becomes more acid quickly and reaches a steady state at about the same time as does the A-horizon organic-matter content (Fig. 8–9). Carbonates, if present, will influence the pH

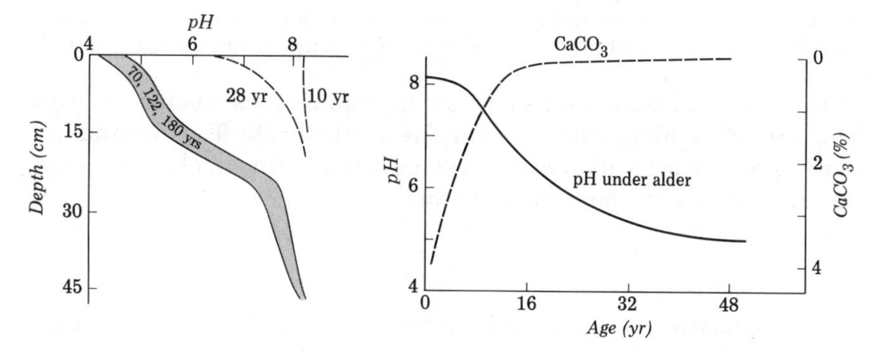

Fig. 8–9 Variation in pH and carbonate content with time, Glacier Bay, Alaska. (Taken from Crocker and Major,[19] Figs. 4 and 11.) Data in the right hand figure are for the top 5.1 cm of the soil.

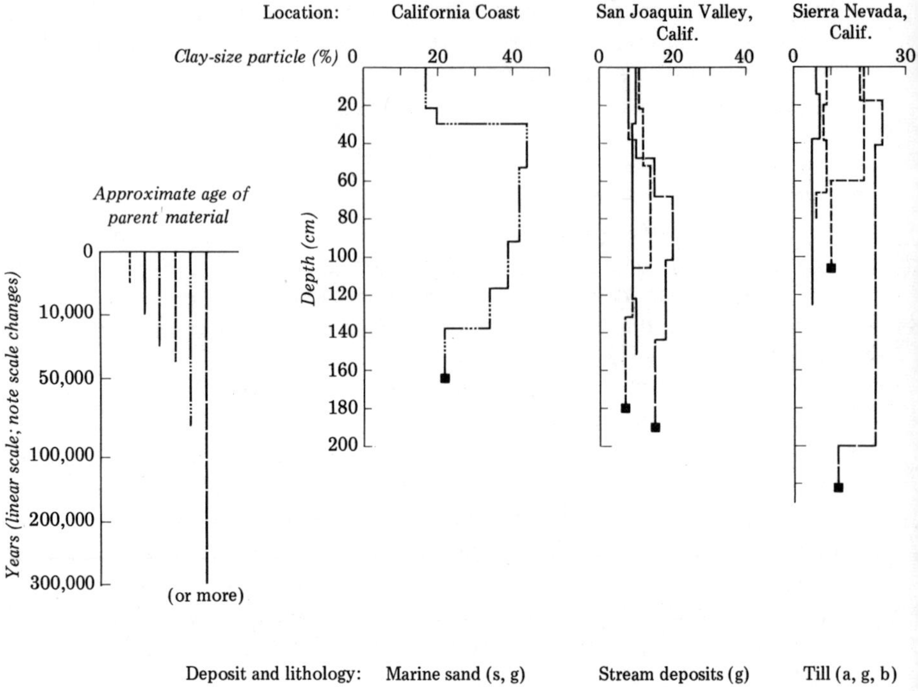

Fig. 8–10 Clay-size-particle variation with depth for deposits of different age from the California coast to Ohio. The indicated age is for the parent material; the age of the soil could be younger because of either erosion between the end of deposition and the beginning of soil formation or burial of the soil by younger deposits. For the Sierra Nevada tills, two analyses of approximately 40,000-year-old soils are shown. Soil without the textural B horizon is usual; soils with textural B horizons are relatively rare, and the one depicted here is buried by a late-Wisconsin till. Although the less than 5000-year-old soil for the Colorado

unless they are leached as rapidly as the organic constituents build up. The rate of leaching will depend on the local climate. The eventual pH at the steady state will depend on some combination of climatic, vegetational, and parent-material influences.

Textural B-horizon development

The formation of a textural B horizon is time-dependent, because clay formation and translocation are relatively slow processes. Although the best approach here would be to calculate the quantities of clay formed and translocated as a function of time, quantitative data

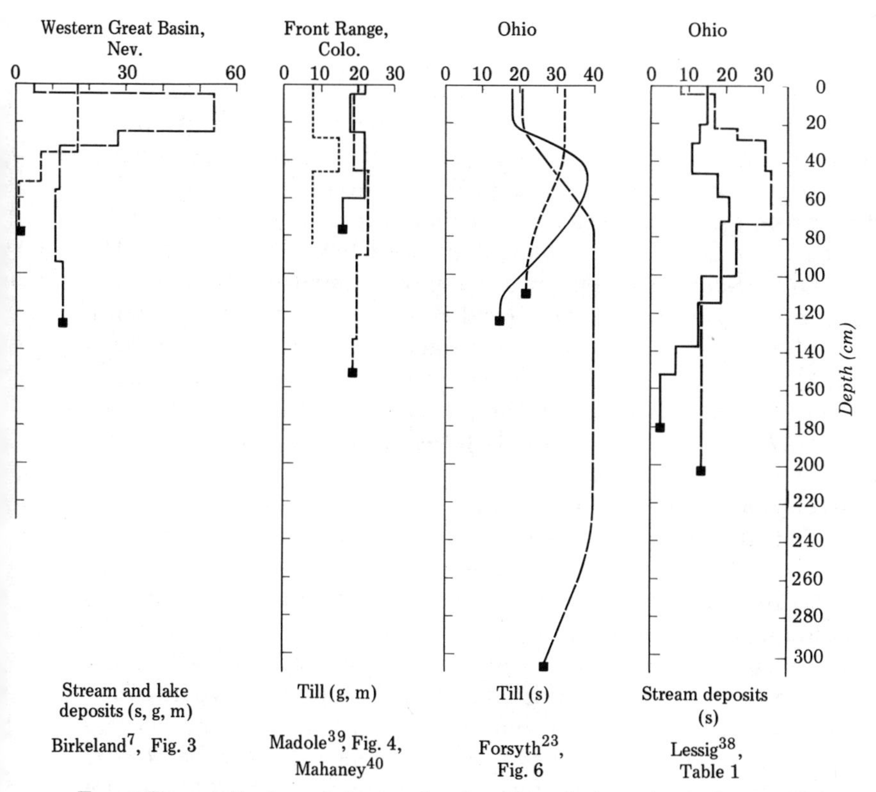

Front Range displays a textural subsurface bulge, the bulge may be due to depositional rather than to pedological processes. The oldest and youngest Ohio tills are composites of several profiles, and the intermediate-age soil is a reconstructed profile; this latter soil is buried by late-Wisconsin till and lacks an A horizon. Key to parent-material lithology: g, granitic rock; a, andesitic rock; b, basaltic rock; m, metamorphic rock; s, sedimentary rock. Filled box at base of clay curve indicates a textural B horizon recognizable in the field.

on rates are quite rare.[32] In lieu of that, I will take the stratigraphic approach and compare clay-content profiles of variously aged deposits from different regions. Another important facet of B-horizon formation is the progressive development of microscopic features seen in thin section. This topic has been reviewed recently by Brewer and Sleeman[15] and will not be covered here.

A plot of clay content against depth usually shows an increase in the amount and in the thickness of B-horizon clay with time. Data on soils from a variety of regions are given in Fig. 8–10; these soils are thought to be representative of the rate of clay buildup for the respective region and for the indicated parent material. Several things are quite evident in the plots. One is that the rate of clay buildup varies remarkably from region to region. Regions with the most rapid rates, the California coast and Ohio, have parent materials characterized by high initial clay contents; hence, one reason for the rate of clay buildup could be translocation of clay already present. Another reason could be that parent materials with high clay content have high water-holding capacities, and these promote more rapid clay formation than would be the case with parent materials of low clay content. The western Great Basin also appears as a region of rapid clay buildup. The origin of the clay in the western Great Basin soils is not known, but if part is eolian, this could account for some of the increase with time. Evidence that eolian influx of solid particles is an important process in desert soil formation is accumulating. In contrast, soils with low initial clay content in the western United States seem to require much more time to develop textural B horizons.

The relationship of clay distribution with depth and time is important in soil-stratigraphic studies, because clay content is one of the basic properties used in the correlation of unconsolidated deposits. One should, therefore, be aware of the regional differences in the rates of clay buildup, or else erroneous correlations might result. To illustrate the problem, one can compare stream deposits in California with those in Ohio (Fig. 8–10). In the California deposits, some 40,000 years of soil formation seem to be required to form a minimal B horizon, whereas soils of younger age in Ohio have well-developed B horizons. Again, one can note the regional variation in the Sangamon soil, a major stratigraphic marker in U.S. Quaternary studies (soils shown as having parent materials about 300,000 years old in Fig. 8–10). This soil is recognized throughout the midcontinent by its thickness, depth of leaching of carbonate minerals, and color. The probable Sangamon soil equivalents in the western United States also are good markers in their respective areas, relative to the other soils in the local succession, but none of the western U.S. Sangamon soils approaches its mid-continental counterpart in the amount of clay and the thickness of the B horizon.

Field studies indicate that although clay distribution in soils is a useful criterion in determining the relative ages of surficial deposits, its usefulness decreases with older deposits. It is commonly observed, for example, that whereas the Sangamon soil is a widespread and easily recognized stratigraphic marker, it is difficult to differentiate older soils of the major interglacials from it on most field soil criteria, including clay content.[45,54] This seems to be true whether the soils are at the surface or buried. Thus, Sangamon-type clay profiles approximate the maximum obtainable for the environment of most regions.

Development of red color

A common observation in many areas is that older soils are redder than younger soils. If the soil redness, determined by the reddest color of the master horizon immediately beneath the A or E horizon, is plotted against the approximate age of the parent material, it is shown that long intervals of time are needed to develop the reddest color in a soil chronosequence (Fig. 8–11). These trends commonly parallel clay-buildup trends, in that the time necessary to attain maximum clay content approximates that necessary to attain maximum redness. The initial buildup of redness, however, seems to be more rapid than the initial buildup of clay. The reason for this discrepancy is unclear, but it might be that the more easily weathered iron-bearing minerals provide the coloring pigment for the soil before the more slowly weathering aluminosilicate minerals alter to clay-size particles. Although parent material can influence color development, its influence in Fig. 8–11 is considered to be slight, because most of the younger soils have 10YR hues. The oldest soils in the chronosequences plotted, however, vary in maximum redness, and the reason for this is not clear; perhaps it is related to differences in climate.

One problem in linking color development with age of parent material is the role of paleoclimate in producing redness. It is known, for example, that red soils form fairly rapidly in areas characterized by high temperatures. Hence, if soil redness increases with time, one or both of two factors may be responsible. One factor is time, and the

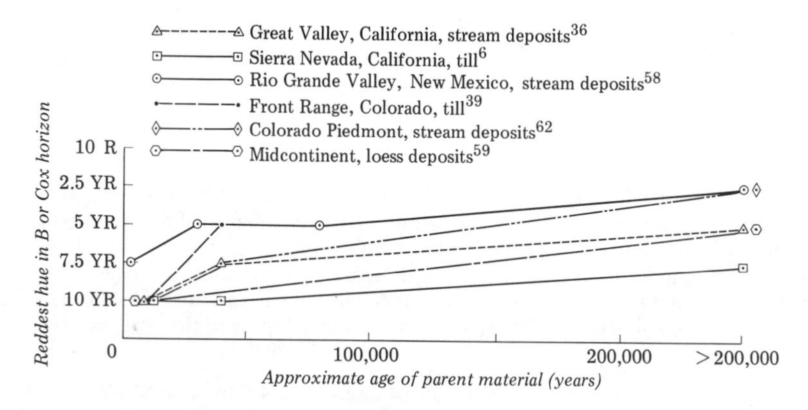

Fig. 8–11 Color hue for the reddest part of the B horizon, or for the Cox horizon if no B has formed, for deposits of different age: central California to mid-continent. Indicated age is for the parent material; the age of the soil could be younger because of erosion between the end of deposition and the beginning of soil formation or burial of the soil by younger deposits. Soils in the western Great Basin of an age similar to those of the Sierra Nevada, California, tills show somewhat similar color trends.[7]

other is warmer past climates. It is difficult to separate the effects of these two factors.[56] Paleoclimatic interpretation of soils, therefore, should not rest solely on color, because other soil properties might give better clues on past climates.

Soil $CaCO_3$ *buildup and removal*

The rate at which carbonate builds up in a soil is dependent on the process of formation as well as the rate of leaching in the soil. If the necessary Ca^{2+} comes from mineral weathering, buildup will be controlled by the rate of weathering of the calcium-bearing minerals. If, however, $CaCO_3$ was originally present in the parent material, the rate of buildup in the soil is a function of the rate at which it can be translocated by leaching waters in the soil profile. If, however, the $CaCO_3$ is of eolian origin, buildup is a function of the rate of both influx and translocation.

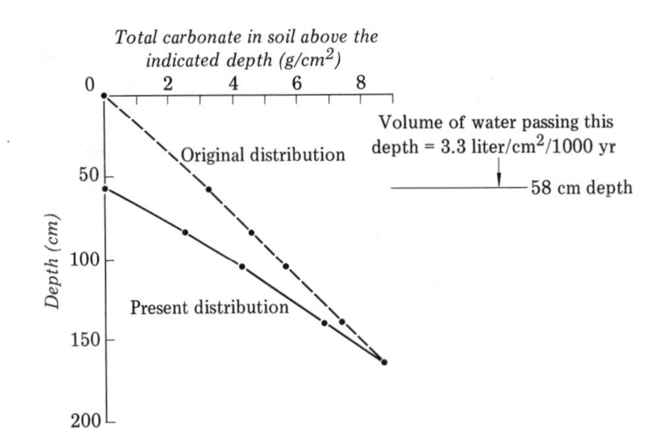

Fig. 8–12 A calculation of the approximate time necessary to redistribute carbonate in a soil.[1] Carbonate is assumed to have been uniformly distributed in the soil, subsequently removed from the uppermost 58 cm, and then precipitated in the 58 to 165 cm interval. It is calculated that about 3.3 liters of water/cm²/1000 yr move by the 58-cm-depth level in the soil, and, from Fig. 5–7, a soil carbonate solubility of 0.1 g/liter is estimated for the prevailing physical and chemical conditions.

$$\text{Time for carbonate translocation} = \frac{\text{Amount of carbonate translocated from uppermost 58 cm of soil}}{\text{Volume of water passing through the 58-cm-depth level of soil} \times \text{Carbonate solubility}}$$

$$= \frac{3.24\ \text{g}}{(3.3\ \text{liter}/1000\ \text{yr}) \times (0.1\ \text{g/liter})} = 9800\ \text{years}$$

Provided the $CaCO_3$ can be shown to have been distributed uniformly in the soil parent material, one can make a rough calculation as to how long it might take to redistribute the $CaCO_3$ by solution, translocation, and re-precipitation. Arkley[1] has presented a method for these calculations, using estimations on the volume of water passing various levels in the soil, as determined from soil and climatic data, and data on $CaCO_3$ solubility. One such calculation is shown in Fig. 8–12, and, although it and other ages he has calculated are fairly rough and only a minimum, the calculated ages are reasonably consistent with the geologic evidence on age.

There are few regional stratigraphic data on the rate of buildup of soil $CaCO_3$. Perhaps the best-studied area in this respect is near Las Cruces, New Mexico; there the Soil Conservation Service has been conducting a long-range project on desert soils. The rate of buildup in nongravelly parent materials is quite rapid (Fig. 8–13), perhaps one of the more rapid ones in the western United States. Of interest here is the observation that a K horizon can form on deposits of about 30,000 years old. In other parts of the western United States, deposits of early-Wisconsin age generally do not have K horizons, and such horizons do not become a prominent soil feature except in soils formed from pre-Wisconsin deposits.

Airborne-sediment traps have been in place in the Las Cruces area for several years, and they and chemical data on rainfall indicate that much of the soil $CaCO_3$ might be of eolian origin.[29,58] The yearly influx of dust approximates 0.1 kg/m^2. Most of this eolian material is non-

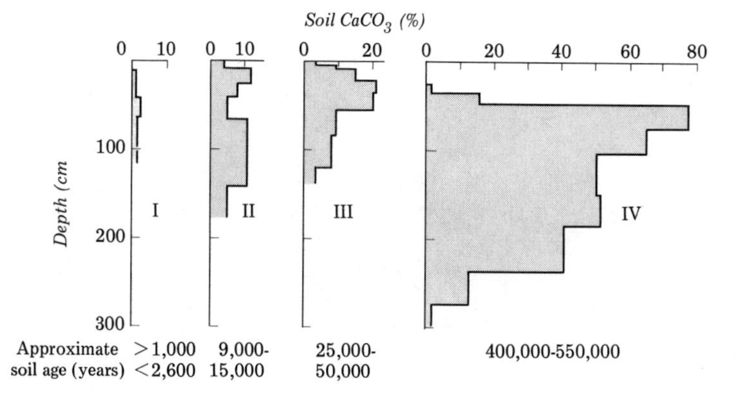

Fig. 8–13 Carbonate distribution with depth for soils of different ages, Las Cruces area, New Mexico.[28,29] The data are for nongravelly parent materials, and the Roman numerals correspond to carbonate buildup stages (Appendix 1). Most ages are approximate, based on data available in March 1972 (L. H. Gile, written commun.).

carbonate clay, silt, and very fine sand; organic carbon makes up less than 10 per cent of the total, and $CaCO_3$ less than 2 per cent.

One might suspect that the age of a soil could be estimated by dividing the $CaCO_3$ influx into the total soil $CaCO_3$. The system seems more complex than that, however. Some complications are (a) part of the water runs off and therefore does not translocate $CaCO_3$ in the soil; (b) some of the eolian $CaCO_3$ may be blown away before the next rainfall can move it into the soil; and (c) the Ca^{2+} that goes to making additional $CaCO_3$ may come from other sources, such as calcium-bearing salts, the Ca^{2+} in the rainfall, and the Ca^{2+} on the exchangeable sites of the colloidal fraction. A rough estimate of the maximum potential $CaCO_3$ influx at Las Cruces from atmospheric sources (dust + precipitation) is 2 $g/m^2/yr$, a figure that is lower than the late-Quaternary rates calculated from the soil data and estimated age of the soils (Fig. 8–14). This suggests that the present measured rate of influx probably is not representative of that over the last 40,000 years.

Parent materials in some humid regions contain original $CaCO_3$, and because of the prevailing climate this material is removed slowly in solution by the percolating waters. Some figures on the rate of $CaCO_3$ depletion from soils are as follows: removal from the top 5 cm of the soil in less than 50 years at Glacier Bay, Alaska,[19] and from the top 2 m in 1000 years in sand dunes along Lake Michigan.[49] Stratigraphic studies in the midcontinent have tended to use the depth of $CaCO_3$ leaching as one criterion for the approximate age of the parent material. Thorp[67]

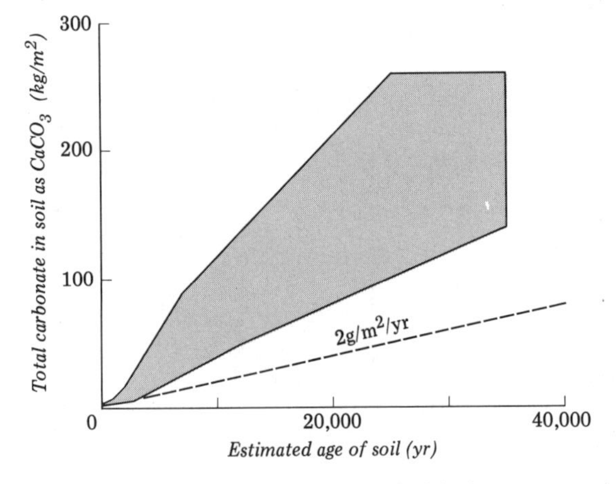

Fig. 8–14 Envelope curve showing the relationship between estimated age of various soils, Las Cruces, New Mexico area, and total carbonate expressed as $CaCO_3$. (Data from Gile and others,[29] Table 20.) For comparison, the approximate present-day maximum potential atmospheric contribution of $CaCO_3$ of 2 $g/m^2/yr$ is plotted.

points out, for example, that Wisconsin drifts in northern Illinois are leached to 0.5 to 1.5 m depth, whereas Illinoian drifts in the southern part of the state are leached to 2.5 to 3.5 m depth. Thus, for any region with a uniform parent material there might be a depth of $CaCO_3$ leaching corresponding to age of the parent material. Many workers however, warn, that this criterion for age should be used with caution because many variables can alter the rate of $CaCO_3$ removal. Furthermore, even if environmental conditions remain constant for long periods of time, the rate of leaching would decrease with depth in the soil because less water is available for leaching at greater depths in the soil, and the water there could be saturated with respect to $CaCO_3$. Hence, one would not expect a linear relationship between depth of leaching and duration of soil formation.

Clay mineralogy

The clay mineralogy of a soil may change with time, and if it does this may confound efforts to use clay minerals as paleoclimatic indicators. Two cases will be explored. In one the initial parent material contains little inherited clay, and clay formation from alteration of nonclay minerals predominates. In the other the parent material contains clay minerals of diverse types, and many of the clays that form in the soil do so by alteration of pre-existing clays. Whether or not a clay mineral will alter is a function of its stability in the soil solution as shown by stability diagrams (Figs. 4–6 and 4–7). If clay minerals form from the weathering of nonclay minerals, the clay produced should be stable in that environment, or else it would not have formed. However, inherited clays may be deposited in an environment in which they are not stable, according to the stability diagrams, and so theoretically they should alter to types stable in that environment.

There is little evidence that clay-mineral content in soils varies with time for parent materials initially low in clay content. A transect along the eastern Sierra Nevada and into the western Great Basin demonstrates the point.[8,9] In this region, a large number of tills and outwash deposits of diverse lithology were sampled in a variety of environments. Soils on deposits of different ages throughout the region were sampled, and their ages range from postglacial to probably several 10^5 years. Very seldom did the data on clay minerals indicate a variation mainly with age. Deposits of several ages in the same drainage basin usually have a common clay-mineral assemblage in the soils. Other factors, specifically parent material and climate, seem to explain most clay-mineral variations. The only effect that time could have had on these soils would be if paleoclimates were markedly different from current climates, but even this effect seems to have been slight. Time alone, therefore, does not seem to cause soil clay minerals of pedogenic origin to alter successively to other clays.

If, however, the properties of a soil change with time so that the internal environment of the soil changes, the originally formed clays may alter. For example, the buildup of clay-size particles in a soil may change the leaching environment enough that (a) existing clay minerals are unstable, or (b) clay minerals that form in the future differ from those that formed in the past, yet both are in equilibrium with the water chemistry at the time of their respective formation. Another example of a pure time control on clay-mineral alteration would be if the weathering of the nonclay minerals was so intense that certain minerals that delivered weathering products to the solution were depleted, and that their depletion so altered the water chemistry that the previously formed clays then became unstable. This may happen in areas of intense weathering, such as the tropics or areas of podzolization, or in stable landscapes in which weathering dates back to the early Quaternary and Tertiary.

In places where a variety of clay minerals are present in the original parent material, changes in mineral species may take place for any mineral not stable in the environment. Jackson and others[34] ranked the clay minerals by relative resistance to weathering and assigned a number to each mineral characteristic of a weathering stage (Fig. 8–15). Three of the more stable minerals are not given in Fig. 8–15; they are allophane (stage 11), hematite (stage 12), and anatase (stage 13). For most of the minerals, a decrease in iron, magnesium, and silicon is associated with increased mineral stability. The arrows in the figure indicate some possible paths for alteration from one mineral to another. In soils with a combination of original clay minerals in the parent material any alteration due to weathering will be regulated by the water-mineral equilibria. The extent of the alteration, or in other words the weathering stage reached, will be a function of time, because the alterations involve slow reactions and water-mineral equilibria; that

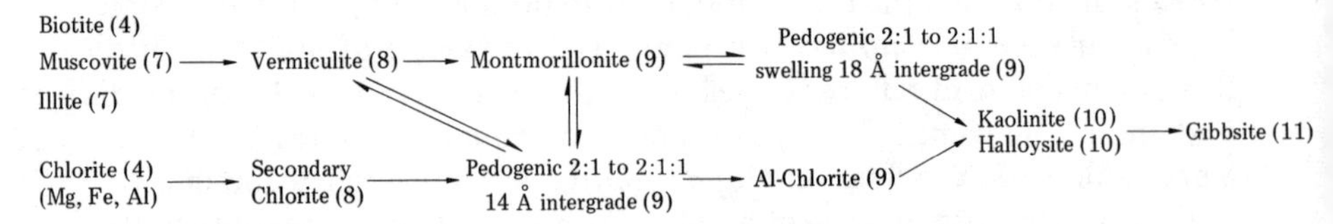

Fig. 8–15 Various clay-mineral reactions in soil. Parenthetical numbers are weathering-stability index numbers for the clay minerals; mineral stability increases with number. (From Jackson,[33] p. 124 *in* Chemistry of the Soil by F. E. Bear, ed., © 1964 by Litton Educational Publishing, Inc. Reprinted by permission of Van Nostrand Reinhold Company.)

is, alteration takes place until an equilibrium mineral assemblage is formed, after which time further alteration probably does not occur.

The midcontinent is a good region to study the effect of time on clay mineral alteration. Tills and loesses of several ages are present in many environments, and parent materials contain a variety of clay minerals derived from bedrock units over which the glaciers advanced. Clay-mineral transformations are judged from the vertical sequence of clay minerals in a soil and by comparison of soils of different ages. Many data have been accumulated on the region, but only two example areas will be discussed.

Clay minerals formed in soils from tills in Illinois demonstrate the changes possible with advancing age (Fig. 8–16). The soil on early-Wisconsin till shows only an alteration of chlorite to vermiculite-chlorite in the B and leached Cox horizons. Some soils on late-Wisconsin till display similar alteration.[25,27] This similarity between clay-mineral alteration in soils formed from early- and late-Wisconsin tills does not always hold true. In some places, illite depletion is recognized in soils on early-Wisconsin tills, but not in soils on late-Wisconsin tills[71]; in addition, some of the early-Wisconsin tills with original montmorillonite show an increase in montmorillonite in the soil. In contrast to the soils on early-Wisconsin tills, Sangamon soils have more altered clays, and the alteration goes to greater depth[16,71] (Fig. 8–16). Chlorite is altered through a vermiculite-chlorite stage to material termed "heterogeneous-swelling material" that probably is either montmorillonite or mixed-layer mineral.[71] Illite alters to illite-montmorillonite and finally to montmorillonite. Yarmouth soils seem to be altered to about the same stage.

Soil formed from late-Wisconsin loess in Illinois shows slight clay mineral alteration at most.[26] The major alteration is that of montmorillonite to the heterogeneous-swelling material. Other changes are slight depletion of illite, modification of chlorite, and possibly formation of vermiculite and kaolinite.

In Illinois, therefore, the stable clay mineral seems to be stage 9 montmorillonite, but of the heterogeneous-swelling material variety. This material can form in the time span represented by the weathering of late-Wisconsin deposits. More intense alteration to great depths, however, require weathering durations on the order of Sangamon time. A different clay-mineral pattern is seen in the accretion-gley deposits, and this is discussed in Chapter 9.

Soils in Indiana show somewhat similar trends.[4] Illite and chlorite are present in the parent tills. The alteration goes to vermiculite and illite-montmorillonite in post-Wisconsin soils (between stages 8 and 9) and to montmorillonite (stage 9) in Sangamon soils. Kaolinite (stage

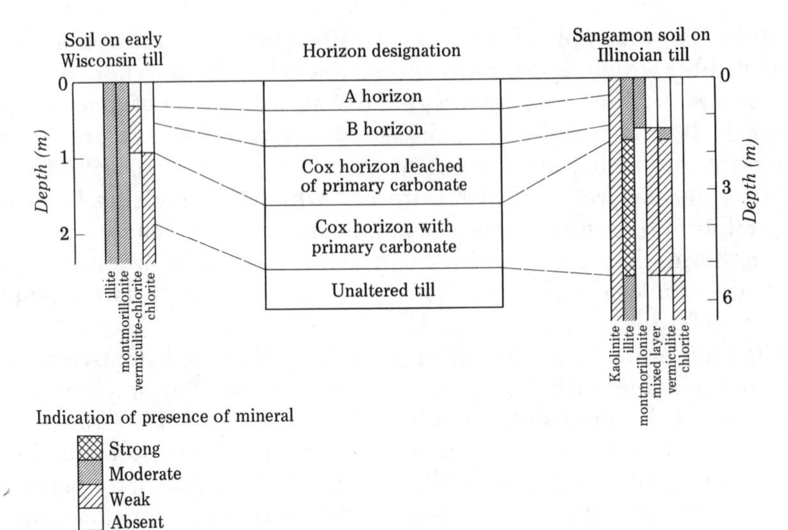

Fig. 8–16 Clay-mineral variation with depth in soils of different ages formed from till, Illinois. (Taken from Frye and others,[25] Fig. 3.) The expandable vermiculite in the original publication has subsequently been shown to be heterogeneous-swelling material[71]; it is here shown as montmorillonite in the Sangamon soil. Note the difference in vertical scale for the two profiles.

10) is found in the parent material of the Yarmouth soils, but also some seems to have formed in the soils.

In summary, it seems that the following generalizations are valid. Clay minerals that form in soils from parent materials low in clay content probably form mineral assemblages stable in that environment, and therefore variation in the minerals with age generally is not found and not expected. Soils formed from parent materials high in clay content, and with a variety of clay minerals, are another matter. Some clays in the assemblage may be unstable and gradually change over to more stable forms. The change will progress from the surface downward, and complete change to a stable assemblage may take considerable time. In these cases, the presence of clay mineral assemblages may be an aid in identifying either the age of a surface soil or the occurrence of a period of weathering and/or soil formation within a thick section of sediments.

Relationship between soil orders and time

There is a fairly good correlation between the soil orders and the age of the underlying deposits or landscapes in some regions. A striking example in the United States is the widespread occurrence of

Ultisols in the midcontinent and in the east (Fig. 2–5). These soils have formed on deposits and landscapes of pre-Wisconsin age. Although it is admittedly difficult to decide whether the main factor involved in their formation is primarily time or paleoclimate,[56] at least some investigators feel that the soils required long periods of time to form.[47] Another example is the band of Ultisols along the west slope of the Sierra Nevada and in the Cascade Range of California[30] (Fig. 2–5). Detailed stratigraphic studies in the region indicate that these soils primarily are restricted to pre-Wisconsin deposits.[18,35] In northwestern Oregon, Trimble[68] suggests that 10^6 years or more were required to form some of the deeply weathered Ultisols.

The Oxisols of the world also seem to have required long periods of time to form. Maignien[41] reviews most of the data on them and shows that many Oxisols date from the Tertiary or early Quaternary. Although the climate in some regions, such as Australia, may have been different during the time of Oxisol formation, these soils are so highly weathered that formation times of tens to hundreds of thousands of years or more do not seem unreasonable. Preliminary stratigraphic work in Hawaii indicates that under moist conditions ($>$ 250 cm rainfall) Oxisols may form from volcanic ash in about 10^5 years (S.C. Porter, pers. commun., 1972). More examples could be given. However, the point is made that some soils classified at the order level require exceedingly long periods of time to form, and therefore some maps of soil distribution at the order level can be used to suggest broad groupings of deposits or landscapes on the basis of age.

In contrast to soil orders that seem to require long periods of time to form, many form in a short time. An example of this is the Spodosols. Tamm (cited in Jenny[37]) reports that Spodosols in northern Sweden require only 1000 to 1500 years to reach a steady-state condition. In northern Michigan, well-expressed Spodosols take more than 3000 years but less than 8000 years to form.[24]

Time necessary to attain the steady-state condition

That soils do reach a steady-state condition is shown by plots of the variation in particular soil properties with time, as seen in many of the figures in this chapter. Curves for the buildup of most properties initially are fairly steep, but after some time the curves flatten, indicating little visible change thereafter with time.

The time necessary to reach the steady state will vary with the soil property being studied, the parent material, and the particular kind of soil profile that forms in a particular environment.[72] Thus, A-horizon properties form rapidly whereas B-horizon properties form rather

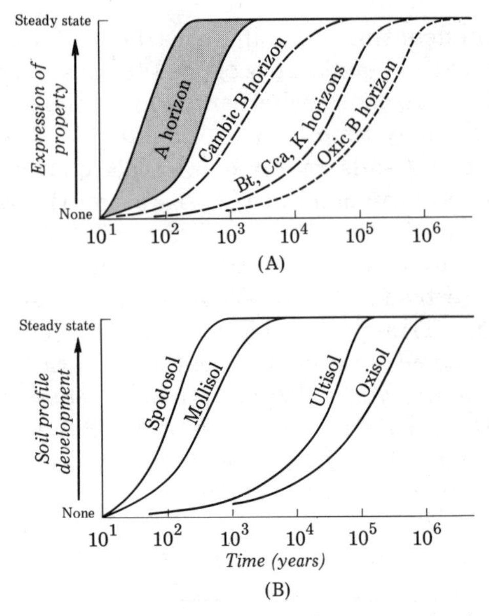

Fig. 8–17 Diagram showing the variations in time to attain the steady state for (A) various soil properties and (B) various soil orders.

slowly (Fig. 8–17, A). The variation in rates of formation of the various kinds of B horizons is a function of the processes responsible for each kind of B horizon. Because a soil profile is the sum total of many soil properties, a profile can be said to be in a steady state only when most of its diagnostic properties are in a steady state. Hence, soil orders will vary in the time necessary to reach a steady state (Fig. 8–17, B).

The steady-state condition is reached for the different properties by different processes. For example, at the steady state for organic matter in the A horizon, the gain of undecomposed organic matter to the soil is balanced by the loss of organic matter due to decomposition. The on-going processes at the steady state for the Bt, Cca, and K horizons are another matter. I doubt if the growth of clay minerals is balanced by their chemical destruction, or that the precipitation of $CaCO_3$ is balanced by its solution and removal. Rather, these steady states might involve a delicate balance between removal of surficial material by erosion and the slow extension of the soil profile at depth into unweathered material.[46] With oxic B horizons, the steady state might be reached only when most of the weatherable minerals have been depleted by weathering, thus rendering further change unlikely.

REFERENCES

1. Arkley, R. J., 1963, Calculation of carbonate and water movement in soil from climatic data: Soil Sci., v. 96, p. 239–248.
2. —— 1964, Soil survey of the eastern Stanislaus area, California: U.S. Dept. Agri., Soil surv. series 1957, no. 20, 160 p.
3. Barton, D. C., 1916, The disintegration of granite in Egypt: Jour. Geol., v. 24, p. 382–393.
4. Bhattacharya, N., 1962, Weathering of glacial tills in Indiana: I. Clay minerals: Geol. Soc. Amer. Bull., v. 73, p. 1007–1020.
5. —— 1963, Weathering of glacial tills in Indiana: II. Heavy minerals: Jour. Sed. Petrol., v. 33, p. 789–794.
6. Birkeland, P. W., 1964, Pleistocene glaciation of the northern Sierra Nevada north of Lake Tahoe, California: Jour. Geol., v. 72, p. 810–825.
7. —— 1968, Correlation of Quaternary stratigraphy of the Sierra Nevada with that of the Lake Lahontan area, p. 469–500 *in* R. B. Morrison and H. E. Wright, Jr., eds., Means of correlation of Quaternary successions: Internat. Assoc. Quaternary Res., VII Cong., Proc. v. 8, 631 p.
8. —— 1969, Quaternary paleoclimatic implications of soil clay mineral distribution in a Sierra Nevada–Great Basin transect: Jour. Geol., v. 77, p. 289–302.
9. —— and Janda, R. J., 1971, Clay mineralogy of soils developed from Quaternary deposits of the eastern Sierra Nevada, California: Geol. Soc. Amer. Bull., v. 82, p. 2495–2514.
10. ——, Crandell, D. R., and Richmond, G. M., 1971, Status of correlation of Quaternary stratigraphic units in the western conterminous United States: Quaternary Res., v. 1. p. 208–227.
11. Birman, J. H., 1964, Glacial geology across the crest of the Sierra Nevada, California: Geol. Soc. Amer. Spec. Pap. 75, 80 p.
12. Blackwelder, E. B., 1931, Pleistocene glaciation of the Sierra Nevada and Basin Ranges: Geol. Soc. Amer. Bull., v. 42, p. 865–922.
13. Bradley, W. C., 1957, Origin of marine-terrace deposits in the Santa Cruz area, California: Geol. Soc. Amer. Bull., v. 68, p. 421–444.
14. —— 1965, Marine terraces on Ben Lomond Mountain, California, p. 148–150 *in* Guidebook for Field Conf. I, Northern Great Basin and California: Internat. Assoc. Quaternary Res., VII Cong., Nebraska Acad. Sci., Lincoln.
15. Brewer, R., and Sleeman, J. R., 1969, The arrangement of constituents in Quaternary soils: Soil Sci., v. 107, p. 435–441.
16. Brophy, J. A., 1959, Heavy mineral ratios of Sangamon weathering profiles in Illinois: Illinois State Geol. Surv. Circ. 273, 22 p.
17. Crandell, D. R., 1967, Glaciation at Wallowa Lake, Oregon: U.S. Geol. Survey Prof. Pap. 575-C, p. C145–C153.
18. —— 1972, Glaciation near Lassen Peak, northern California: U.S. Geol. Surv. Prof. Pap. 800-C, p. C181–C190.
19. Crocker, R. L., and Major, J., 1955, Soil development in relation to vegetation and surface age at Glacier Bay, Alaska: Jour. Ecology, v. 43, p. 427–448.

20. ___ and Dickson, B. A., 1957, Soil development on the recessional moraines of the Herbert and Mendenhall glaciers, southeastern Alaska: Jour. Ecology, v. 45, p. 169–185.
21. Dickson, B. A., and Crocker, R. L., 1953, A chronosequence of soils and vegetation near Mt. Shasta, California. II. The development of the forest floors and the carbon and nitrogen profiles of the soils: Jour. Soil Sci., v. 4, 142–154.
22. Flint, R. F., 1971, Glacial and Quaternary geology: John Wiley and Sons, New York, 892 p.
23. Forsyth, J. L., 1965, Age of the buried soil in the Sidney, Ohio, area: Amer. Jour. Sci., v. 263, p. 571–597.
24. Franzmeier, D. P., and Whiteside, E. P., 1963, A chronosequence of podzols in northern Michigan. I. Physical and chemical properties: Mich. State Univ. Agri. Exp. Sta. Quat. Bull., v. 46, p. 21–36.
25. Frye, J. C., Willman, H. B., and Glass, H. D., 1960, Gumbotil, accretion-gley, and the weathering profile: Illinois State Geol. Surv. Circ. 295, 39 p.
26. ___, Glass, H. D., and Willman, H. B., 1968, Mineral zonation of Woodfordian loesses of Illinois: Illinois State Geol. Surv. Circ. 427, 44 p.
27. ___, ___, Kempton, J. P., and Willman, H. B., 1969, Glacial tills of northwestern Illinois; Illinois State Geol. Surv. Circ. 437, 47 p.
28. Gile, L. H., 1970, Soils of the Rio Grande Valley border in southern New Mexico: Soil Sci. Soc. Amer. Proc., v. 34, p. 465–472.
29. ___, Hawley, J. W., and Grossman, R. B., 1971, The identification, occurrence and genesis of soils in an arid region of southern New Mexico: Training sessions, desert soil-geomorphology project, Dona Ana County, N. M., Soil Conservation Service, 177 p.
30. Harradine, F., 1966, Comparative morphology of lateritic and podzolic soils in California: Soil Sci., v. 101, p. 142–151.
31. Hay, R. L., 1959, Origin and weathering of late Pleistocene ash deposits on St. Vincent. B. W. I.: Jour. Geol., v. 67, p. 65–87.
32. ___ 1960, Rate of clay formation and mineral alteration in a 4000-year-old volcanic ash soil on St. Vincent, B.W.I.: Amer. Jour. Sci., v. 258, p. 354–368.
33. Jackson, M. L., 1964, Chemical composition of the soil, p. 71–141 *in* F. E. Bear, ed., Chemistry of the soil: Reinhold Publ. Corp., New York, 515 p.
34. ___, Tyler, S. A., Willis, A. L., Bourbeau, G. A., and Pennington, R. P., 1948, Weathering sequence of clay-size minerals in soils and sediments. I. Fundamental generalizations: Jour. Phys. Colloid. Chem., v. 52, p. 1237–1260.
35. Janda, R. J., 1966, Pleistocene history and hydrology of the upper San Joaquin River, California: Unpub. Ph.D. thesis, Univ. Calif. (Berkeley), 425 p.
36. ___ and Croft, M. G., 1967, Stratigraphic significance of a sequence of Noncalcic Brown soils formed on Quaternary alluvium in the northeastern San Joaquin Valley, California, p. 157–190 *in* R. B. Morrison and H. E. Wright, Jr., eds., Quaternary soils: Internat. Assoc. Quaternary Res., VII Cong., Proc. v. 9, 338 p.

37. Jenny, H., 1941, Factors of soil formation: McGraw-Hill, New York, 281 p.
38. Lessig, H. D., 1961, The soils developed on Wisconsin and Illinoian-age glacial outwash terraces along Little Beaver Creek and the adjoining upper Ohio Valley, Columbiana County, Ohio: Ohio Jour. Sci., v. 61, p. 286–294.
39. Madole, R. F., 1969, Pinedale and Bull Lake glaciation in upper St. Vrain drainage basin, Boulder County, Colorado: Arctic and Alpine Res., v. 1, p. 279–287.
40. Mahaney, W. C., 1970, Soil genesis on deposits of neoglacial and late Pleistocene age in the Indian Peaks of the Colorado Front Range: Unpub. Ph.D. thesis, Univ. of Colorado, 246 p.
41. Maignien, R., 1966, Review of research on laterites: UNESCO, natural resources research, IV, 148 p.
42. Matelski, R. P., and Turk, L. M., 1947, Heavy minerals in some podzol soil profiles in Michigan: Soil Sci., v. 64, p. 469–487.
43. Meyer, B. and Kalk, E., 1964, Verwitterungs-mikromorphologie der mineralspezies in mittel-Europäischen Holozän-böden aus Pleistozänen und Holozänen lockersedimenten, p. 109–129 *in* A. Jongerius, ed., Soil micromorphology: Elsevier Publ. Co., New York, 540 p.
44. Miller, C. D., 1971, Quaternary glacial events in the northern Sawatch Range, Colorado: Unpub. Ph.D. thesis, Univ. of Colorado, Boulder, 86 p.
45. Morrison, R. B., 1964, Lake Lahontan: Geology of the southern Carson Desert: U.S. Geol. Surv. Prof. Pap. 401, 156 p.
46. Nikiforoff, C. C., 1949, Weathering and soil evolution: Soil Sci., v. 67, p. 219–230.
47. Novak, R. J., Motto, H. L., and Douglas, L. A., 1971, The effect of time and particle size on mineral alteration in several Quaternary soils in New Jersey and Pennsylvania, U.S.A., p. 211–224 *in* D. H. Yaalon, ed., Paleopedology: Israel Univ. Press, Jerusalem, 350 p.
48. Ollier, C. D., 1969, Weathering: Oliver and Boyd, Edinburgh, 304 p.
49. Olson, J. S., 1958, Rates of succession and soil changes on southern Lake Michigan sand dunes: Botanical Gazette, v. 119, p. 125–170.
50. Parsons, R. B., Scholtes, W. H., and Riecken, F. F., 1962, Soils on Indian mounds in northeastern Iowa as benchmarks for studies of soil genesis: Soil Sci. Soc. Amer. Proc., v. 26, p. 491–496.
51. ——, Balster, C. A., and Ness, A. O., 1970, Soil development and geomorphic surfaces, Willamette Valley, Oregon: Soil Sci. Soc. Amer. Proc., v. 34, p. 485–491.
52. Porter, S. C., 1969, Pleistocene geology of the east-central Cascade Range, Washington: Guidebook for third Pacific Coast Friends of the Pleistocene field conference, Sept. 27–28, 1969, 54 p.
53. Rahn, P. H., 1971, The weathering of tombstones and its relationship to the topography of New England: Jour. Geol. Educ., v. 19, p. 112–118.
54. Richmond, G. M., 1962, Quaternary stratigraphy of the La Sal Mountains, Utah: U.S. Geol. Surv. Prof. Pap. 324, 135 p.
55. Ruhe, R. V., 1956, Geomorphic surfaces and the nature of soils: Soil Sci., v. 82, p. 441–455.

56. ____ 1965, Quaternary paleopedology, p. 755–764 *in* H. E. Wright, Jr., and D. G. Frey, eds., The Quaternary of the United States: Princeton Univ. Press, Princeton, 922 p.
57. ____ 1967a, Geomorphology of parts of the Greenfield quadrangle, Adair County, Iowa: U.S. Dept. Agri. Tech. Bull. 1349, p. 93–161.
58. ____ 1967b, Geomorphic surfaces and surficial deposits in southern New Mexico: New Mex. Bur. Mines and Mineral Resources, Mem. 18, 66 p.
59. ____ 1968, Identification of paleosols in loess deposits in the United States, p. 49–65 *in* C. B. Schultz and J. C. Frye, eds., Loess and related eolian deposits of the world: Internat. Assoc. Quaternary Res., VII Cong., Proc. v. 12, 369 p.
60. ____ 1969, Quaternary landscapes in Iowa: Iowa St. Univ. Press, Ames, 225 p.
61. Schmidt, D. L., and Mackin, J. H., 1970, Quaternary geology of the Long and Bear valleys, west-central Idaho: U.S. Geol. Surv. Bull. 1311-A, 22 p.
62. Scott, G. R., 1963, Quaternary geology and geomorphic history of the Kassler Quadrangle, Colorado: U.S. Geol. Surv. Prof. Pap. 421-A, 70 p.
63. Sharp, R. P., and Birman, J. H., 1963, Additions to classical sequence of Pleistocene glaciations, Sierra Nevada, California: Geol. Soc. Amer. Bull., v. 74, p. 1079–1086.
64. ____ 1969, Semiquantitative differentiation of glacial moraines near Convict Lake, Sierra Nevada, California: Jour. Geol., v. 77, p. 68–91.
65. Stork, A., 1963, Plant immigration in front of retreating glaciers, with examples from the Kebnekajse area, northern Sweden: Geografiska Annaler, v. XLV, p. 1–22.
66. Syers, J. K., Adams, J. A., and Walker, T. W., 1970, Accumulation of organic matter in a chronosequence of soils developed on wind-blown sand in New Zealand: Jour. Soil Sci., v. 21, p. 146–153.
67. Thorp, J., 1968, The soil—a reflection of Quaternary environments in Illinois, p. 48–55 *in* R. E. Bergstrom, ed., The Quaternary of Illinois: Univ. of Illinois, College of Agri., Spec. Publ. no. 14, 179 p.
68. Trimble, D. E., 1963, Geology of Portland, Oregon and adjacent areas: U.S. Geol. Surv. Bull. 1119, 119 p.
69. Wagner, R. J., and Nelson, R. E., 1961, Soil survey of the San Mateo area, California: U.S. Dept. Agri., Soil surv. series 1954, no. 13, 111 p.
70. Wahrhaftig, C., 1965, Stepped topography of the southern Sierra Nevada, California: Geol. Soc. Amer. Bull., v. 76, p. 1165–1190.
71. Willman, H. B., Glass, H. D., and Frye, J. C., 1966, Mineralogy of glacial tills and their weathering profiles in Illinois: Part II. Weathering profiles: Illinois State Geol. Surv. Circ. 400, 76 p.
72. Yaalon, D. H., 1971, Soil-forming processes in time and space, p. 29–39 *in* D. H. Yaalon, ed., Paleopedology: Israel Univ. Press, Jerusalem, 350 p.
73. Yeend, W. E., 1969, Quaternary geology of the Grand and Battlement Mesas area, Colorado: U. S. Geol. Surv. Prof. Pap. 617, 50 p.

CHAPTER 9

Topography–soil relationships

Topography, or local relief, controls much of the distribution of soils in the landscape, to such an extent that soils of markedly contrasting morphologies and properties can merge laterally with one another and yet be in equilibrium under existing local conditions (Fig. 9–1). Many of the differences in soils that vary with topography are due to some combination of microclimate, pedogenesis, and geologic surficial processes, and the sorting out of the effects of each on soil distribution is difficult. The fields of pedology and geomorphology probably overlap here more than with any other pedologic factor.

Soil properties vary laterally with topography. One reason for this is the orientation of the hillslopes on which soils form; this affects the microclimate and, hence, the soil. Another is the steepness of the slope; this affects soil properties because the rates of surface-water runoff and erosion vary with slope. In areas of rolling terrain soil properties vary because lower areas are likely to be areas of accumulation of water runoff and sediment derived from surrounding higher-lying areas. Also, low areas might be influenced by a high water table, which could have a considerable effect on the soil. In this chapter, I will discuss examples of these processes. Of course, we assume other factors to be nearly constant in this analysis; this might be true for some examples, but perhaps not for all.

Influence of slope orientation on pedogenesis

Slope orientation results in microclimatic and vegetation differences, and thus in soil differences. Jenny[7,8] argues that topography is the primary factor in explaining soil variation in these field situations. Here we are concerned with topographic relief of meters, or tens of meters, and therefore regional climate can be considered a constant.

An instructive example of the effect of slope orientation on soil properties is the study of Finney and others[3] in southeastern Ohio (Fig. 9–2). Several NW–SE-trending valleys were studied. The parent material is mostly colluvium derived from sandstone. The microclimate varies quite markedly with orientation, with the SW-facing slopes displaying higher temperatures on the leaf litter, as well as greater annual fluctuations. Temperatures beneath the leaf litter show the same trends with orientation, but the differences between maximum and minimum values are less. Soil-moisture values follow the temperature differences, with soils on NE-facing slopes generally being more moist than those on SW-facing slopes. Vegetation correlates with the

Fig. 9–1 Variation in soils with topographic position in Indiana. Poorly drained, dark-colored Aquolls occur in the lowlands, and somewhat poorly drained Aqualfs occur in the higher parts of the landscape. These latter soils are light colored here because erosion has exposed the light-colored B horizon. [Photo from Marbut Memorial Slide Collection, prepared and published by the Soil Science Society of America (Madison, Wisconsin) in 1968.]

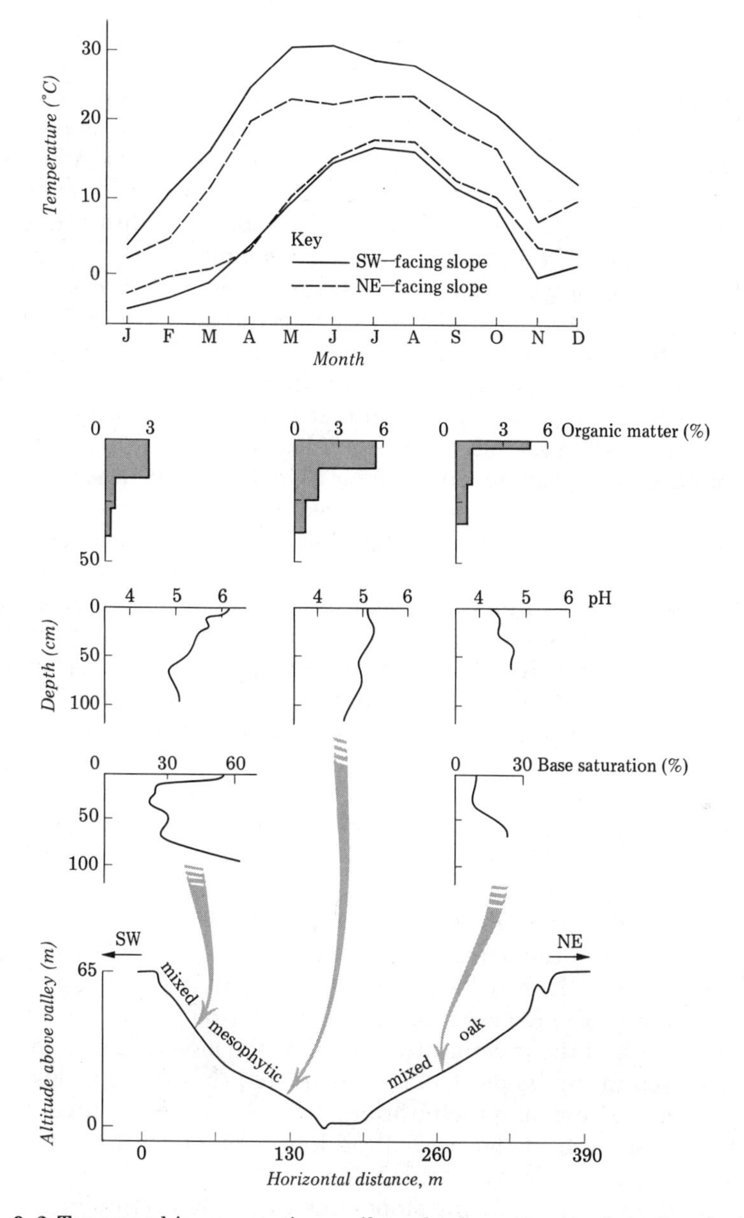

Fig. 9–2 Topographic, vegetative, soil, and microclimatic data for slopes of different orientation, southeastern Ohio. (Taken from Finney and others,[3] Figs. 1, 3, 5, 6, and 7.) Temperatures are maximum and minimum monthly averages on leaf litter, 1956.

moisture-temperature trends. A mixed-oak association is dominant on the SW-facing slopes, whereas a mixed mesophytic plant association is dominant on the NE-facing slopes. The microclimatic-vegetation differences produce soil suborder differences. NE-facing slopes are characterized by Ochrepts with thick A horizons on slopes with gradients over 40 per cent and Udalfs with relatively thick A horizons on the rest of the slopes. However, Udalfs with thin A horizons predominate on the SW-facing slopes, and Ochrepts with thin A horizons are present locally (H. R. Finney, written commun., 1972).

Slope orientation greatly affects soil organic carbon distribution with depth, the presence or absence of an E horizon, pH, and per cent exchangeable bases (Fig. 9–2). Organic matter differences probably result from greater moisture and vegetation cover on the NE-facing slopes, combined with greater organic matter decomposition rates on the SW-facing slopes along with some loss due to surface runoff. E horizons are more common on SW-facing slopes, and the reason for this is not clear. If anything, one would predict that they would be more common on the moister NE-facing slopes. Base saturation and pH trends are somewhat parallel with depth, and these can be explained in terms of vegetation differences and fire. The vegetation assemblage on the NE-facing slopes has a higher base content, and bases are returned to the soil surface as litterfall. However, the mixed-oak association has a lower base content and probably a higher incidence of fire. Fire not only burns off some of the soil organic matter, but also brings about the loss of bases released from the organic matter by either runoff or deep leaching.

A somewhat similar relationship of soils to slope orientation is seen in the rolling hills of eastern Washington.[13] Parent material is Wisconsin loess deposited in NW–SE-trending ridges that have about 65 m of relief (Fig. 9–3). Although the regional precipitation is 53 cm/yr, most of which occurs in the winter, distinct microclimates are produced by the orientation of the hillslopes. It is estimated that the S-facing slopes receive close to the average annual precipitation. However, because about one-fifth of the precipitation is snow, the moisture can be easily redistributed in the landscape. It is estimated that the hilltops might receive only 25 cm of precipitation, with losses attributed to the removal of snow by wind as well as high evaporation rates on the exposed ridges. In contrast, the N-facing slopes have more effective moisture than do the S-facing slopes due to a combination of accumulated drifting snow and lower evaporation rates. The virgin vegetation follows these trends, with the more mesic shrubs, grasses, and forbs on the N-facing slopes and the more xeric forbs and grasses on the S-facing slopes.

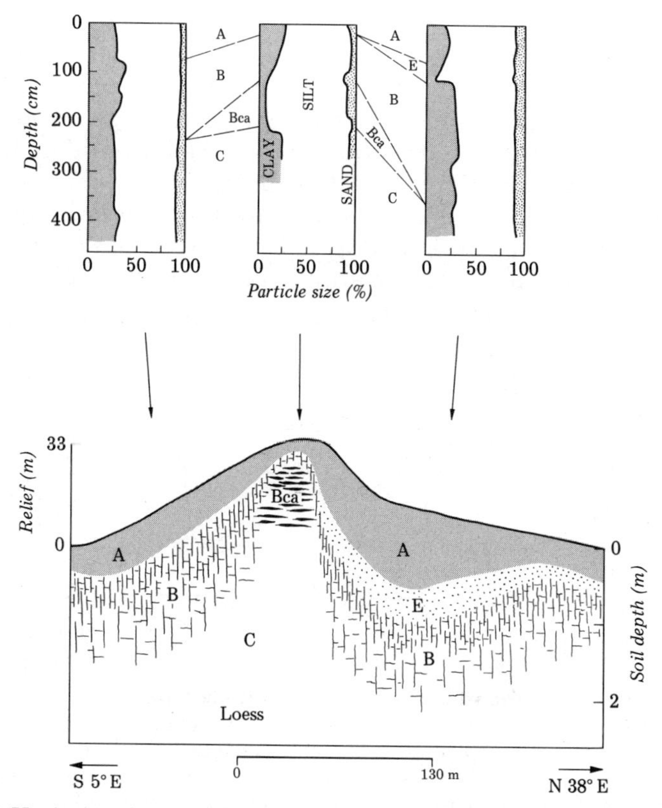

Fig. 9–3 Variation in soil properties with slope orientation in loess hills, eastern Washington. (Taken from Lotspeich and Smith,[13] Figs. 2 and 3, © 1953, The Williams & Wilkins Co., Baltimore.)

Soils are closely related to the microclimate (Fig. 9–3). The S-facing slopes are characterized by noncalcareous Mollisols that are thought to be normal for the region. In contrast, the N-facing slopes, largely because of the additional moisture, have more strongly developed Mollisol profiles and greater amounts of organic matter in the A horizon. The soils on the ridgetops are thin Mollisols, characterized by weak B-horizon development and carbonate accumulation at depth, both properties that are clearly the result of a more arid microenvironment. Thus, it is seen that fairly large differences in soil properties and profiles can be closely associated with each other because of microclimatic differences associated with slope orientation.

Soil properties related to position on a slope

Numerous studies have shown that many soil properties are related to the position of a particular soil on the slope. Nettleton and others[15] have described soils formed from tonalite at various positions on a slope in southern California (Fig. 9–4). They assume that the three soils have formed on slopes of about the same geomorphic age, and therefore the soil differences can be attributed to topographic position. Because of the warm climate, with precipitation mainly in the winter and early spring (38 cm/yr), the soils have ochric A horizons. However, B-horizon properties vary markedly downslope. The Vista soil has only a cambic B, whereas the Fallbrook has an argillic B, and the Bonsall a natric B. The C horizons are tonalite grus.

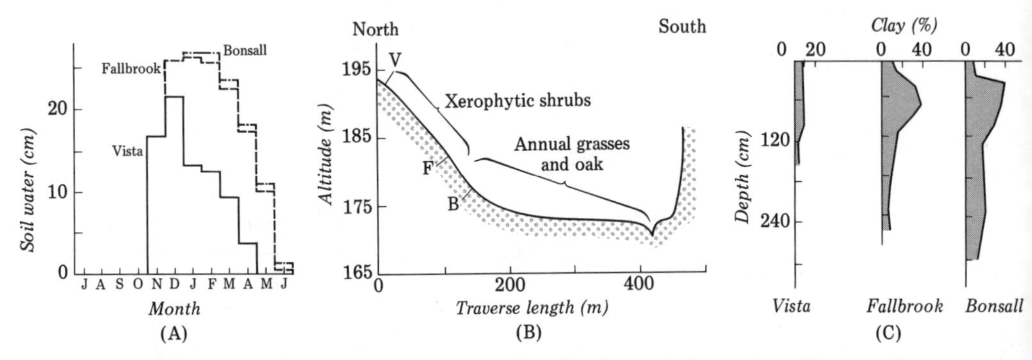

Fig. 9–4 Data for a toposequence of soils formed from tonalite, southern California. (A) estimated soil water for 1965–66; (B) topographic and vegetative relationships [at the sampled sites the slopes are 8% for Vista (V), 12% for Fallbrook (F), and 8% for Bonsall (B)]; (C) distribution of clay with depth. (Taken from Nettleton and others,[15] Figs. 1 and 12, and Table 3.)

Soil-moisture measurements were taken at various times of the year, and from these data and precipitation data for 1965–66, the moisture in the soils was estimated on a monthly basis (Fig. 9–4). These data show that the soils that lie downslope (Fallbrook and Bonsall) receive more soil moisture than the upslope soil (Vista), and that they retain their moisture longer.

A marked feature of these soils is that the clay content increases downslope (Fig. 9–4). Weathering of the primary minerals is greater downslope, and thus most of the clay present can be attributed to weathering of the underlying rock at that site. Weathering and clay-formation differences downslope are most likely due to soil moisture as determined by slope position. Soils in lower slope positions can receive more moisture than those in upslope positions because of lateral movement, either at the surface or within the soil. Furthermore, once clay formation begins, the soil has a higher water-holding capacity that results in accelerated clay formation as compared to soils in which clay contents are low.

A somewhat similar relationship of soils with slope position has been described in southeastern Nebraska.[1] Wisconsin loess of uniform particle size is the parent material for the soils. Soils formed in the downslope positions are more strongly developed than are soils in upslope positions and on interfluves (Fig. 9–5). Resistant to non-resistant mineral ratios also vary markedly with slope position, and this suggests that much of the clay content variation is due to mineral weathering within each profile. Soil-moisture values in the C horizon for the summer of 1966 averaged 21 per cent (by volume) at site A compared with 25 per cent at site B. Although this slight difference in moisture might explain the differences in weathering and clay formation, it is possible that past conditions of greater moisture in the downslope soils, suggested by gleyed features, may have accelerated the trends.

The southeastern Nebraska data can be compared with soil data in well-drained sites in other areas for an appreciation of how rapidly the soil-forming processes are sped up in these moister downslope positions. For example, the clay-distribution profile at site B (Fig. 9–5) is

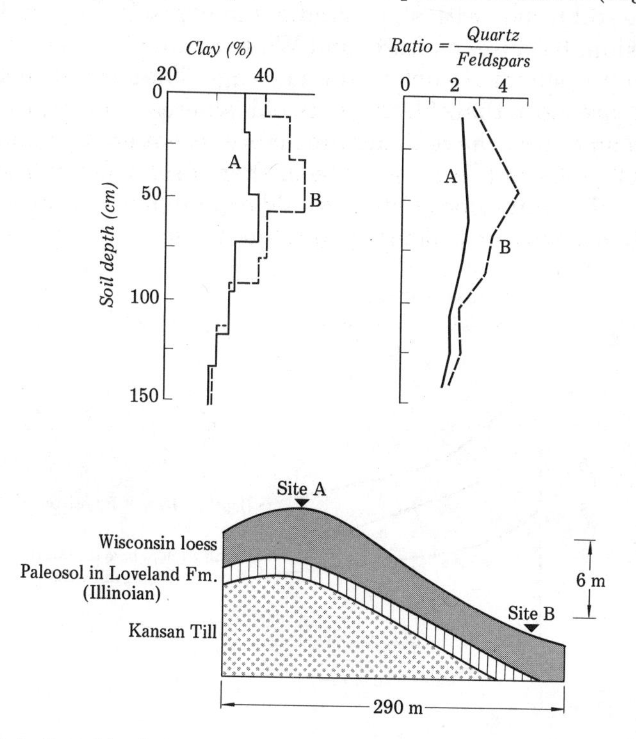

Fig. 9–5 Relationship between topography, stratigraphy, and soils, southeastern Nebraska. (Taken from Al-Janabi and Drew,[1] Figs. 3 and 4, Table 2.)

somewhat similar to that on the first marine terrace in California (Fig. 8–10), yet the latter soil may be seven times older or more. The quartz to feldspars ratio with depth can be compared with those of Ruhe (Fig. 8–4). The ratio at site A is typical for Wisconsin-age deposits, but that at site B compares with those obtained from soils formed on Yarmouth-Sangamon surfaces. Thus, in this case there seems to be about a tenfold acceleration of weathering and clay formation with slope position.

One problem with studying soils on slopes is that the slopes may differ in age with position, and therefore the soils may not be the same age from place to place. Ruhe and Walker[18] point out such a case for soils formed on rolling loess topography in Iowa (Fig. 9–6). Compared here are three slope components in order of increasing steepness in going downslope: the summit, shoulder, and backslope. Many soil properties correlate well with the slope steepness. In particular, with increasing steepness the soils are thinner, and there is less organic matter in the A horizon; also the depths to a pH of 6, to base saturation of 80 per cent, and to a greater than 1 per cent carbonate all decrease. Thus, the soils are shallower and less well developed on steeper slopes.

These soil trends with slope gradient may result, in part, from surface erosion, because, as Ruhe and Walker[18] and Walker and Ruhe[21] point out, the slopes are not all the same age. The summit surfaces, for example, are more than 14,000 years old, whereas the slopes adjacent to the summit areas have undergone more recent erosion and are less than 6800 years old. Age of surface, therefore, a factor not always appreciated in soil-slope studies, would explain most of the soil differences in a comparison of summit and slope areas.

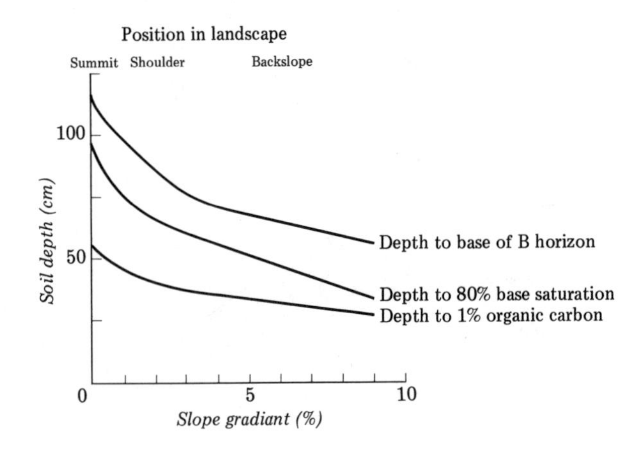

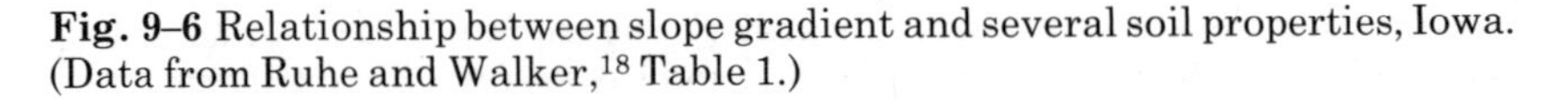
Fig. 9–6 Relationship between slope gradient and several soil properties, Iowa. (Data from Ruhe and Walker,[18] Table 1.)

Slopes are difficult to date. However, alluvium derived from slope erosion commonly has been deposited at the foot of slopes. Therefore, by radiocarbon dating of the alluvium one obtains dates on the times of stability and erosion of the slopes from which the alluvium came.

Soil variation in rolling topography with closed depressions

It is common to find markedly different soils in juxtaposition in rolling topography in which rounded hills rise above closed depressions. Soils on the uplands commonly are well drained, whereas those in the depressions are poorly drained and rich in clay and organic matter, with signs of various degrees of gleying. In dry climates, saline and alkaline soils occupy the depressions, and better-leached soils the hills. The differences in soil properties with position could be due to pedogenesis in place, resulting from differences in moisture, leaching, and vegetation over the rolling landscape. In this case the various parts of the landscape are assumed to be approximately the same age, and soil differences are attributed to the topographic factor. Recent work, however, has cast some doubt on this simple model, because it fails to take into account the fact that some material in the depressions could be derived from erosion of the landscape that slopes into the depression. With this model it is unlikely that the parent material of the soils formed on the slopes is the same as that in the depressions. The differences, however, have their primary origin in topographic position.

Sand dunes along the Mediterranean coast in Israel demonstrate these variations.[2] Soils on the tops and sides of the sand dunes have well-developed Bt horizons, with color hues of 2.5YR, overlying a C horizon of low clay content (Fig. 9–7). In the depressions, however, the soil for 3 m depth is close to 60 per cent clay, and the soil color meets the criteria for a gleyed horizon. Other notable differences below the surface from ridgetop to depression are decrease in content of free iron oxide, change from predominantly kaolinite to montmorillonite, and increase in pH.

Soil variation of the Israel sand dunes is explained by eolian influx combined with slope processes and pedogenesis. Much of the clay in these soils results from eolian influx, some of which, after reaching the surface, is translocated downward in the soil. During pedogenesis the clay minerals are thought to have altered from montmorillonite to kaolinite under favorable chemical and leaching conditions. The soil in the depressions is thought to have been mechanically transported into those positions and not to have formed in place, mainly because there are few weatherable mineral grains in the landscape that can alter to clay. The main process envisaged is downslope transfer of the

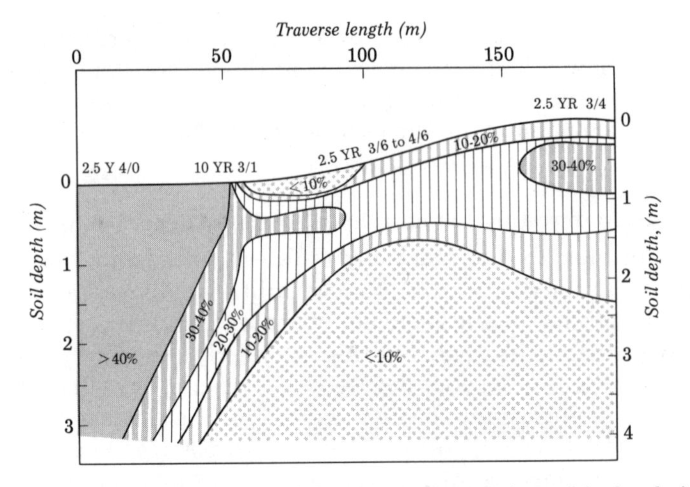

Fig. 9–7 Generalized distribution of clay-sized particles with depth in soils developed from sand-dune deposits, Israel. (Modified from Dan and others,[2] Fig. 9.) The difference in altitude from the ridge top to the depression is 6 m. Munsell color notations are for Bt horizons of soils formed above the depression and for soil material at a comparable depth within the depression.

eolian dust on loose organic litter by running water. It is also suggested that some clay transfer could take place within the soil by laterally moving soil water at the top of the strongly developed Bt horizon (about 20 cm depth). It is not felt that the evidence supports the transfer of sand-dune sediment from the slopes to the depression, because so little sand is found in the depression. Leaching conditions in the depressions are slight and thereby favor retention of montmorillonite as the main clay mineral. Pedogenesis is going on in these different parent materials during slope erosion and deposition in the depressions.

The till landscapes of the midcontinent are good examples of the interplay of pedological, sedimentological, and erosional processes in areas characterized by closed depressions. We will first consider processes operative on the present landscape and then discuss the well-known problem of gumbotil.

Walker[20] has made a detailed study of closed depressions and associated hillslopes in Iowa. The parent material is till dated at about 13,000 BP. The soils vary from Mollisols in well-drained positions to Aquolls in poorly drained closed depressions. Major differences in the soils are increase in clay and organic matter toward the center of the depression, along with a decrease in the gravel content (Fig. 9–8).

The stratigraphy and radiocarbon dating of several bogs indicates that the sedimentation rate in the bogs and the erosion rate of the surrounding slopes has varied in the past. The uppermost 60 cm of material in the bog shown in Fig. 9–8 is rich in organic matter and was deposited during a period of relatively slow slope erosion dating from about 3000 BP to the present. Material below 60 cm, however, dates back to 8000 BP; it is relatively low in organic matter and was deposited during an interval of relatively rapid slope erosion. Older deposits, indicating additional periods of slope erosion and stability, are found deeper in the bog.

The bog, therefore, is mostly postglacial sediment, and its properties directly influence the soil properties. These sediments were derived from the adjacent slopes, and the operative processes brought about the lateral separation of size fractions, so that gravel sizes were not moved into the centers of the depressions, whereas clay sizes were. Thus, the

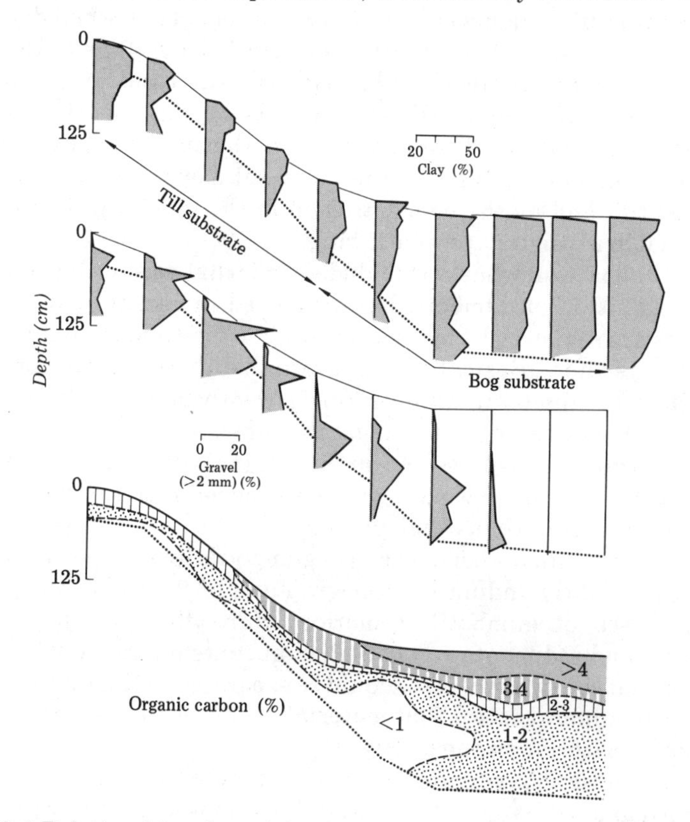

Fig. 9–8 Relationship of particle size and organic carbon with topographic position in a closed depression and on surrounding hillslopes, Iowa. (Taken from Walker,[20] Figs. 19, 30 and 31.) The dotted line separates hillslope sediments from till on the hillslopes, and younger from older bog deposits in the low-lying area.

major lateral differences in soil-particle sizes are sedimentological, not pedologic, in origin. Add to this primary textural control the variation in soil moisture due to variation in internal drainage, and most of the lateral variation in soils is adequately explained. Finally, the stratigraphic relationships indicate that most of the soils in the area formed over the past 3000 years under grassland vegetation. Hillslope soils that formed over a previous period under postglacial forest vegetation were essentially removed by erosion before 3000 BP.

The formation of the well-known gumbotils of the midcontinental Pleistocene record probably results from a combination of the topographic factor in an undulating landscape and the time factor. Kay[9] defined gumbotil as "... a gray to dark-colored, thoroughly leached, non-laminated, deoxidized clay, very sticky and breaking with a starchlike fracture when wet, very hard and tenaceous when dry, and ... chiefly, the result of weathering of drift." Kay[9,10] and Kay and Pearce[11] considered gumbotil to be primarily the result of prolonged chemical weathering of pre-Wisconsin tills on flat, poorly drained plains; it is a gleyed soil. That chemical weathering had occurred was shown by chemical analyses of the gumbotil compared with the unweathered till, and by the concentration in the gumbotil of siliceous pebbles that are resistant to weathering.

A lively debate on what was included and what was excluded in the original definition of gumbotil has developed in recent years. These arguments are reviewed by many workers,[4,5,12,17,19] and the details will not be repeated here. Basically, it is agreed that there are two contrasting origins for gumbotil. One origin is weathering in place under conditions of poor internal drainage; the other origin is that they are deposits laid down in depressions on the original till surface. Although Kay considered chemical weathering to be the main factor in gumbotil origin, he did agree that other processes, such as slope wash, were operative. Ruhe,[16] in his comparison of gumbotils located on swells and in swales in slightly undulating topography (Fig. 9–9), also recognized both processes of gumbotil formation. Ruhe thought that either weathering and sedimentation could go on contemporaneously in the swales, or that the two processes could be separated in time, and demonstrated from mineral ratios that material in the swales is more weathered than is material on the swells.

Frye and others[4] have called attention to these two different origins of gumbotil and have suggested that, if the term is kept, it be restricted to profiles that have weathered in place from till. They suggest that gleyed material that accumulates in depressions be called accretion gley. It is important to recognize both of these materials and probably to use separate terms for both, because the profiles formed in place

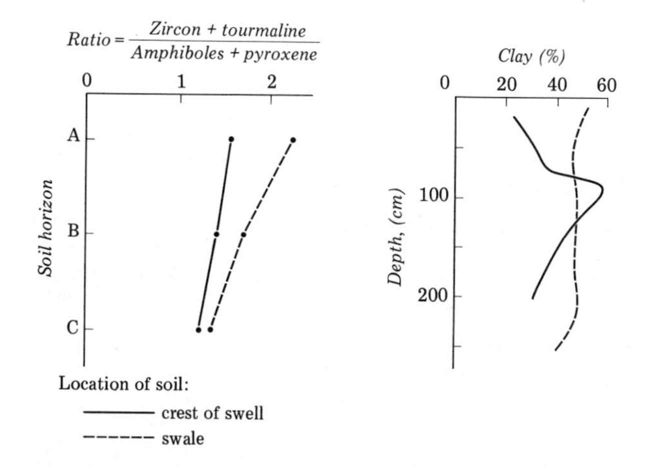

Fig. 9–9 Comparison of pre-Wisconsin soils formed in different parts of an undulating landscape, Iowa. (Taken from Ruhe,[16] Figs. 4 and 5, Table 1, © 1956, The Williams & Wilkins Co., Baltimore.) Parts of both profiles would be classed as gumbotil.

attain their main characteristics by pedogenesis in poorly drained areas, whereas accretion gleys attain their main characteristics by sedimentary processes, as well as pedogenesis, in poorly drained areas. Some criteria found to be helpful in differentiating gumbotil from accretion gley are listed in Table 9–1.

The relationship of the gumbotils to the adjacent accretion gleys is analogous to that of deposits and soils with topography described by Walker,[20] mentioned above. This relationship has been pointed out by Ruhe,[17] and the lateral variation in gumbotils and accretion gleys appears to be due to the topographic factor. Time is also involved here, however, because gumbotils are found only on pre-Wisconsin surfaces. Thus, given enough time, the materials and soils described by Walker might progress to the gumbotil and accretion gley stage of development (compare Figs. 9–8 and 9–9 and Table 9–1).

Variation in soils with topographic position in the humid tropics

Topographic position controls the formation of some Oxisols in tropical regions.[6,14] The origin of Oxisols has been shown to differ between upland sites and sites located on slopes. Oxisols form by the usual weathering reactions and soil-forming processes in the uplands. The slopes, however, are characterized by fairly large quantities of laterally moving soil moisture derived mostly from upslope. Mobile constituents released by weathering can move with these waters (Fig. 9–10). Because iron and manganese are more mobile than aluminum,

Table 9–1

Some criteria for differentiating gumbotil from accretion gley[4,5,19]

CRITERION	GUMBOTIL (formed in place by weathering)	ACCRETION GLEY (deposited in lows in the landscape)
Stratification	Massive, not stratified, leached of carbonates	Stratification can be discerned as humus-rich layers toward the top of the deposit and as layers of contrasting particle size
Contact with fresh till	Gradational contact with fresh till includes a leached and oxidized zone over an unleached and oxidized zone	Contact with underlying fresh till can be sharp; configuration of the contact might suggest that the deposit occupies a depression on an old paleo-landscape
Distribution of resistant pebbles	Pebbles of resistant lithologies show an orderly increase in relative abundance upward	Some deposits may have few resistant pebbles, relative to the till, due to sorting on adjacent hillslopes
Silicate mineral weathering	Silicate minerals are decomposed in order of weatherability; decomposition shows an orderly increase toward surface	Weatherable silicate minerals may be depleted; depletion toward the surface need not be systematic
Clay mineral alteration	Clay minerals show an orderly arrangement of weathering stage with position in the profile	Clay minerals are a mixture of minerals of different stages of weathering; mechanical mixing is suggested

one sees a lateral segregation of these elements; aluminum is relatively concentrated in the upland areas, iron in the downslope areas, and manganese, because it is more mobile than iron, moves farthest down-

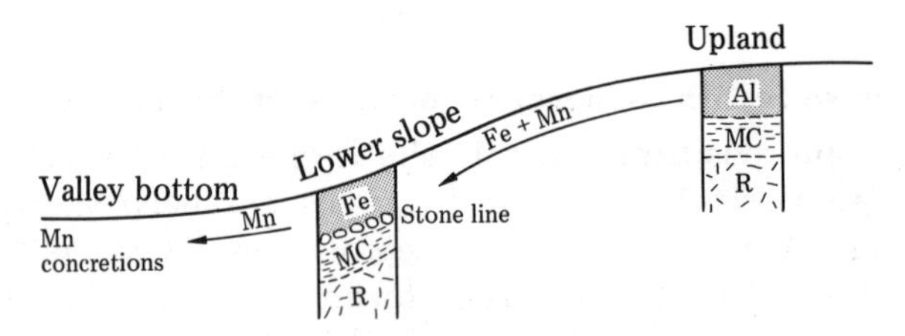

Fig. 9–10 Variation in morphology and chemistry of Oxisols with topographic position; the arrows depict general movement of soil moisture, iron, and manganese. (Taken from Hamilton,[6] Fig. 1.) Key: Al, bauxitic horizon; Fe, ferruginous horizon; MC, mottled clay; R, parent material.

slope. Oxisols on the slopes, therefore, owe much of their iron content to precipitation as crystalline and amorphous iron compounds in voids in the soils or in the colluvial deposits on the slopes.

With dissection of the landscape, the previously deposited iron can be remobilized and moved into younger soils and deposits downslope. For example, if the landscape in Fig. 9–10 were to be dissected, a younger landscape would form downslope from the previously formed iron-rich Oxisol. Because of dissection, the iron-rich soil would be relatively well drained and waters from it could move laterally. The iron could, therefore, if re-mobilized, move to lower parts of the landscape and cement those soils and deposits. This process could continue with each new cycle of downcutting in the landscape.

REFERENCES

1. Al-Janabi, A. M., and Drew, J. V., 1967, Characterization and genesis of a Sharpsburg-Wymore soil sequence in southeastern Nebraska: Soil Sci. Soc. Amer. Proc., v. 31, p. 238–244.
2. Dan, J., Yaalon, D. H., and Koyumdjisky, H., 1968, Catenary soil relationships in Israel, 1. The Netanya catena on coastal dunes of the Sharon: Geoderma, v. 2, p. 95–120.
3. Finney, H. R., Holowaychuk, N., and Heddleson, M. R., 1962, The influence of microclimate on the morphology of certain soils of the Allegheny Plateau of Ohio: Soil Sci. Soc. Amer. Proc., v. 26, p. 287–292.
4. Frye, J. C., Shaffer, P. R., Willman, H. B., and Ekblaw, G. E., 1960a, Accretion gley and the gumbotil dilemma: Amer. Jour. Sci., v. 258, p. 185–190.
5. ——, Willman, H. B., and Glass, H. D., 1960b, Gumbotil, accretion-gley, and the weathering profile: Ill. State Geol. Surv. Circ. 295, 39 p.
6. Hamilton, R., 1964, Microscopic studies on laterite formation, p. 269–276 *in* A. Jongerius, ed., Soil micromorphology: Elsevier Publ. Co., New York, 540 p.
7. Jenny, H., 1941, Factors of soil formation: McGraw-Hill, New York, 281 p.
8. —— 1958, Role of the plant factor in the pedogenic functions: Ecology, v. 39, p. 5–16.
9. Kay, G. F., 1916, Gumbotil, a new term in Pleistocene geology: Science, v. 44, p. 637–638.
10. —— 1931, Classification and duration of the Pleistocene period: Geol. Soc. Amer. Bull., v. 42, p. 425–466.
11. —— and Pearce, J. N., 1920, The origin of gumbotil: Jour. Geol., v. 28, p. 89–125.
12. Leighton, M. M., and MacClintock, P., 1962, The weathered mantle of glacial tills beneath original surfaces in north-central United States: Jour. Geol., v. 70, p. 267–293.

13. Lotspeich, F. B., and Smith, H. W., 1953, Soils of the Palouse loess: I. The Palouse catena: Soil Sci., v. 76, p. 467–480.
14. Maignien, R., 1960, Review of research on laterites: UNESCO, Natural Resources Res. IV, 148 p.
15. Nettleton, W. D., Flach, K. W., and Borst, G., 1968, A toposequence of soils on tonalite grus in the southern California Peninsular Range: U.S. Dept. Agri., Soil Cons. Serv., Soil Surv. Invest. Rep. no. 21, 41 p.
16. Ruhe, R. V., 1956, Geomorphic surfaces and the nature of soils: Soil Sci., v. 82, p. 441–455.
17. ____ 1965, Paleopedology, p. 755–764 *in* H. E. Wright, Jr., and D. G. Frey, eds., The Quaternary of the United States: Princeton Univ. Press, Princeton, 922 p.
18. ____ and Walker, P. H., 1968, Hillslope models and soil formation. I. Open systems: 9th Internat. Cong. Soil Sci. Trans., v. 4, p. 551–560.
19. Trowbridge, A. C., 1961, Discussion: accretion-gley and the gumbotil dilemma: Amer. Jour. Sci., v. 259, p. 154–157.
20. Walker, P. H., 1966, Postglacial environments in relation to landscape and soils on the Cary drift, Iowa: Iowa State Univ., Agri. and Home Econ. Exp. Sta. Res. Bull. 549, p. 835–875.
21. ____ and Ruhe, R. V., 1968, Hillslope models and soil formation. II. Closed systems: 9th Internat. Cong. Soil Sci. Trans., v. 4, p. 561–568.

CHAPTER 10

Vegetation–soil relationships

The biotic factor in pedogenesis is difficult to assess because of the dependence of both vegetation and soil on climate and the interaction of soil and vegetation. Jenny[15] depicts the interrelationship of these three factors thus

$$\begin{array}{ccc} & \text{Climate} & \\ \swarrow & & \searrow \\ \text{Vegetation} & \rightleftharpoons & \text{Soil} \end{array}$$

Here we are concerned with the lower part of the triangle, specifically the influence of vegetation on the soil. Field sites can be found where vegetation is the most important variable producing differences in soil properties. In many places these vegetational effects could be interpreted as microclimatological influences brought about by vegetation differences. As in the discussion of the other factors, all factors except vegetation will be kept constant. A constant climate is assumed by considering only the regional climate; in areas where this remains constant, we can determine the influence of vegetation on the soils. Two aspects of the vegetation-soil relationship will be considered here: vegetation and soil morphology and vegetation and soil chemistry. In addition, the influence of former vegetation on a soil will be reviewed.

Soil variation at the forest–prairie boundary

An often-cited example of vegetation-soil relationships is the comparison of forested and grassland soils at the forest-prairie boundary of the midcontinent[1,14,25,28] and of Canada.[24] The study of White and Riecken[34] in the midcontinent is used here (Fig. 10–1). Alfisols are present in the deciduous forests, and they possess an A/E/B/C soil-horizon sequence. Mollisols in the grassland possess an A/B/C soil-horizon sequence. Transitional soils commonly occur between these two kinds of soils in the field, and their properties are intermediate between the two end members.

Several diagnostic soil properties are closely related to differences in vegetation. The most obvious one is organic-matter distribution. Although all the soils have an equally high content of organic matter at the surface, the distribution with depth varies with vegetation. Forested A horizons are relatively thin, and the organic-matter content decreases rapidly with depth, whereas grassland A horizons are thick, and the organic-matter content remains high for a considerable depth. These differences occur partly because of the manner in which organic matter enters the soil. In forested soils the main input of

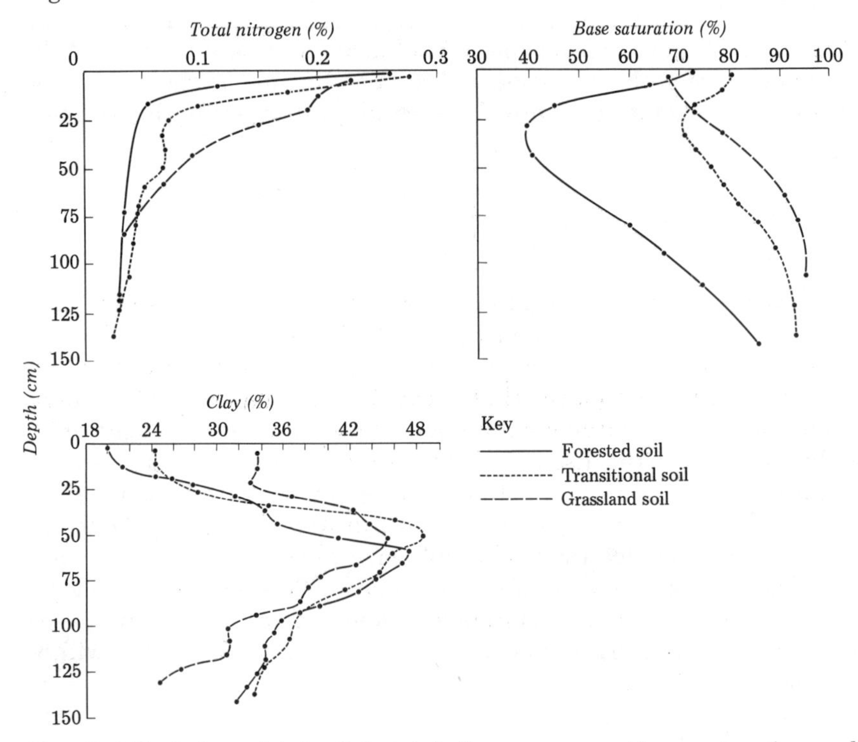

Fig. 10–1 Variation with depth in total nitrogen, per cent base saturation, and clay content for soils formed under different vegetation covers. (Taken from White and Riecken,[34] Figs. 1, 3, and 5.)

organic matter is by litterfall to the soil surface, whereas in grassland soils the organic matter input is both by litterfall and by root decay at depth. The overall percentage base saturation tends to be higher in grassland than in forests. This difference could be a function of greater leaching in the forests along with greater annual biomass production and cycling of cations in grassland. The greater leaching in the forests relative to the grasslands could be due, in part, to lower evapotranspiration rates and the presence of more chelating agents and more acid leaching waters under a forest canopy. In addition, the ratio of free Fe_2O_3 in the A and/or E horizons relative to that in the B horizons is lower in forested than in grassland soils. The origin of this difference probably lies in the differences in chelating ability of the organic compounds formed during decomposition in the two contrasting environments. Another striking morphological feature is clay distribution with depth. The surface layers of grassland soils have higher clay contents than do comparable layers in forested soils, indicating that the environment in the forests is such that clay particles can be translocated to a greater extent than is the case in the grasslands.

Thus, differences in soils at the forest-prairie boundary are striking. To review, forested sites commonly show greater leaching of cations, correspondingly lower pH's, and greater clay translocation than do adjacent grassland sites. In addition, organic-matter content is higher at greater depths in the grassland sites.

If the forest-prairie boundary shifts, then the soil properties under the new vegetation will be somewhat out of phase with the vegetation; this leads to soils with properties transitional between the end members discussed above[34] (Fig. 10–1). It is conceivable that organic-matter distribution and base saturation with depth in the transitional soils would be the same if the change at the site was either forest to grassland or the reverse. The time necessary to reach the new steady state for these properties probably is in the order of 10^3 years or more. If the change were grassland to forest, iron and clay particles would start to move, and the time necessary to reach profiles characteristic of the forest is not known. If, on the other hand, the change were forest to grassland, the iron and clay profiles might persist, and these can be cautiously used to indicate former vegetational influences.

Variation in soil with distance from trees and with tree species

Within a constant regional climate tree species will vary from site to site, and thus soil properties may vary from site to site. Some soil variations are quite subtle and can be observed only as changes in soil chemistry; these include pH or exchangeable cations. Other soil variations are quite striking and can be observed as changes in soil morphology. An extreme example of the latter variation is found in New Zealand, where local podzolization occurs under *Podocarpus sp.*[12] and Kauri pine (*Agathis australis*).[7] The influence of the tree is such that the Spodosols occur only under each tree and not beyond the influence of the tree. Here we will present data on soil properties with distance outward from several tree species.

Zinke[38] has demonstrated the effect of a single shore pine tree, *Pinus contorta*, on surface soil properties. The tree was 45 years old and growing in a sand dune containing shell material. Some marked changes in soil properties are associated with distance from the tree and whether the sample is taken under or beyond the tree crown (Fig. 10–2). The soil is much more acid beneath the tree relative to the non-vegetated sand, the difference in pH being 1.5 or more. Nitrogen is almost an order of magnitude greater under the tree than it is beyond the influence of the tree. Because cation-exchange capacity is derived mostly from the organic matter, it follows the nitrogen trends, being about twice as great beneath the tree (approximately 28 me/100 g soil) as beyond the tree (approximately 16 me/100 g soil).

Zinke[38] also studied surface-soil variation with radial distance out from several different tree species and found that all the species studied had similar trends (Fig. 10–3). Differences between species, although

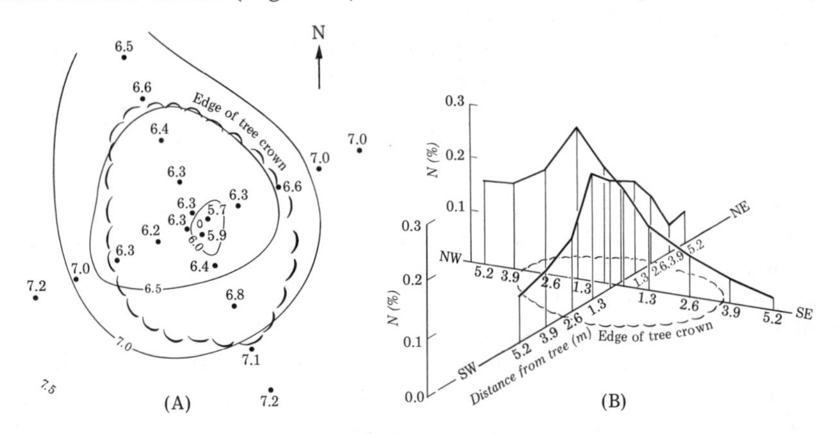

Fig. 10–2 Variation in surface-soil properties with distance from a pine tree (*Pinus contorta*). (A) Recorded pH values and isolines of approximately equivalent pH values. The NW-SE diameter is approximately 8 m. (B) Nitrogen content in per cent by weight. Samples were taken from the 0 to 6.4 cm depth layer. High values extend NW beyond the tree crown edge probably because of the prevailing winds. (Taken from Zinke,[38] The pattern of individual forest trees on soil properties, Figs. 1 and 2, by permission of the Duke University Press, Durham, North Carolina.)

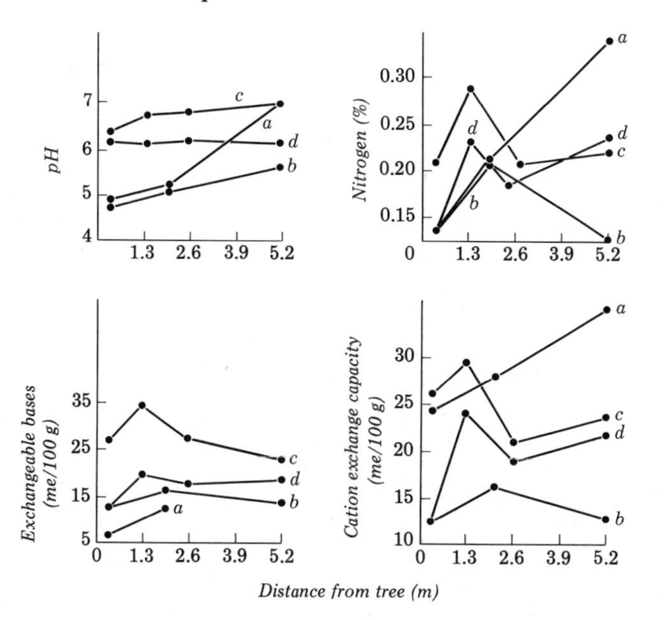

Fig. 10–3 Variation in surface soil properties with distance from old forest trees of different species in California. Samples were taken from the 0 to 6.4 cm depth layer. The outer sampling point always extends beyond the influence of the crown canopy, whose edge is 2.6 to 5.2 m from the trunk. The tree species are (a) and (b) ponderosa pine (*Pinus ponderosa*), (c) incense cedar (*Libocedrus decurrens*), and (d) Douglas fir (*Pseudotsuga menziesii*). The high nitrogen value for (a) is due to sampling near a nitrogen-fixing plant. (Taken from Zinke,[38] The pattern of individual forest trees on soil properties, Fig. 4, by permission of the Duke University Press, Durham, North Carolina.)

evident, may not reflect only a species effect because samples were taken from sites throughout California and other soil-forming factors may not have been the same. The general trend, however, is decreasing pH toward the trees. Nitrogen content, exchangeable bases, and cation-exchange capacity all are low near the tree stems, increase to a maximum some distance from the stem, but within the area covered by the crown, and generally decline outward.

The above trends are interpreted as being due primarily to the influence on the soil of bark and leaf fall near the tree trunk, relative to no tree litter in the opening between the trees.[38] Bark litter predominates in a ring surrounding the tree closest to the stem, and leaf litter predominates in a ring out to about the edge of the crown. The trends in soil properties therefore are a function of the predominance either of bark or of leaf litter. Of the two kinds of litter, bark litter is the more acid and has a lower cation and nitrogen content. Thus the soils directly reflect the composition of the litter. Zinke also suggests that

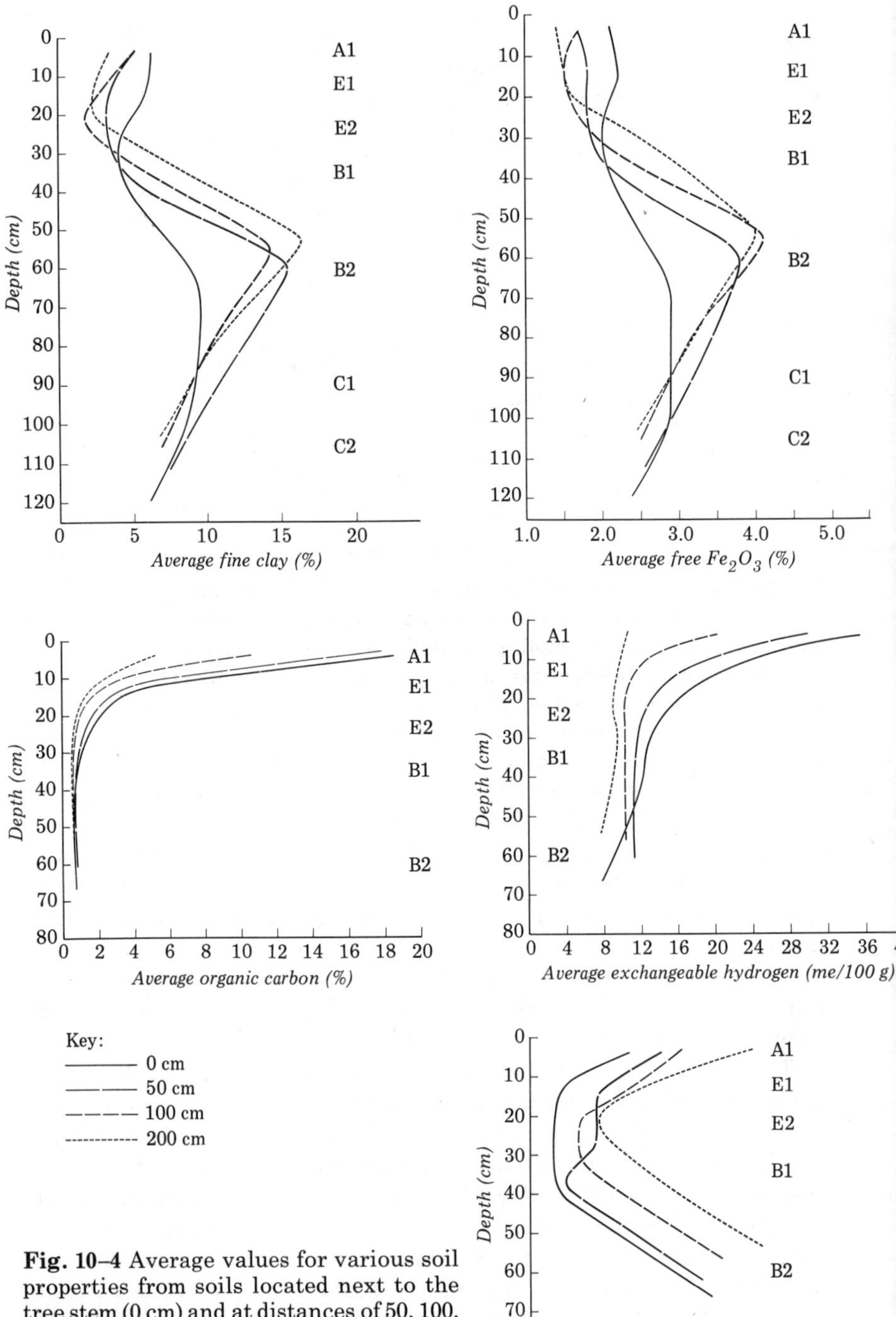

Fig. 10–4 Average values for various soil properties from soils located next to the tree stem (0 cm) and at distances of 50, 100, and 200 cm outward. (Taken from Gersper and Holowaychuk,[9] Fig. 8, and Gersper and Holowaychuk,[10] Figs. 3, 5, 10, and 11.)

the water that reaches the soil, whether by stemflow, drip from foliage, or without vegetation interception, may exert some control on the soil-property trends, because these waters differ in their chemical composition. No data, however, were reported on the composition of these waters.

Variation in soil properties outward from a beech tree, *Fagus grandifolia*, in the midcontinent provides data on the influence of stemflow water on pedogenesis.[9,10] The parent material is till containing $CaCO_3$, and the soil formed is a Udalf. Trends with distance from the stem are quite apparent and vary more for chemical properties than for physical properties (Fig. 10–4). Close to the tree the B horizons are more mottled, and the E horizons are thicker; there is less fine clay in the B horizon relative to sites 200 cm from the tree. The main trends in chemical properties in going away from the tree are decreasing amounts of total nitrogen and organic carbon, increasing percentage base saturation, increasing pH and decreasing exchangeable H^+, and higher free Fe_2O_3 along with greater differentiation of Fe_2O_3 with depth.

The above trends reflect the amount and composition of stemflow water. Stemflow water is that water derived from rainfall that moves along the branches to the tree stem and then down the stem to the ground. Thus, a larger quantity of water can enter the soil near a tree stem than can enter a soil located beyond tree-stem influences. Gersper and Holowaychuk[10] estimate that the soil at the tree stem can, by this means, receive as much as five times the quantity of water than can a soil in the open that receives its water from rainfall alone. Furthermore, they cite literature indicating that water in contact with vegetation has a much higher content of chemical elements than has rainfall, and that stemflow water has a higher content of elements than water that drips from the crown. The estimated amount of material added to the soil each year is quite large (Table 10–1). A major source of the elements in stemflow water is thought to be dust and other matter adsorbed on tree surfaces as well as insects and bark washed from the tree. Gersper and Holowaychuk feel that the main factors responsible for soil trends are variation in influx of water and chemical elements delivered to the soil next to the tree stem relative to influx some distance from the tree stem. Designated "biohydrologic" soil-forming factor, it is important in those soils in which the interaction of precipitation with vegetation affects soil formation.

The relative importance of stemflow or of litterfall on soil variation with distance from a specific tree seems to vary with the nature of the bark.[11] Trees with smooth bark have high rates of stemflow, and thus soil-property variations are largely controlled by stemflow. However,

Table 10–1

Estimated annual quantities of elements added to a Udalf through stemflow near the stem of an American beech tree *(Fagus grandifolia)*, expressed on a per hectare basis

Element	*Quantity added (kg/ha)*
C	10,385
Ca	1,033
K	842
Na	202
Mg	191
P	112
Fe	22

(Taken from Gersper and Holowaychuk,[10] Table 2)

trees with rough bark (such as those reported by Zinke, above) have low rates of stemflow, and soil-property variation with distance from the tree depends on the amount and composition of the litterfall.

Many workers have recognized that some soil properties vary with tree species. Perhaps the data with the best environmental control are those from plantations in which various tree species have been growing for several decades.[6,20–23] Organic carbon and total nitrogen, pH, and the exchangeable cations commonly vary with tree species. Ovington's work shows, in a very general way, that soils under conifers have greater amounts of organic carbon and total nitrogen, lower pH values, greater amounts of sodium, potassium, and phosphorus, and lower amounts of calcium and magnesium, relative to the soils under deciduous trees. These soil properties are among those that can change most rapidly with changing environmental conditions. Thus, if a new tree species occupies the soil site formerly occupied by another tree species, the soil properties probably would alter rather quickly (10^2–10^3 years) to values determined by the new tree species. In short, the influence of former tree species on the soil site may not linger long after a change in species.

Deciphering past vegetational change by soil properties

One goal of the study of Quaternary soils is to assess past vegetational change by the examination of soil properties. One way of doing this is to establish the relationship between present vegetation and soils and then to examine carefully both surface and buried soils to determine if their properties might indicate formation under a past vegetation different from the present vegetation. One would hope to go one step further in the interpretation, and that is to correlate past changes in vegetation with past changes in climate.

One of the areas in which distinctly different soil patterns are closely linked with vegetation is at the forest-grassland boundary. As discussed earlier, soil properties in many places change markedly across the vegetational boundary. Under ideal conditions, Spodosols or Alfisols with an E horizon form in the forest, and Mollisols or Inceptisols form in the grasslands. Whether or not a change in past vegetation could be determined by soil properties would depend on the time elapsed since the vegetational change and on the ability of a specific soil property to persist relatively unchanged under the new conditions.

The time required for adjustment to a new environment depends upon the specific surface-soil property. Properties like pH and exchangeable bases change quickly and thus carry little if any legacy of past vegetation. The amount and distribution of organic matter with depth probably would persist for longer periods of time but perhaps not as long as it would take to reach steady-state values for that environment beginning with an unvegetated, unweathered surface (10^2+ to 10^3+ years). For example, in those ecosystems where the rate of turnover of organic matter is rapid, the change to new steady-state values would be rapid. An example of this might be the grassland areas of the humid midcontinent. If, however, the rate of turnover of organic matter is slow, such as in the forested and tundra ecosystems of the northern latitudes or of high altitudes, changes with new vegetation would be quite slow, and properties related to the former vegetation might persist for thousands of years. Furthermore, if the change is from forest to grassland, the E horizon may persist for long periods of time if the grassland A horizon is comparable in thickness to the forest A horizon or if turnover rates of organic matter are slow in the grassland ecosystem. Jungerius[17] discusses some of these changes in vegetation and soils in southern Alberta and points out that in places the platy structure of the E horizon of the former forested soil can persist in present-day Mollisols and thus provide a clue to past vegetational patterns. In contrast, a change from grassland to forest can be accompanied by the formation of an incipient E horizon at the base of the grassland A horizon.[27]

Chances for deciphering past vegetational trends farther back in time ($> 10^4$ years) probably are best for buried soils, because these can be compared with the present-day surface soils in the same region. Soil morphological features seem to provide the best data for these interpretations. Ruhe and Cady[26] compared the morphology of the buried Sangamon and late-Sangamon soils in Iowa with that of the surface

Mollisols and concluded that the former, because they contain an E horizon, had formed under a forest cover. Jungerius[17] used comparisons of surface- and buried-soil morphology to estimate treeline fluctuations in Alberta since the Wisconsin glaciation. Sorenson and others[29] used the distribution and morphology of surface and buried soils west of Hudson Bay to demonstrate fluctuations in the forest-tundra boundary over the past 6000 years (Fig. 10–5). Spodosols are formed in the forests, whereas Inceptisols with A/C profiles are formed on well-drained sites in the tundra, a distribution similar to that seen in Alaska.[32] The latitudinal distribution of these contrasting soil morphologies therefore provides data on the former positions of the forest-tundra boundary.

Although the interpretation that non-forested soils with E horizons north of present treeline were forested in the past probably is correct for many places, some E horizons in well-drained soils do not seem to require a forest cover for their formation. For example, some surface soils in the tundra zone just north of the treeline show evidence of weak podzolization.[31,32] In these cases, either the forest-tundra border was farther north in the recent past or podzolization can take place in well-drained sites in the southern part of the tundra zone in the absence of trees.

Other soils with E horizons in northern regions seem to have formed under non-forested conditions. For example, James[13] reports on soils with an E horizon 480 km north of the present treeline along the west coast of Hudson Bay that probably formed under the present-day, lichen-heath vegetation. Spodosol-like soils with E horizons are present in Greenland,[33] again far north of treeline. In addition, Spodosol-like soils are found on the north slope of the Brooks Range under

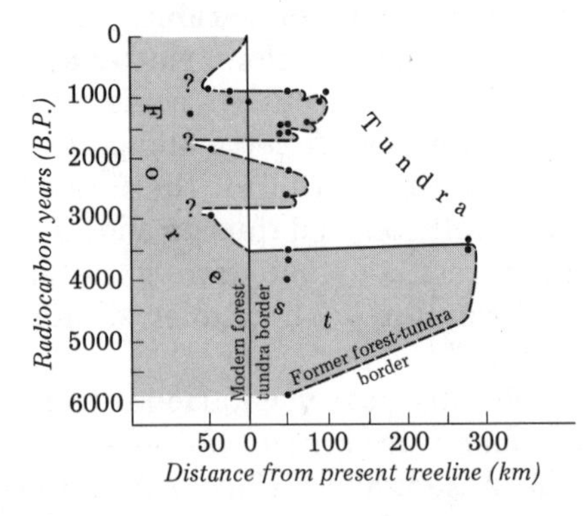

Fig. 10–5 Tentative Holocene migrations of forest-tundra boundary, southwest Keewatin, Canada. (Taken from Sorenson and others.[29]) Filled circles denote radiocarbon dates.

dwarf heath vegetation.[4,5] Thus, one should be cautious in relating all non-forested, Spodosol-like soils to formation under a forest vegetation.

It might be possible to relate variations in soil morphology to fluctuation in the forest-tundra boundary in the high mountains. Bliss and Woodwell[2] report on the presence of Spodosols both under fir-spruce (*Abies-Picea*) krummholz and at a nearby site under sedge-heath vegetation. Spodosols also occur above treeline in the Alp[3,18] and in Scotland.[30] These alpine Spodosols apparently formed under the present-day vegetation. Reconnaissance work by myself and students in various parts of the Rocky Mountains, however, has failed to demonstrate the presence of E horizons in surface soils above the present treeline; a suggested interpretation is that treeline has not gone to altitudes higher than the present in the recent past. Thus, although soils above treeline generally lack E horizons, they have been reported in certain environments, and one has to be cautious in using only the presence or absence of an E horizon in the reconstruction of former treelines.

Biogenic opal in soils as an indicator of past vegetation

Many plants secrete opal that conforms to the shape of the cell from which it forms; upon decomposition the opal remains in the soil.[19,36] These opal forms are called phytoliths. Although their morphologies vary with plant species, little work has been done on the taxonomy of these forms. They appear to persist for long periods of time, however, as shown by radiocarbon dates of about 13,000 years for phytoliths in Ohio.[35]

Forested and grassland soils differ in their phytolith content, and this difference in content can be used to reconstruct past vegetation. Grasslands produce more phytoliths per unit area than do forests, and these differences show up in the phytolith contents of the surface soils. To decipher vegetational change, therefore, one has to sample a transect across the two vegetation types. If the change in phytolith content coincides with the present forest-grassland boundary, the interpretation is that the boundary has been stable for some time.[37] If, however, both the forested and grassland soils have similar phytolith contents, the boundary apparently has shifted. In some areas the vegetational change seems to have been from grassland to forests,[16] whereas in others the evidence favors a shift from forests to grassland.[36]

Phytoliths are also useful in deciphering the environment of formation for buried soils. Dormaar and Lutwick[8] used phytoliths in combination with the infrared spectra of humic acids to identify buried soils as such and to reconstruct their vegetational history. Because it is difficult to reconstruct the environment of formation for buried soils from morphology alone, phytoliths, which can persist long after the soil organic matter has been removed by processes following burial, may provide important clues to past vegetation.

REFERENCES

1. Bailey, L. W., Odell, R. T., and Boggess, W. R., 1964, Properties of selected soils developed near the forest-prairie border in east-central Illinois: Soil Sci. Soc. Amer. Proc., v. 28, p. 257–263.
2. Bliss, L. C., and Woodwell, G. M., 1965, An alpine podzol on Mount Katahdin, Maine: Soil Sci., v. 100, p. 274–279.
3. Bouma, J., Hoeks, J., van der Plas, L., and van Scherrenburg, B., 1969, Genesis and morphology of some alpine podzol profiles: Jour. Soil Sci., v. 20., p. 384–398.
4. Brown, J., 1966, Soils of the Okpilak River region, Alaska: Cold Regions Res. and Engr. Lab. (Hanover, N. H.), Res. Report 118, 49 p.
5. ___ and Tedrow, J. C. F., 1964, Soils of the northern Brooks Range, Alaska: 4. Well-drained soils of the glaciated valleys: Soil Sci., v. 97, p. 187–195.
6. Challinor, D., 1968, Alteration of surface soil characteristics by four tree species: Ecol., v. 49, p. 286–290.
7. Crocker, R. L., 1952, Soil genesis and the pedogenic factors: Quat. Rev. Biol., v. 27, p. 139–168.
8. Dormaar, J. F., and Lutwick, L. E., 1969, Infrared spectra of humic acids and opal phytoliths as indicators of palaeosols: Can. Jour. Soil Sci., v. 49, p. 29–37.
9. Gersper, P. L., and Holowaychuk, N., 1970a, Effects of stemflow water on a Miami soil under a beech tree: I. Morphological and physical properties: Soil Sci. Soc. Amer. Proc., v. 34, p. 779–786.
10. ___ and ___ 1970b, Effects of stemflow water on a Miami soil under a beech tree: II. Chemical properties: Soil Sci. Soc. Amer. Proc., v. 34, p. 786–794.
11. ___ and ___ 1971, Some effects of stem flow from forest canopy trees on chemical properties of soils: Ecol., v. 52, p. 691–702.
12. Jackson, M. L., and Sherman, G. D., 1953, Chemical weathering of minerals in soils: Advances in Agron., v. 5, p. 219–318.
13. James, P. A., 1970, The soils of the Rankin Inlet area, Keewatin, N.W.T., Canada: Arctic and Alpine Res., v. 2, p. 293–302.
14. Jenny, H., 1941, Factors of soil formation: McGraw-Hill, New York, 281 p.
15. ___ 1958, Role of the plant factor in the pedogenic functions: Ecol., v. 39, p. 5–16.
16. Jones, R. L., and Beavers, A. H., 1964, Variation of opal phytolith content among some great soil groups of Illinois: Soil Sci. Soc. Amer. Proc., v. 28, p. 711–712.

17. Jungerius, P. D., 1969, Soil evidence of postglacial tree line fluctuations in the Cypress Hills area, Alberta, Canada: Arctic and Alpine Res., v. 1, p. 235–245.
18. Kubiëna, W. L., 1953, The soils of Europe: Thomas Murby and Co., London, 314 p.
19. Lutwick, L. E., 1969, Identification of phytoliths in soils, p. 77–82 *in* S. Pawluk, ed., Pedology and Quaternary Research: Univ. Alberta Printing Dept., Edmonton, 218 p.
20. Ovington, J. D., 1953, Studies of the development of woodland conditions under different trees. I. Soils pH: Jour. Ecol, v. 41, p. 13–34.
21. —— 1956, Studies of the development of woodland conditions under different trees. IV. The ignition loss, water, carbon and nitrogen content of the mineral soil: Jour. Ecol., v. 44, p. 171–179.
22. —— 1958a, Studies of the development of woodland conditions under different trees. VI. Soil sodium, potassium and phosphorous: Jour. Ecol., v. 46, p. 127–142.
23. —— 1958b, Studies of the development of woodland conditions under different trees. VII. Soil calcium and magnesium: Jour. Ecol., v. 46, p. 391–406.
24. Pettapiece, W. W., 1969, The forest grassland transition, p. 103–113 *in* S. Pawluk, ed., Pedology and Quaternary Research: Univ. Alberta Printing Dept., Edmonton, 218 p.
25. Ruhe, R. V., 1969, Soils, paleosols, and environment, p. 37–52 *in* W. Dort, Jr., and J. K. Jones, Jr., eds., Pleistocene and Recent environments of the central Great Plains: Univ. Press of Kansas, Lawrence, 433 p.
26. —— and Cady, J. C., 1969, The relation of Pleistocene geology and soils between Bentley and Adair in southwestern Iowa: U.S. Dept. Agri. Tech. Bull. 1349, p. 1–92.
27. Sawyer, C. D., and Pawluk, S., 1963, Characteristics of organic matter in degrading chernozemic surface soils: Can. Jour. Soil Sci., v. 43, p. 275–286.
28. Smith, G. D., Allaway, W. H., and Riecken, F. F., 1950, Prairie soils of the upper Mississippi Valley: Advances in Agron., v. 2, p. 157–205.
29. Sorenson, C. J., Knox, J. C., Larsen, J. A., and Bryson, R. A., 1971, Paleosols and the forest border in Keewatin, N.W.T.: Quaternary Res., v. 1, p. 468–473.
30. Stevens, J. H., and Wilson, M. J., 1970, Alpine podzol soils on the Ben Lawers massif, Perthshire: Jour. Soil Sci., v. 21, p. 85–95.
31. Tedrow, J. C. F., 1968, Pedogenic gradients of the polar regions: Jour. Soil Sci., v. 19, p. 197–204.
32. ——, Drew, J. V., Hill, D. E., and Douglas, L. A., 1958, Major genetic soils of the arctic slope of Alaska: Jour. Soil Sci., v. 9, p. 33–45.
33. Ugolini, F. C., 1966, Soils of the Mesters Vig District, northeast Greenland. 1. The arctic brown and related soils: Meddelelser om Grønland, Bd. 176, no. 1, p. 1–22.
34. White, E. M., and Riecken, F. F., 1955, Brunizem-gray brown podsolic soil biosequences: Soil Sci. Soc. Amer. Proc., v. 19, p. 504–509.

35. Wilding, L. P., 1967, Radiocarbon dating of biogenic opal: Science, v. 156, p. 66–67.
36. ____ and Drees, L. R., 1969, Biogenic opal in soils as an index of vegetative history in the Prairie Peninsula, p. 96–103 *in* R. E. Bergstrom, ed., The Quaternary of Illinois: Univ. Illinois Coll. Agri. Spec. Publ. no. 14, 179 p.
37. Witty, J. E., and Knox, E. G., 1964, Grass opal in some chestnut and forested soils in north central Oregon: Soil Sci. Soc. Amer. Proc., v. 28, p. 685–688.
38. Zinke, P. J., 1962, The pattern of individual forest trees on soil properties: Ecol., v. 43, p. 130–133.

CHAPTER 11

Climate-soil relationships

The climatic factor is considered by many to be the most important factor in determining the properties of many soils. This climate-soil relationship can be seen by comparing the U.S. soil map (Fig. 2–5) with a map of precipitation and temperature (Appendix 2); most soil orders and suborders are restricted to certain climatic regions. In this chapter we will look into some of these large-scale relationships, mention specific soil properties that are related to climate, and discuss those properties that might be most useful in reconstructing past climates.

Moisture and temperature are the two aspects of climate most important in controlling soil properties. Moisture is important because water is involved in most of the physical, chemical, and biochemical processes that go on in a soil, and the amount of moisture delivered to the soil surface influences the weathering and leaching conditions with depth in the soil. Temperature influences the rate of chemical and biochemical processes. The temperature rule of van't Hoff is often mentioned to demonstrate the influence of temperature on such reactions: for every 10°C rise in temperature the velocity of a chemical reaction increases by a factor of two or three.[23] Jenny[23] has been able to derive separate functions for precipitation and for temperature in some areas, provided other aspects of the climatic factor and the other soil-forming factors could be considered constant.

Climatic parameters

A numerical value for the climate can be used to demonstrate quantitatively the functional relationships between climate and the various soil properties. In places the mean annual values of either precipitation or temperature can be used as an approximation of the climate. The use of mean annual values, however, fails to take into account the monthly distribution of precipitation and temperature (Fig. 11–1). It is important to know, for example, whether or not precipitation is seasonal, and, if it is, whether the precipitation maximum coincides with the annual temperature maximum or minimum, because these climatic variables strongly influence soil leaching and soil-water chemistry, both of which are important in determining

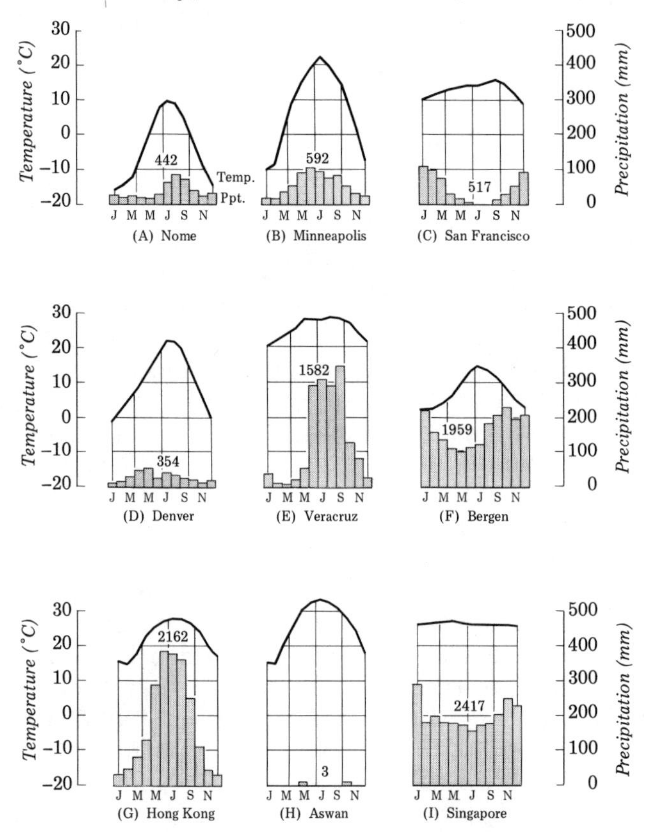

Fig. 11–1 Monthly variations in precipitation and temperature for various climatic stations around the world. Precipitation given on each graph (mm) is total mean annual. (Reproduced by kind permission of John Bartholomew & Son Ltd. from *The Times Atlas of the World*,[40] pp. XXVIII–XXIX.) Climate and altitude of each station as follows. (A) Tundra, 7 m; (B) continental, 285 m; (C) mediterranean, 47 m; (D) steppe, 1613 m; (E) tropical savanna, 16 m; (F) marine, 43 m; (G) humid subtropical, 33 m; (H) desert, 111 m; (I) tropical rain forest, 5 m. Upper curve, temperature; histogram, precipitation.

key soil properties. A fair amount of work has gone into calculating a single number that adequately describes those aspects of climate for which data are readily available and which relate to soil properties. These are discussed by Jenny[23] and will not be repeated here.

Water balance can be used to describe some of the climatic characteristics of a region that are important to soil formation.[1,2,3] This balance represents the gains and losses of soil moisture over a certain interval of time (week, month, year). Gains are from precipitation (P); these data come from climatological stations. Losses, however, are from evapotranspiration; values for potential evapotranspiration (ET_p) are calculated by the method of Thornthwaite and Mather.[41] If Manhattan, Kansas, is used as an example (Table 11–1), the water balance shows the approximate monthly variation in soil moisture. For those months with positive values for ($P-ET_p$), water is stored in the soil at moisture values above permanent wilting point; this water is available for plant growth, weathering reactions, and, if the water is

Table 11–1

Water balance at Manhattan, Kansas, for a soil with 10.2-cm available water-holding capacity

	MONTH												ANNUAL
	J	F	M	A	M	J	J	A	S	O	N	D	
Potential evapotranspiration (ET_p)	0.0	0.0	1.9	5.1	9.3	13.9	16.4	15.0	10.1	5.2	1.4	0.0	78.3 cm
Precipitation (P)	2.0	3.0	3.8	7.1	11.1	11.7	11.5	9.5	8.6	5.8	3.8	2.2	80.1 cm
Soil-moisture gains, $P-ET_p=(+)$	2.0	3.0	1.9	2.0	1.8					0.6	2.4	2.2	15.9 cm*
Soil-moisture losses, $P-ET_p=(-)$						2.2	4.9	5.5	1.5				14.1 cm
Soil-moisture storage at end of month (A and B horizons)	7.2	10.2	10.2	10.2	10.2	8.0	3.1	0.0	0.0	0.6	3.0	5.2	
Possible deep percolation below B horizon**			1.9	2.0	1.8								5.7 cm

*This annual value is the leaching index of Arkley.[2]

**In this case it is assumed that the water-holding capacity of the A and B horizons is 10.2 cm. Therefore any surplus of water over 10.2 cm can move to greater depths in the soil.

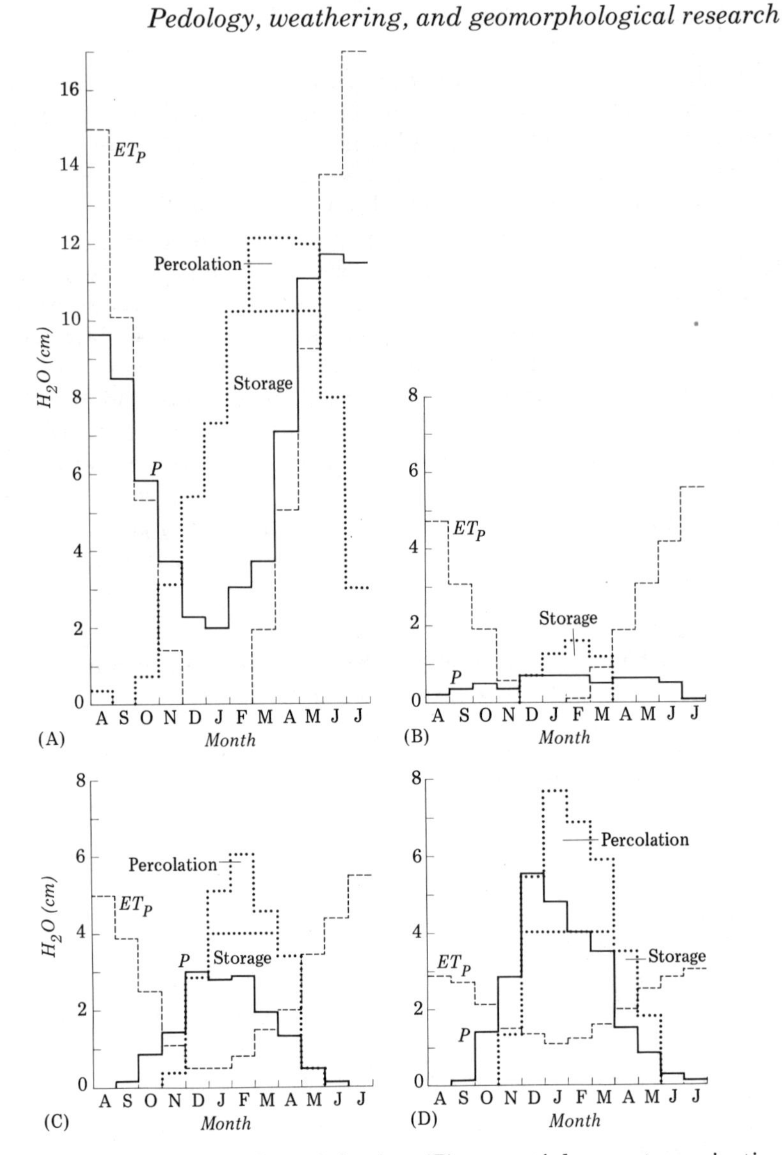

Fig. 11–2 Monthly values of precipitation (*P*), potential evapotranspiration (ET_p), soil-moisture storage in the A and B horizons (10.2-cm available water-holding capacity), and soil-moisture percolation below the B horizon for various climatic stations. Mean temperature as follows:

Station	*Mean temperature (°C)* Jan.	July
(A) Manhattan, Kan.	−1.6	26.6
(B) Fallon, Nev.	−1.2	22.9
(C) Sacramento, Calif.	7.5	23.2
(D) Half Moon Bay, Calif.	9.9	17.0

moving, it can translocate material within the soil. In contrast, months with negative values of $(P - ET_p)$ are marked by the removal of water from the soil. Water removal can go on until the soil reaches water contents approaching permanent wilting point. If the water-holding capacity of the soil is known, the amount of water percolating to depths beneath the B horizon can be calculated.

Plots of water balances for several different climatic regions indicate significant soil-moisture differencies of pedologic importance (Fig. 11–2). For example, both the California coast (Half Moon Bay) and the California Great Valley (Sacramento) are characterized by soil-moisture buildup for about the same months, but a greater amount of water leaches through the soil at the coastal site than at the inland site. Fallon, Nevada, located in the rain shadow of the Sierra Nevada, is characterized by a short period of soil-moisture storage and slight leaching; thus, soil profiles there are shallow and slightly leached and contain Cca horizons. However, Manhattan, Kansas, is in an area of summer precipitation and thus is characterized by fairly high soil-moisture contents throughout much of the year, including some of the warm summer months. These data point out the basic differences between the climate of the midcontinent and that of California and the Basin and Range Province; that is, in the midcontinent the soils are moist during much of the warm season, whereas the western areas are winter wet and summer dry. From this it can be predicted that the rates of biological and chemical processes will vary between these regions. Compared with western sites, sites in the midcontinent should produce more above-ground organic matter annually, and they should have higher rates of organic-matter decomposition and of mineral weathering.

The data on water balance and water-holding capacity of the soil can be combined to approximate water movement with depth in soils.[2] Knowing the frequency of wettings per year, and assuming that downward movement of water takes place only when field capacity for that part of the soil has been reached and that moisture can be removed from the soil by evapotranspiration down to permanent wilting point, one can construct curves depicting water movement (Fig. 11–3). These curves then can be compared with soil data, such as clay-mineral variation with depth, or the top of the Cca horizon, in order to determine which soil features can be attributed to present-day water movement and which may have formed under some past water-movement regime.

Although a single number cannot be derived from the water-balance data to adequately characterize a particular climate, water balances do provide data from which one can rank soils from various

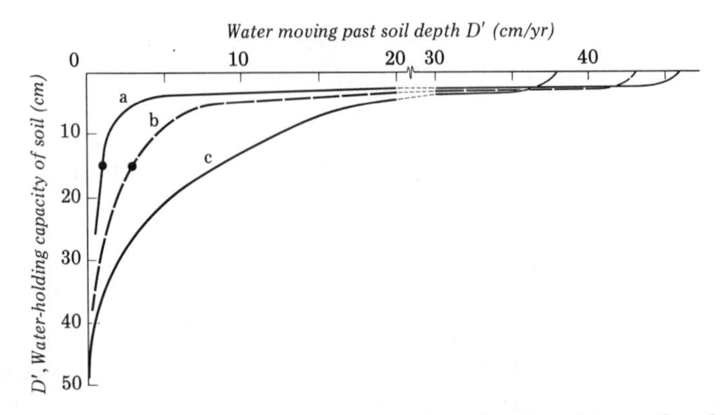

Fig. 11–3 Amount of water moving past various depths of the soil, calculated from water-balance data. Depth is in water-holding-capacity units (D′), but actual soil depths can be calculated if data on water-holding capacities for specific soil profiles are known. (Part of data from Arkley,[2] Fig. 4, © 1963, The Williams & Wilkins Co., Baltimore.) Filled circles, depth to the top of the $CaCO_3$ accumulation layer for nearby soils. Key for curves: a, Clovis, N.M.; b, Boise City, Okla.; c, Santa Monica, Calif.

regions by the amount of water leaching through the soil. Arkley,[2] for example, has derived a number called the leaching index, which is the mean seasonal excess of $(P - ET_p)$ for those months of the year in which $P > ET_p$ (Table 11–1). Soils located in areas characterized by a high leaching index commonly have properties associated with large amounts of water percolating through the soil, such as 1:1 clays, whereas soils in regions with a low leaching index have properties associated with slight leaching, such as 2:1 clay minerals or a Cca horizon close to the surface.

Because most data in the literature on climate-soil relationships use mean annual values of precipitation and temperature, these will generally be used here. Water-balance data might be more meaningful, however.

Regional soil trends related to climate

Many soil properties show distinct trends with regional climate in going from the equator to poles, as is shown by the diagram of Strakhov[38] (Fig. 11–4). These variations in the soils originate in such processes as organic-matter influx and decomposition, presence or absence of chelating agents, soil-water chemistry, and the depth and rate of leaching of water through the soil. These processes, in turn, are controlled by the climate. The tropical forest regions are characterized by intense, deep weathering, with iron and aluminum oxides and hydroxides predominant close to the surface. With depth, clay minerals

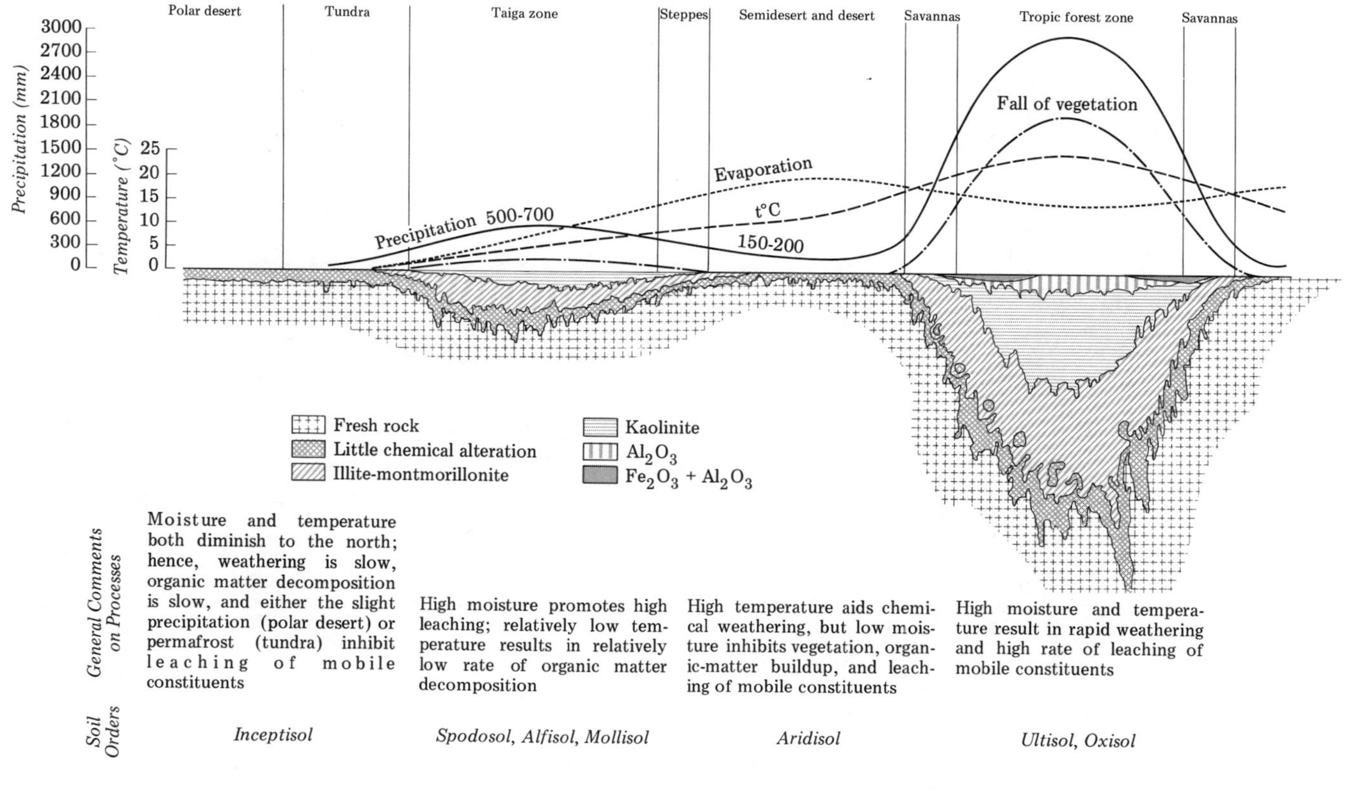

Fig. 11–4 Diagram of relative depth of weathering and weathering products as they relate to some environmental factors in a transect from the equator into the north polar region. (Taken mostly from Strakhov,[38] Fig. 2.)

of the 1:1 and finally the 2:1 varieties are found. Organic matter in these soils varies, because even though the amount of organic matter annually added to the soil is high, the decomposition rate of organic matter varies. The above trends diminish to the north in the savanna region. The deserts are characterized by low organic-matter input relative to the rate of decomposition, and low organic-matter content in the soil results. Slight leaching produces 2:1 clays and $CaCO_3$ in the soil. Increased precipitation and decreased evapotranspiration characterize the steppes compared to conditions in the deserts; the result is a fairly thick cover of vegetation, thick A horizons rich in organic matter, with moderately leached subsurface horizons. We find $CaCO_3$ in soils only in the more arid parts of the steppes. North of the steppes is the taiga, a region of fairly high soil leaching. The low temperatures of

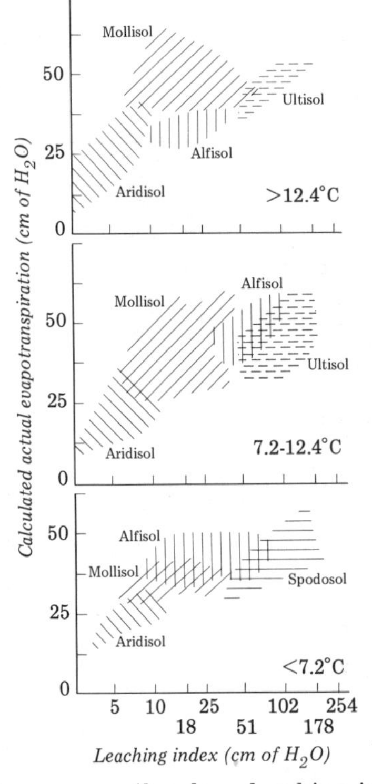

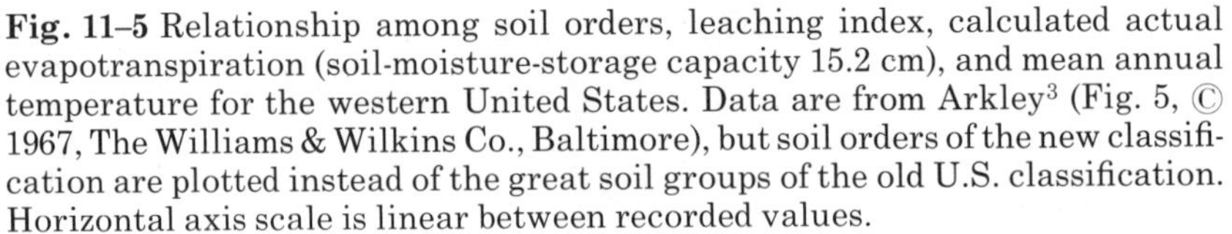
Fig. 11–5 Relationship among soil orders, leaching index, calculated actual evapotranspiration (soil-moisture-storage capacity 15.2 cm), and mean annual temperature for the western United States. Data are from Arkley[3] (Fig. 5, © 1967, The Williams & Wilkins Co., Baltimore), but soil orders of the new classification are plotted instead of the great soil groups of the old U.S. classification. Horizontal axis scale is linear between recorded values.

the taiga result in fairly low rates of organic-matter decomposition and a prominent A horizon forms. Where conditions favor iron movement, an E horizon can form and, in areas of greatest leaching, 1:1 clays can form. The tundra lies north of the taiga, and the combination of fairly low precipitation, low temperature, and permafrost close to the surface in places result in a moist soil with relatively slow rates of organic-matter influx and decomposition. These conditions produce a soil in which an A horizon with a fairly high organic content commonly overlies a gleyed horizon. Continuing northward, the tundra gives way to the polar desert, a region of low precipitation and temperature.[12,39] Because vascular plants are nearly absent in this desert, organic-matter content in the soil is fairly low and provided mainly by lichens, algae, and diatoms. Although these polar desert soils generally are permeable, the absence of appreciable moisture and water movement in the soil causes them to be saline and alkaline, with very little weathering. In many respects these soils are similar to those of some hot deserts.

Very little has been done on the relationship between climatic parameters and large-scale soil-classification units, probably because the relationship is an extremely complex one. Arkley[3] has attempted such a correlation for the western United States, in which soils shown on a regional soil map were related to water-balance climatic parameters obtained from climatic stations located within the mapped soil units (Fig. 11–5). The parameters chosen were calculated actual evapotranspiration, based on a soil-moisture storage capacity of 15.2 cm, leaching index, and mean annual temperature. These parameters were considered to relate best to the major processes responsible for the various soil-classification units. The plot of data indicates that, although there is considerable overlap of some soil orders, and although some orders range widely in values for the climatic parameters, most soil orders seem to fall within well-defined values for the climatic parameters. In some places the overlap might occur because the climatic parameters chosen were not those most highly correlated with the particular orders; in other places the overlap might occur because both orders are stable in that climatic regime, and one soil order may grade into another with time. These plots also bring up a major problem in working with soil morphology for paleoclimatic reconstruction. That is that each order is represented by a large variation in climatic parameters, and thus soil classification units do not give very precise data on paleoclimate, except in those transitional regions that separate soils of markedly different morphologies. For more precise information it might be best to construct figures for soil suborders rather than orders.

Variation in specific soil properties with climate

The main soil morphological and mineralogical properties that correlate with climate are organic-matter content, clay content, kind of clay minerals, color, the presence or absence of $CaCO_3$, and depth to the top of the $CaCO_3$-bearing horizon. One can quantify much of the data on soil-climate relationships for some properties; this is not possible for other properties and all one can do, given the present state of knowledge, is make some qualitative observations. Because the main thrust in this direction has been by Jenny and his associates, the data here will rely heavily on their work.

Organic-matter content

Jenny[23,26] has studied many climatic transects to determine the trends in organic-matter constituents in the soil (organic carbon and total nitrogen). In general, he finds that soil nitrogen increases logarithmically with increasing moisture and decreases exponentially with rising temperature. These relationships hold for such diverse climatic regions as the Great Plains, India, and California (Figs. 11–6

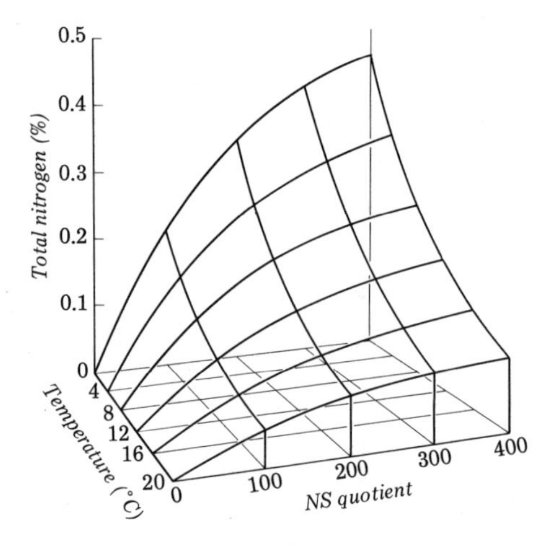

Fig. 11–6 Relationship between total nitrogen content of the surface soil and climate for grassland soils of the Great Plains. (Taken from Jenny,[23] Fig. 92.) Moisture is given as the NS quotient

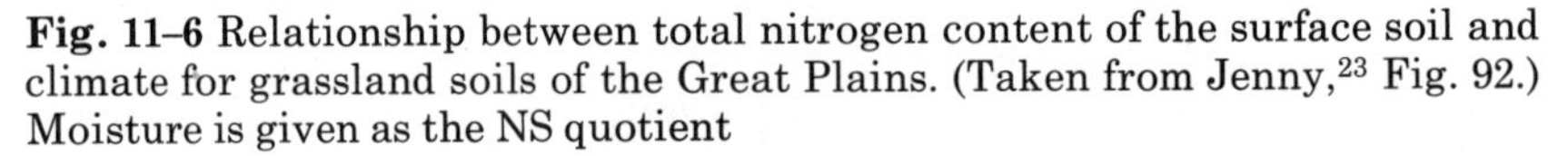

$$m = \frac{\text{Precipitation (mm)}}{\text{Absolute saturation deficit of air (mm Hg)}}$$

The equation that describes the surface is

$$N = 0.55\, e^{-0{\cdot}08\, T}\,(1 - e^{-0{\cdot}005\, m})$$

where N is total nitrogen, T, mean annual air temperature, and m the NS quotient.

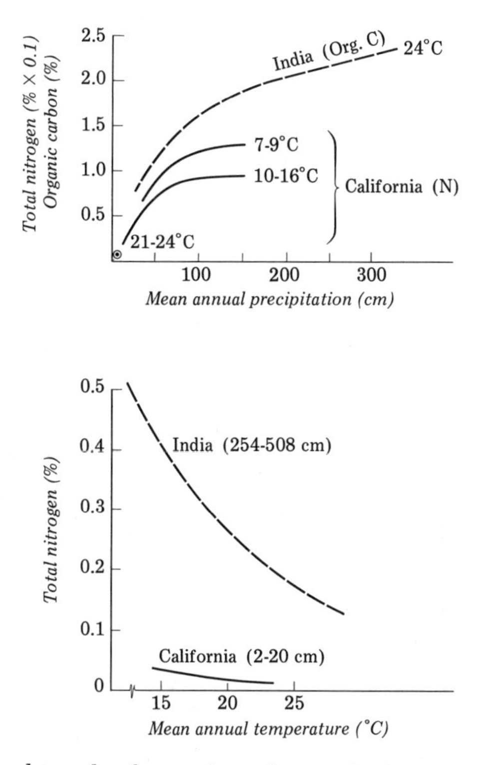

Fig. 11–7 Idealized trends of organic carbon and nitrogen with mean annual precipitation and temperature: India and California. [Taken from Jenny,[26] Missouri Agri. Exp. Sta. Res. Bull. 765 (1961), and Harradine and Jenny,[18] Figs. 5 and 6, © 1958, The William & Wilkins Co., Baltimore.] Temperature values are mean annual (upper graph); the numbers in parentheses (lower graph) are mean annual precipitation values.

and 11–7). In general, this means that at fairly low values of either precipitation or temperature each increment of change in either climatic parameter has a much greater effect on the amount of organic carbon or of nitrogen in the soil than has an identical increment of change at higher values of precipitation and temperature. The figures also indicate the obvious general organic-matter content trends with climatic region; that is, low in deserts, intermediate in temperate regions and high in some tropical regions. Total organic carbon and nitrogen contents per unit surface area in soils of the temperate and tropical regions also follow these trends (Table 11–2).

The above climatic trends are related to yearly gains and losses of organic matter. Jenny[25,26] compared the dynamic nature of several tropical and temperate ecosystems to find some basic differences (Table 11–2). The tropical soils he studied contain more organic carbon and

Table 11–2
Annual gains, losses, and decomposition constants for several forest ecosystems

	TROPICAL FORESTS OF COLUMBIA		CALIFORNIA		
	Sealevel	*1540 m*	*Temperate forest (1640 m)*		*Cold forest (3280 m)*
			Oak	*Pine*	*Pine*
Litterfall (g/m²)					
Weight*	730	935	149	305	101
Total nitrogen	10.4	15.7	1.27	1.54	—
Organic carbon	391	510	74.6	164	—
*Forest floor** (g/m²)*					
Weight*	432	1,455	2,517	12,635	11,081
Total nitrogen	8.8	35.4	31.1	117	—
Organic carbon	225	795	1,224	6,463	—
Decomposition constant of litterfall and forest floor (annual percent loss)					
Weight*	62.8	39.1	5.6	2.4	0.90
Total nitrogen	54.2	30.7	3.9	1.3	—
Organic carbon	63.5	39.1	5.7	2.5	—
Total weights in soil profile † (g/m²)					
Total nitrogen	2,502	3,521	633	650	—
Organic carbon	36,681	45,196	11,606	17,317	—

*Volatile weight only, because the forest floor was contaminated with sand grains.
**Forest floor is the fresh and partially decomposed plant debris overlying the mineral soil.
†Includes forest floor and amounts in mineral soil.

(Taken from Jenny,[25] Table 1, © 1950, The Williams & Wilkins Co., Baltimore, and Jenny,[26] Table 2, Missouri Agri. Exp. Sta. Res. Bull. 765)

total nitrogen than the temperate soils, but the latter have a higher proportion of organic constituents in the forest floor. This is the case in spite of data indicating that litterfall is much greater in tropical forests. The annual rates of decomposition of the litterfall and forest floor material explain this apparent discrepancy, for the rates are much higher in the tropical region. In California, the rates of decomposition also have been shown to decrease with elevation and thus with climate[27] (Table 11–2).

These studies can serve as models for further study of ecosystems in other climates. Thus, to understand the relationship between climate and the organic-matter constituents in a soil to the extent that Jenny has, we must know the gains and losses involved, the decomposition constants, and the total weight of the organic-matter constituents, as well as the climatic parameters.

Clay content

It is commonly reported that clay content in soils varies with climate, but precise data on this relationship are not often available because in many places factors other than climate have influenced clay production, and their influence is hard to quantify.

Two of the early quantitative relationships between soil clay content and climate are the moisture and temperature functions of Jenny,[22] in which he found a linear relationship with moisture and an exponential relationship with temperature (Fig. 11–8). Thus, if all other factors are constant, one would expect low rates of clay production in cold-dry cold-wet, and hot-dry environments, increasing rates with increasing moisture contents, and highest rates in hot-humid environments. These data were obtained before we knew much about Quaternary stratigraphy and the relationships between soil morphology and age, so, although Jenny's generalized trends probably still hold, the absolute amounts of clay for any climate might not coincide too precisely with the data of Fig. 11–8.

To determine the precise relationship between clay production and climate, it would be best to have data on total clay present per unit area of the ground surface, data similar to that available for organic matter (Table 11–2). Percentage values, although based on the less than 2 mm fraction, do not reflect the true amounts of clay in the profile, because variations in gravel content have a marked effect on absolute clay

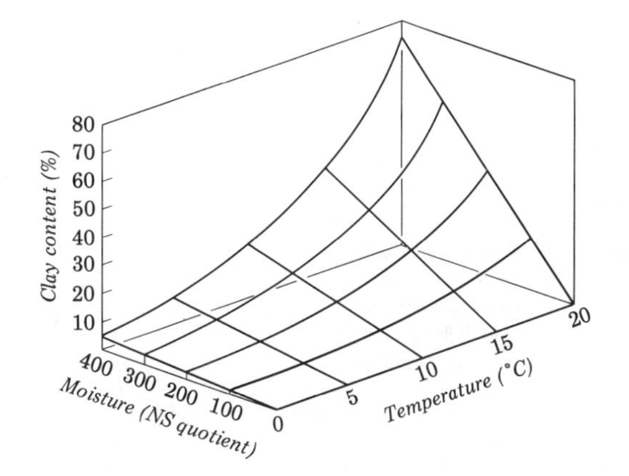

Fig. 11–8 Idealized relationship between per cent clay in the top 92 to 102 cm of soil derived from granites and gneisses and mean annual moisture and temperature. Moisture is given as the NS quotient (see Fig. 11–6 caption). (Taken from Jenny,[22] Fig. 8, © 1935, The Williams & Wilkins Co., Baltimore.)

content (see Fig. 1–7). Until such data are collected we will not be able to quantify clay production as a function of any of the factors.

One way of qualitatively assessing clay production with climate is to compare stratigraphically dated soils along a transect covering various climatic regions. One such transect is the Truckee River stream-terrace deposits that can be traced from the cool, humid Sierra Nevada of eastern California to the warm, dry Basin and Range Province in Nevada.[6] There the clay content increases in the direction of increasing temperature and decreasing moisture (Fig. 11–9). The explanation for this behavior might lie in the variation in leaching index along the transect, the index being high in the humid area and low in the dry area. At the humid end of the transect, with high rates of leaching, many constituents released by weathering might be leached from the soil, whereas at the drier end, constituents released by weathering might remain in the soil and react to form clays. Thus, the explanation for these relationships might depend more on soil-leaching conditions than on moisture alone. It would have been difficult to predict this trend from Fig. 11–8. Perhaps if enough clay data on different climatic regions are collected, one could attempt to reconstruct a graph like that of Fig. 11–8.

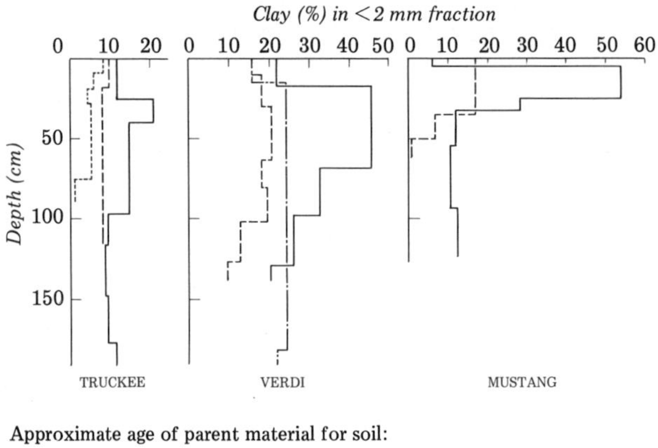

Fig. 11–9 Clay distribution in soils formed from stream terrace deposits and a lacustrine sand (younger soil at Mustang) on the Truckee River, California and Nevada. (Taken from Birkeland,[6] Fig. 3.) Truckee lies at the foot of the eastern Sierra Nevada; climatic parameters are 80-cm mean annual precipitation, 6° C mean annual temperature, and a leaching index of 65.8 cm. Comparable climatic parameters for Reno, Nevada (between Verdi and Mustang) are 18 cm, 9.5° C, and 7.4 cm. Summers are dry at all locations.

Clay production can be qualitatively related to climate by comparing stratigraphically dated deposits and their respective soils in many different environments. One useful age datum is soil formed on late-Wisconsin till. Alpine tills in many parts of the Cordilleran Region, in areas characterized by igneous and metamorphic rocks of granitic composition, have only A/Cox soil profiles; textural B horizons usually are absent. This relationship holds for soils in the Sierra Nevada,[8] the eastern Cascade Range,[33] and the Rocky Mountains.[35] Soils of similar age formed from parent materials of at least partly volcanic rock in the high-precipitation areas of the Puget Sound Lowland, Washington, or on Mt. Rainier, Washington, also lack a textural B horizon.[13,14] These latter data come somewhat as a surprise because one would predict a greater rate of clay production in high-rainfall areas with volcanic parent materials, because these materials are less resistant to alteration relative to parent materials derived from granitic rock. The conclusion to be drawn from these data is that soils formed from tills of somewhat similar texture look alike after 10^4 years of soil formation, in spite of large variations in climate and parent materials.

Moderately developed soils can be found on early-Wisconsin deposits at some of the localities mentioned above (see Fig. 8–10). In many places, however, the soils on these deposits are only weakly developed, perhaps because erosion occurred at some time subsequent to initial soil development. So again, at these localities there does not seem to be an obvious textural relationship attributable solely to climate. Pre-Wisconsin soils are not too abundant, and textural data are so few that no valid conclusions on regional rates of clay production can be drawn.

Other areas in the United States are characterized by much more rapid clay production than are the mountains of the Cordilleran Region (Fig. 8–10). One such area is the California Coast; here the textural profile for the first marine terrace (pre-Wisconsin) indicates more clay production relative to older soils in other parts of the Cordillera. The midcontinent is another area of seemingly rapid clay production. Soils formed on late- and early-Wisconsin tills and on pre-Wisconsin tills have more strongly developed B horizons than their Cordilleran counterparts. Although part of this variation could result from parent materials with relatively high clay content in the midcontinent, at least some of the greater clay production of the midcontinent could result from the climate, which is characterized by significant rainfall during the warm summers. Soils suitable for a comparison of soil development on parent materials of similar textures include those on outwash deposits of the eastern Sierra Nevada (Fig. 11–9) and of Ohio (Fig. 8–10). Again, the midcontinent shows up as a region of rapid clay production, relative to the Sierra Nevada, but the rates in the midcontinent and at the western edge of the Basin and Range Province (Verdi, Fig. 11–9) are comparable, to a certain extent, at least for soils formed on deposits of early-Wisconsin age.

Many workers have observed that the arctic regions have a characteristic low clay production. Recent reconnaissance work with stratigraphically dated soils on Baffin Island seems to bear this out.[9] Deposits of Wisconsin and pre-Wisconsin age have been recognized, and textural B horizons have yet to be found in the area. Parent materials are igneous and metamorphic rocks, granitic in composition, and therefore are similar to the parent materials of many soils formed on till in the Cordilleran Region.

One problem in determining clay production in arid regions so that it can be compared with rates of other regions is that of airfall contamination. For example, Gile and others[17] report significant influx of fine-grained noncarbonate and carbonate material into artificial traps in part of the Rio Grande Valley of New Mexico. From this it would seem that one should be careful not to ascribe all clay in aridland soils to pedogenesis without data on the weathering of the nonclay minerals. Of the aridland soils at Mustang, Nevada (Fig. 11–9), for example, those soils formed on the early-Wisconsin deposit have etched pyroxene grains (J. G. LaFleur, unpub. report, 1972), and this would suggest that at least some of the clay could have formed within the profile.

In summary, although many clay data are available for soils, we still do not have a very precise idea on the variation in clay content with climate, when all other factors are kept constant. As Quaternary stratigraphic studies proceed in many diverse climates, soils of different ages can be collected and their clay production analyzed. We must improve our techniques, however, to get the best data. Bulk densities should be taken in the field so that the rates can be based not on per cents but on weight of clay formed per 100 g parent material or per unit surface area for the entire soil. In addition, detailed studies of the nonclay fraction to determine the relative degree of weathering in the soil as well as the absolute amounts of weathering of specific minerals can be undertaken; these should show some relationship to the amounts of clay produced.

Clay mineralogy

Clay minerals formed in the soil vary with the water chemistry and the rate of leaching, and thus with the climate. Many transects relating clay mineralogy to climate have been described, but only a few will be mentioned here. In any climatic transect, however, the clay mineralogy seems best correlated with precipitation or leaching index, and, in going toward areas of greater precipitation, the clay mineralogy trend usually is toward species containing progessively less silica. The reason for this trend is that soil leaching increases with increasing precipitation, and therefore more silica is lost from the soil during the course of weathering (Fig. 11–10). Data are given here for well-drained soils so that the mineralogy reflects regional climate. This is important because it is commonly reported, even in tropical regions where kaolinite has formed in well-drained sites, that montmorillonite can form in soils in topographic positions in which internal soil drainage is restricted. The discussion here will focus mostly on soil clay minerals formed from crystalline-rock parent materials. Clay minerals formed from sedimentary rocks or deposits derived largely from sedimentary rocks will not be discussed, because some of their clay minerals are inherited.

Even though tropical regions typically are used to characterize the end products of intense weathering (aluminum and iron compounds), there is a definite mineralogical trend with climate.[28] Commonly, montmorillonite forms at low amounts of precipitation, kaolinite at higher amounts, and oxides and hydroxides of iron and aluminum at still higher amounts. In Hawaii, Sherman[37] reports that montmorillonite predominates below about 100-cm precipitation, kaolinite between about 100 and 200, and the iron and aluminum compounds above 200 (Fig. 11–11). In an ash layer in Hawaii dated at 10,000 to 17,000 years, Hay and Jones[19] found montmorillonite at 25- to 65-cm precipitation, and gibbsite at greater than 370-cm precipitation. In soils formed on mafic lavas and tuffs in the Caribbean, Beaven and Dumbleton[5] reported montmorillonite predominant at less then 150-cm precipitation, kaolinite predominant at greater than 200-cm precipitation, and mixtures of the two at intermediate values of precipitation.

Barshad[4] reported on the clay mineralogy of several hundred

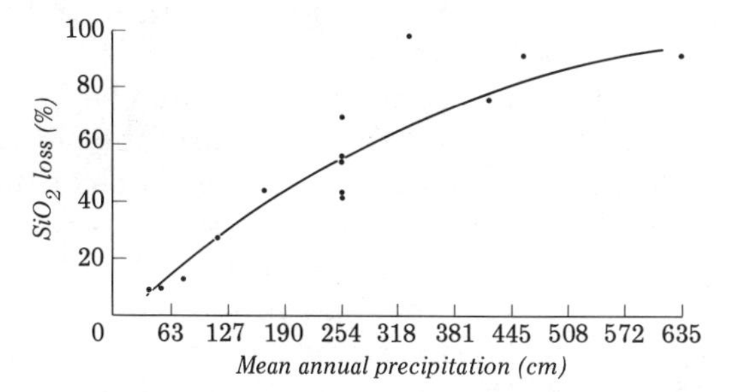

Fig. 11–10 Overall trend of silica loss with precipitation in a 10,000- to 17,000-year-old volcanic ash, Hawaii. (Taken from Hay and Jones,[19] Fig. 2, published by the Geological Society of America.) This is the same ash for which mineral weathering data are presented in Fig. 11–14.

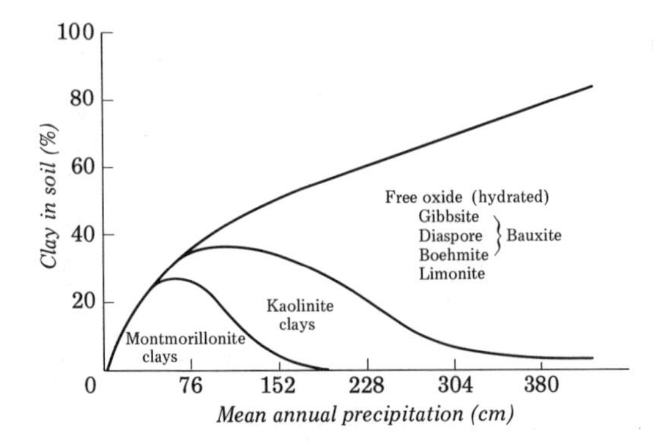

Fig. 11–11 Clay mineralogy as a function of precipitation under a continuously wet climate, Hawaii. (Taken from Sherman,[37] Fig. 3 in *Problems of Clay and Laterite Genesis*, published by AIME.) Sherman also presents data for an alternating wet and dry climate with a somewhat similar relationship of precipitation and soil clay minerals.

surface-soil samples (0 to 15 cm depth) in California and found a close correspondence of clay mineral with precipitation (Fig. 11–12). Generally, montmorillonite is only found at less than about 100-cm precipitation, and gibbsite at greater than about 100. Kaolinite and/or halloysite are present over a wide range of precipitations and are predominant above about 50-cm precipitation. Illite and vermiculite are also present, the former only in felsic igneous rocks and the latter in both felsic and mafic igneous rocks. The abundances of the various minerals varies somewhat for the same amount of precipitation between soils formed from felsic igneous rocks and those formed from mafic igneous rocks. The main reason for this shift in clay-mineral abundances with parent material is the cation content of the parent material. Thus, for example, montmorillonite can form under higher precipitation from mafic igneous rocks than from felsic igneous rocks because the former have a higher cation content, and high cation content in the soil solution favors formation of montmorillonite.

Clay-mineral studies of soils from a large number of till and outwash deposits of many ages along the east side of the Sierra Nevada in California and Nevada substantiate the conclusions of Barshad.[7,8] In general, in the northern part of the range in deposits with a mixed lithology of granitic, andesitic, and basaltic rocks, the change from high halloysite content to high montmorillonite content takes place at the pine-sagebrush vegetation boundary at about 40-cm annual precipitation (Fig. 11–13). However, near the southern end of the study area soils formed from granitic till presently located below the pine-sagebrush boundary, in what is probably a drier climate, do not contain montmorillonite. Again, this variation with parent material can be predicted from the data of Barshad, although the exact climatic values at the transition from one predominant clay mineral to another may vary with the precise environmental conditions, including any paleoclimatic influences.

In summary, each area has a unique environment and therefore a characteristic clay-mineral assemblage. If, however, there is a sufficiently large variation in the amount of precipitation, clay minerals with less silica will predominate at higher precipitations. If precipitation is high enough and leaching extensive enough, most silica is removed and only oxides and hydroxides of aluminum and iron form. The amounts of precipitation at which one clay mineral or oxide and/or hydroxide predominate over others will vary with the environment of the region. Looking at climatic relationships alone, one would predict that the boundary between predominant 1:1 clay and 2:1 clay would be at higher precipitations in a climate characterized by a summer rainfall than in one in which rainfall is concentrated primarily in the winter.

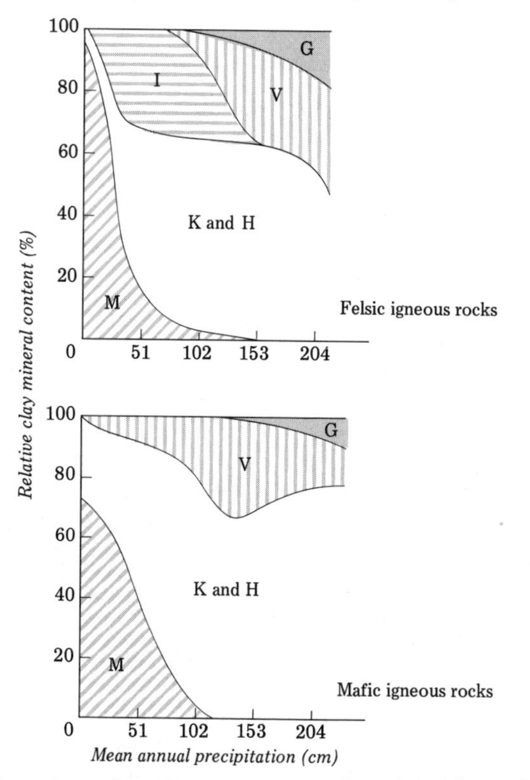

Fig. 11–12 Relative clay-mineral content as a function of precipitation for surface-soil samples, California. (Taken from Barshad,[4] Figs. 1 and 2.) Mean annual temperatures range from 10° to 15.6° C.

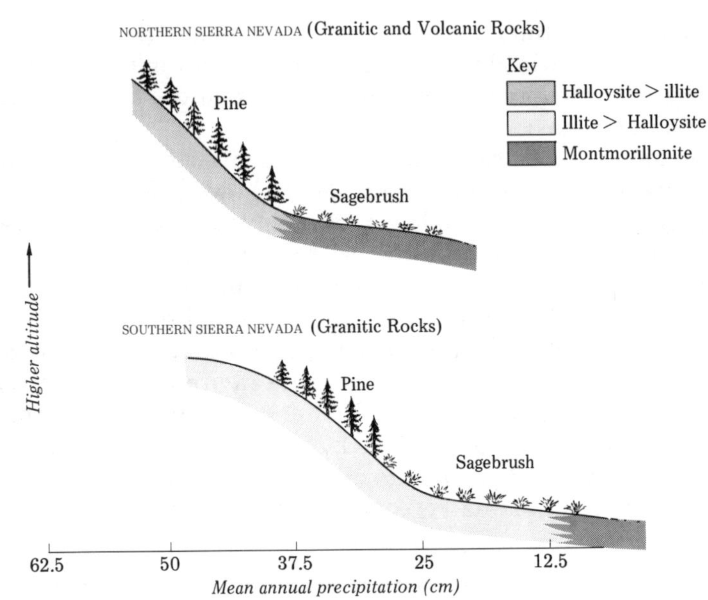

Fig. 11–13 Diagram of the relationship of clay mineralogy and climate with different parent materials along the eastern Sierra Nevada. (Data from Birkeland and Janda,[8] Table 3.)

Mineral weathering

The variation in rate of mineral etching with time for different regions (Fig. 8–7) is attributed, in part, to differences in climate. If weathering in soils alone is compared, one obtains the following ranking, from regions of greatest amount of weathering to least amount: St. Vincent = central Europe = Michigan > Colorado Piedmont > eastern Sierra Nevada. This suggests that hot-humid and cool-humid climates are those most conducive to mineral and rock alteration. It would not necessarily follow that clay-production rates follow these trends, because clays are produced only when the correct constituents in the correct proportions are present and precipitate in the soil.

Recent work on a volcanic ash unit in a climatic transect in Hawaii clearly demonstrates the predominant influence of rainfall on mineral weathering.[19] The ash, originally about 95 per cent vitric ash, was laid down from 17,000 to 10,000 years ago; it has been weathering under a precipitation that ranges from about 25 to 635 cm/yr. At less than 115-cm annual precipitation only glass has altered, between 115- and 255-cm annual precipitation plagioclase is the only mineral to show slight alteration, above 225-cm annual precipitation all minerals are altered to various degrees, and at 570- to 635-cm annual precipitation all except olivine are completely altered (Fig. 11–14). These are really exceptional rates of weathering, which again demonstrate that rapid weathering takes place in humid tropical regions.

Soil redness

The color of a soil, in particular its degree of redness, is generally related to climate. One has to be very cautious in this regard, because at least part of the redness of old soils is a function of time of soil formation (Fig. 8–11). Part of the differences in soil color with climatic region, however, is a function of climate. Temperature probably is more important than precipitation in producing redness, because red soils are common in the humid southeastern United States, the humid tropics, and hot deserts of the southwestern United States and Baja California, Mexico; red soils are not common in areas of low temperatures and varying amounts of precipitation (e.g., along the northern borders of the United States and in polar regions).

Calcium carbonate

Several features of $CaCO_3$ accumulation in soils are related to the climate, more specifically to that fraction of the precipitation that leaches downward in the soil. Here we will briefly discuss the amount of

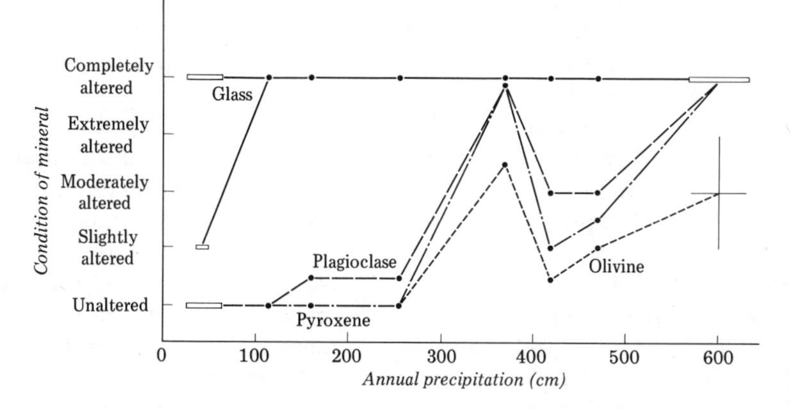

Fig. 11–14 Mineral weathering in volcanic ash as a function of annual precipitation, Hawaii. (Data from Hay and Jones,[19] Table 4.) The plagioclase is labradorite, and the pyroxene clinopyroxene. Bars indicate ranges in either precipitation or mineral condition. Weathering of olivine varies because only when the glass coatings on the grains have dissolved does the mineral begin to weather. Mean annual temperatures range from 23° C near sea level to 16° C at 1202 m altitude. The more rapid weathering at 370-cm precipitation relative to that at 255 and 420 is unexplained; it could be due to a higher mean annual temperature, a finer grain size, or higher precipitation in the late Quaternary (R. L. Hay, written commun., 1972).

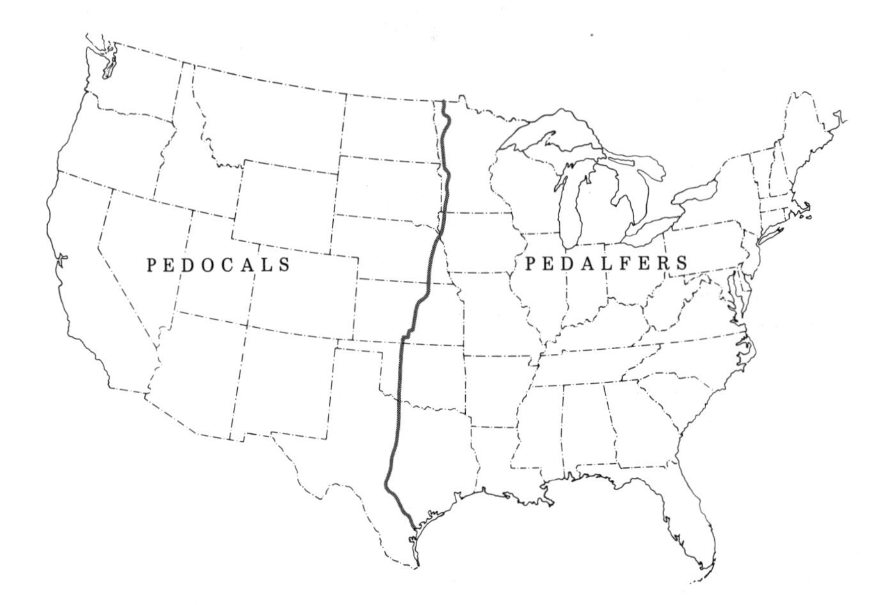

Fig. 11–15 Approximate position of the pedalfer-pedocal boundary for the midcontinent. (Taken from Jenny,[23] Fig. 99.)

precipitation at which $CaCO_3$ begins to appear in soils and the relationship between the amount of precipitation and the depth to the top of the $CaCO_3$-bearing horizon.

In any climatic transect running into a sufficiently dry region, $CaCO_3$ usually appears in the more arid soils. The climatic value at which it first appears is obviously a function of the distribution of moisture and temperature throughout the year. In the midcontinent, for example, an area of significant summer precipitation, the boundary between the pedocals and the pedalfers (Fig. 11–15) is at about 50-cm mean annual precipitation and 5–6°C mean annual temperature in northern Minnesota and at about 60-cm mean annual precipitation and 22°C in southern Texas. However, in going from the northern Sierra Nevada to the Basin and Range Province, $CaCO_3$ first appears in the soils near Reno, Nevada, with a mean annual precipitation of 18 cm and a mean annual temperature of 9.5°C. The pedocal-pedalfer boundary occurs at a lower precipitation at Reno relative to the midcontinent partly because soil leaching per unit of rainfall is more effective in a winter-wet, summer-dry climate such as Reno's. Another factor that might explain the moister climate at the midcontinent pedalfer-pedocal boundary is texture, because midcontinent parent materials commonly are finer textured than are the parent materials near Reno. Still another factor to be considered in explaining this boundary is the calcium content of the parent materials, because Jenny[24] has shown that $CaCO_3$ horizons can persist to higher values of precipitation in soils with calcium-rich parent materials relative to parent materials low in calcium.

Although Fig. 11–15 accurately depicts the pedalfer-pedocal boundary for the midcontinent, it is not accurate for the western United States because here the mountains and intermontane basins have markedly different climates and soils. Nearly every mountain-basin transect is characterized by a relatively moist-cool climate in the mountains and a relatively dry-warm climate in the basins. Under appropriate conditions, therefore, soils on the lower slopes of the mountains and in the basins will contain $CaCO_3$. These relationships are repeated over and over again in the western United States. Richmond[34] has made a particularly detailed study of these relationships in the La Sal Mountains of Utah and finds that the pedalfer-pedocal boundary is related not only to present altitude and climate (Fig. 11–16) but also to past conditions of pedogenesis (Fig. 11–18). He also found

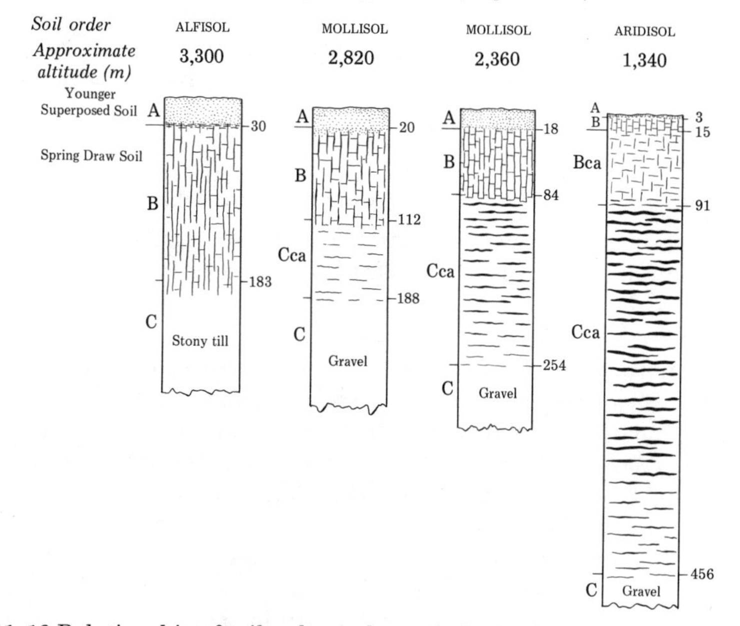

Fig. 11–16 Relationship of soil order and morphology with altitude, the La Sal Mountains, Utah. (Taken from Richmond,[34] Fig. 18.) The soil shown is of pre-Wisconsin age. Numbers beside profiles are soil-horizon boundary depths in centimeters.

that with lower altitude the Cca horizons of all soils of a particular age get thicker, their pH shows a slight increase, and Bca horizons are more common (Fig. 12–9).

Although the amount of $CaCO_3$ in the soil is primarily a function of the duration of pedogenesis, depth to the top of the Cca or K horizon is closely related to the amount of precipitation. This is qualitatively shown by the La Sal Mountains data (Fig. 11–16). Depth relationships have been quantitatively assessed in two regions of the United States. Jenny[23] worked on a transect from the relatively dry climate of the Colorado Great Plains eastward to the more humid regions of Missouri and found the following relationship to be fairly representative

$$D = 2.5(P - 12)$$

where D is the depth to the top of the Cca or K horizon, and P is the mean annual precipitation. Arkley[2] analyzed the same relations for a number of California and Nevada soils and got a different relationship

$$D = 1.63\ (P - 0.45)$$

The difference between the two relationships is ascribed to climate. In California and Nevada, with a predominant winter rainfall, a unit increment of rainfall is much more effective in leaching $CaCO_3$ to a particular depth than it is in the midcontinent, where significant rainfall comes during the season of high evapotranspiration.

Reconstruction of past climates from pedological data

Morphology and various other properties of soils can be used to infer past climates. This is important to Quaternary research because in many places the soils represent hiatuses in the depositional record and thus are the only record left of certain time intervals. Before paleoclimatic interpretations can be made, however, one has to be sure that the observed feature was truly imparted on the soil by a past climate and is not related to other factors of soil formation. It has commonly been recognized, for example, that the effect of a longer interval of soil formation can give the same pedological result as can a climatic change. Moreover, although many workers have used soil features to infer past climates, studies that have backed up such interpretations with quantitative soil data are relatively few. It is hoped that work on this aspect of pedology will be more quantitative in the future.

Soil properties vary in their usefulness as tools for paleoclimatic interpretation. If the soil has remained at the surface since the climatic change, the property or properties used to decipher changes in climate must have been resistant enough to persist in the soil and not be altered entirely during subsequent pedogenesis. Obviously, those soil properties that alter readily with changing environmental conditions, such as pH, cannot be used as indicators of past conditions. The same is true for buried soils. Here, however, the properties imparted to the soil during pedogenesis must be resistant to subsequent diagenetic alteration to be useful in paleoclimatic reconstruction.

Success in deciphering paleoclimate from soils evidence will depend partly on the field area chosen. The most promising areas are those of rather sharp transition from one climatic region to another. As an illustration, if one is working in the center of a large region of Mollisols that are characterized by a relatively wide range in climate, a past change in climate may have left the area of interest still within an area of Mollisol formation. It is possible that no diagnostic soil feature would record this climatic change. In contrast, if one worked at the margins of the Mollisol area, where they grade into either the Aridisols or the Alfisols, and if a past shift in climate were accompanied by a shift in the geographic position in the soil orders, then marked changes in the soil properties would have accompanied the change in climate, and one could hope to read these properties as due to a past different climate in either surface or buried soils.

Overall soil morphology

Some soils in a region, either at the surface or buried, might have a soil morphology quite different from those currently being formed in the region. If they have, one might be able to reconstruct the past climate by comparing the present-day climate and morphology of soils in the region with the climate and morphology of the soil or soils in question. As a guideline, one might use data like that in Fig. 11–5 to determine how much of a change is suggested by the differences in soil morphologies. This figure also illustrates the importance of working with soils in regions along the boundary between two contrasting soil orders. Again, for soils located in the middle of a large soil-order region, a fair amount of climatic change could result in little morphological change. An example of this approach is seen in the work of Janda and Croft[21] on soil formation and Quaternary climates in the northeastern San Joaquin Valley of California. Soils on terraces of all ages are Xeralfs, and the field evidence suggests that they all have formed under a climate conducive to Xeralf formation. Janda and Croft then looked at the west-coast distribution of Xeralfs and noted that they presently lie in areas characterized by 16- to 84-cm mean annual precipitation, falling mainly in the winter, and 11.3 to 18.3°C mean annual temperature. They concluded that this wide range in climatic parameters probably would result in Xeralf formation, and thus rather large changes in the climate could take place without those changes being registered in soil morphological features. About all they could conclude was that the Quaternary climates in that area probably had not deviated beyond the above precipitation and temperature values.

Southwestern Australia is an example of an area where a fairly large climatic change occurred, as shown by the morphologies of different-aged soils. Deep Oxisols lie on old dissected plateau remnants, whereas younger soils in the river valleys are much less leached, and some contain $CaCO_3$ and are alkaline.[31] The area is characterized by less than 50-cm annual precipitation. Clearly the Oxisols could not form in the present climate because it is doubtful if much water wets as deeply as their thickness. A much wetter former climate, perhaps extending back into the Tertiary, is postulated to explain the presence of the Oxisols.

Organic-matter distribution and content

Soil properties associated with the organic matter of the soil are not too useful in reconstructing paleoclimate. Because organic-matter properties of soil reach a steady state quite rapidly (Fig. 8–8), with change in environmental conditions the distribution and content of the organic matter quickly adjust to values in equilibrium with the new conditions. This is especially true for surface soils, and if climatic change is to be deciphered from these properties the change would have to have taken place in less time, before the present, than the time necessary to reach the steady state for that environment. An A horizon in a buried soil might be useful, but the organic matter is usually so depleted that the A horizon is seldom recognizable. A paleoclimatic reconstruction might be possible if some of the organic-matter properties could be shown to have been inherited from a former vegetation cover (Ch. 10) and if the change in vegetation was due to a change in climate.

B-horizon properties

The position of the textural B horizon, its color, and amount of clay all might contribute to an understanding of past climates. The interpretations are hard to make, however, because data are few and open to several interpretations.

The position of the textural B horizon should relate to the climate under which the soil formed. It is commonly reported that aridland soils are shallow and that soils in progessively wetter climates are progressively thicker; the B-horizon position would parallel these trends. One might be able to relate the position of the B horizon to the water-holding capacity of the soil and to the water movement within the soil. I know of no work relating these properties, but if it could be shown that a B horizon extends to much greater depths than those indicated by the soil-water-movement calculations, one might be able to conclude that the present climate is drier than that under which the B horizon formed.

A surface soil that is redder than younger surface soils in the area might be used as an indication of a former warmer climate.[15,29] This can only be done, however, with equivalent parent materials. Moreover, the duration of soil formation has to be known fairly certainly so that the time factor can be ruled out as a cause of the redness. If it can be demonstrated that the red soil formed over a time span similar to that over which the younger non-red soils formed, a temperature change might well have occurred.

The amount of clay present in a soil can be used to indicate past climatic change if the duration of soil formation is precisely known. Here, if the clay content of an old soil is greater than that in a soil formed under the present climate, and if both soils formed over a similar duration of time, it can be concluded that the older soil formed in an environment conducive to a more rapid production of clay. To increase the rate of clay production one can postulate an increase in precipitation, an increase in temperature, or both. In order for this approach to really work, we need data on clay production in post-glacial time in many environments, but these data are not too abundant. Furthermore, we need better data on the age of the soils and the duration of their formation, but these data also are difficult to obtain.

The above approach to paleoclimate can be demonstrated with two field examples in the western United States. Hunt and Sokoloff[20] remarked that strongly developed pre-Wisconsin soils of the Rocky Mountain region may have developed in the same duration of time as the weakly developed post-Wisconsin soils, but under different climatic conditions. This may well be true, but I do not know of any climatic region in the United States in which post-Wisconsin soils are anywhere as strongly developed as the pre-Wisconsin soils they describe. As a second example, Morrison and Frye[30] state that the moderately developed mid-Wisconsin soil in the Lake Lahontan Basin probably formed in not much more than 5000 years. This soil formed from a sand (see soil formed on early Wisconsin deposit at Mustang, Fig. 11–9) and so a considerable amount of weathering must have gone on to produce the profile. Again, in reconnaissance studies in many areas I have not seen a 5000-year-old soil anywhere as well developed as that one, given a similar parent material. I would suggest that we postulate more time for these soils to form. This does not negate the approach, however, which I feel is valid. A climate different from the current climate may have been present when the soils formed; this is just very hard to prove by pedological criteria.

Clay mineralogy

Clay-mineral formation and transformation in the soil is a slow process, and therefore clay mineralogy may be a useful tool in assessing past climatic influences on the soil. Here we will simplify the situation and mention only the interpretations that can be made with the 1:1 clays, kaolinite and halloysite, and montmorillonite.

The stability of a clay mineral in a changing, soil-leaching environment is a function of the clay mineral originally formed and the direction of the climatic change.[7,32] As mentioned earlier, kaolinite and halloysite formation is favored by a high leaching environment and montmorillonite by a relatively low leaching environment. In well-drained parent materials, therefore, the clay minerals formed are mostly a function of climate. A change in soil-leaching environment triggered by a change in climate may bring about a change in the clay mineralogy, but the climatic change usually has to be in one direction to be read. For example, many studies have shown kaolinite to be a stable mineral, one that will persist for long intervals of time in a neutral to alkaline environment. Thus, if kaolinite is formed in a soil under a fairly high leaching and relatively acid environment, and a change in climate toward aridity occurred, the kaolinite probably would persist in the soil and serve as an indicator of a former wetter climate. Montmorillonite, on the other hand, is less stable than kaolinite and will change to the latter if greater leaching conditions obtain. In summary, therefore, a climatic change from humid to arid could be read in the clay mineralogy, but one from arid to humid most likely could not be read. One final point should be made and that is that clay-mineral transformation is a slow process, and therefore the climatic change would have had to persist long enough to allow time for slow reactions to take place. If the climatic change were of short duration, it may not be registered in the clay mineralogy.

A transect from the relatively humid east side of the Sierra Nevada into the dry western Great Basin can serve as an example of the use of clay mineralogy to help define Quaternary climatic change.[7] The clay mineralogy of soils developed from tills, outwash deposits, and lacustrine deposits is closely correlated with the present climate (Fig. 11–17). Halloysite is the predominant clay mineral in the humid environment, and montmorillonite is predominent in the arid environment. The change from one mineral assemblage to the other takes place at the present position of the pine-sagebrush vegetation boundary, at about

40-cm mean annual precipitation. The interesting thing about this study is that the clay mineralogy of soils formed on river-terrace deposits of all ages changes at about the same location under similar present-day environmental conditions. This is taken to mean that long-term past climatic change toward a wetter climate in the western Great Basin has not taken place, for if it had it probably would be seen as predominant kaolinite or halloysite in the older soils far east of their easternmost predominant occurrence in the younger soils. Short-term climatic changes could have occurred, however, and not have been registered in the clay mineralogy.

The above conclusions seem to be valid for a large number of sites along the eastern Sierra Nevada.[8] Because there is little clay- mineral variation corresponding to age of the parent materials, it was concluded that past climates probably have had little effect on the clay-mineral distribution. Moreover, if climatic changes had occurred, the soil-water chemistry did not change sufficiently to alter the equilibrium mineral assemblage (see Fig. 4–6). Or, it could be that climatic change may have been significant but not of long enough duration to alter the previously formed clay minerals significantly.

More work must be done on this aspect of paleopedology before all the significant factors are sorted out. Data on clay-mineral stabilities in changing environments are needed, as are data on the time needed to convert originally formed minerals to new species. Furthermore, these data should be looked at in view of independent data on past climates. For example, if the major portion of clay formation takes place in a relatively warm and dry interglacial interval, and the subsequent glacial climate is wetter and cooler, the cooler temperatures may inhibit alteration to a new clay-mineral species if the rate of alteration can be shown to be strongly temperature-dependent. If this is the case, the clay mineralogy of old soils will carry more of the record of the interglacials than of the glacials. A final point about clay-mineral species is that, like so many other soil properties, it is generally not possible to separate the effects of precipitation from those of temperature. For example, in a particular region the climatic change may have been toward both higher precipitation and higher temperature. The overall effect on the soil-leaching environment and on the soil-water chemistry may have been nil, and thus one would not expect a change in clay mineralogy to accompany the climatic change. What probably would happen in this case would be an acceleration in the rate of alteration of primary minerals to clay minerals.

Position of horizon of $CaCO_3$ *accumulation and morphology of the K horizon*

Because the presence or absence of $CaCO_3$ in a soil is related to soil-

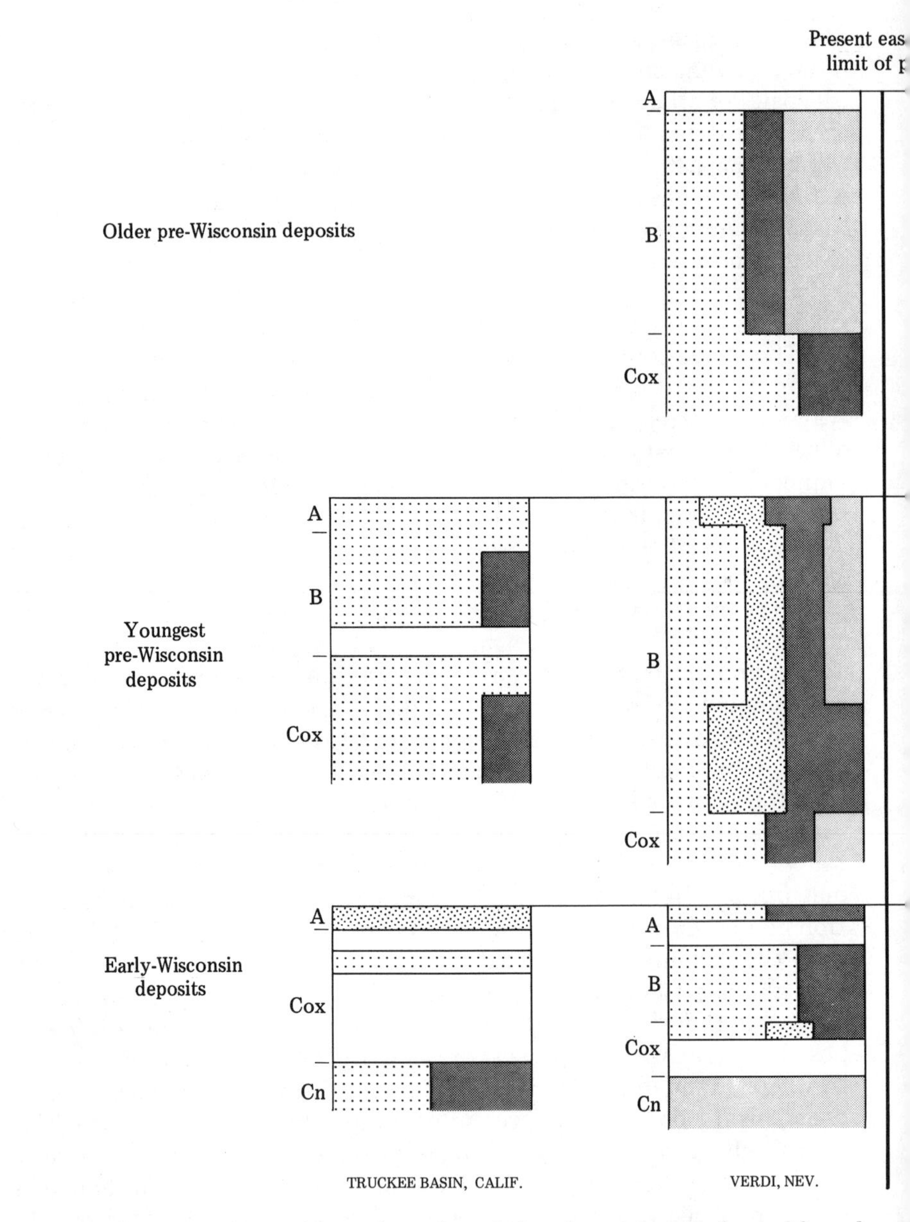

Fig. 11–17 Approximate clay-mineral abundances in soils formed from depo
of different age along the Truckee River, California and Nevada. (Data fr

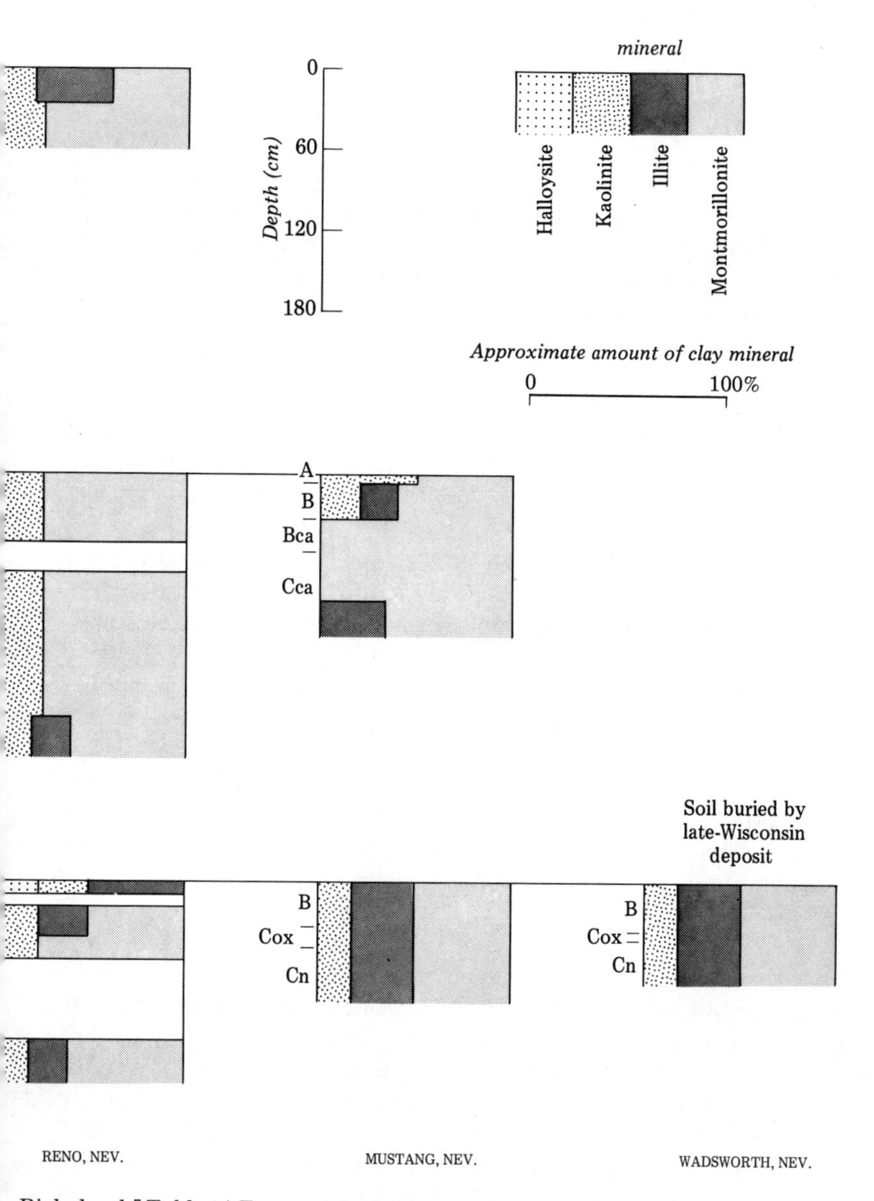

Birkeland,[7] Table 3.) Textural data for some of these soils are given in Fig. 11–9, and climatic data are in that figure legend.

water movement, and hence to climate, the position and morphological features of $CaCO_3$ accumulations may provide insight into past climates.

The position of the top of the zone of $CaCO_3$ accumulation is related to the regional climate, as long as the water-holding capacity of the soil is taken into account. Because Arkley[2] found that the tops of many of the Cca or K horizons plot close to the sharp break in the calculated soil-water-movement curves that represents a rapid decrease in water movement with depth (Fig. 11–3), he concluded that climatic change does not have to be called upon to explain the position of the tops of such horizons. This seems to be a reasonable conclusion. However, before one can explain the lack of such correspondence between climate and the top of the Cca or K horizon as being paleoclimatically significant one would have to investigate thoroughly the possibility of surface erosion or deposition as reasons for the top of such a horizon being either closer to or farther from the surface, respectively, than predicted.

The position of the Cca and/or K horizons is related to that of the textural B horizon; in most places they lie directly beneath the B. If, however, they lie some distance below the base of the B, and this position can be shown not to be of ground-water origin, the possibility of a shift to a wetter climate, driving the $CaCO_3$ to greater depth, should be investigated. However, if the Cca horizon extends up into the base of the textural B horizon, there may have been a climatic change toward aridity as long as no other reasons for the upward movement of the top of the Cca can be demonstrated. Two reasons for non-climatic upward movement of the top of the Cca can be mentioned. One is that as clay accumulates in the B horizon, the water-holding capacity of the horizon increases. The result of this is that more water is held in the B than previously was the case. Hence, the water carrying Ca^{2+} and HCO_3^- through the B does not move so deeply, and these ions no longer move out of the base of the B horizon; instead they precipitate to form a Bca horizon. The other non-climatic reason for upward movement of the top of the $CaCO_3$ accumulation layer involves the K horizon. As the K horizon forms, the pores of the soil are progressively plugged so that deep water penetration is inhibited. The downward moving waters, therefore, are held at the top of the K, and eventually ionic concentrations in the soil solution build up to the point where $CaCO_3$ precipitates out. In this way the top of the K moves toward the surface, engulfing the base of the B horizon in the process.[16]

Lateral and altitudinal displacements of the pedalfer-pedocal boundary along climatic transects in soils of different age can also be used as indicators of climatic change. Richmond[34,36] reports such dis-

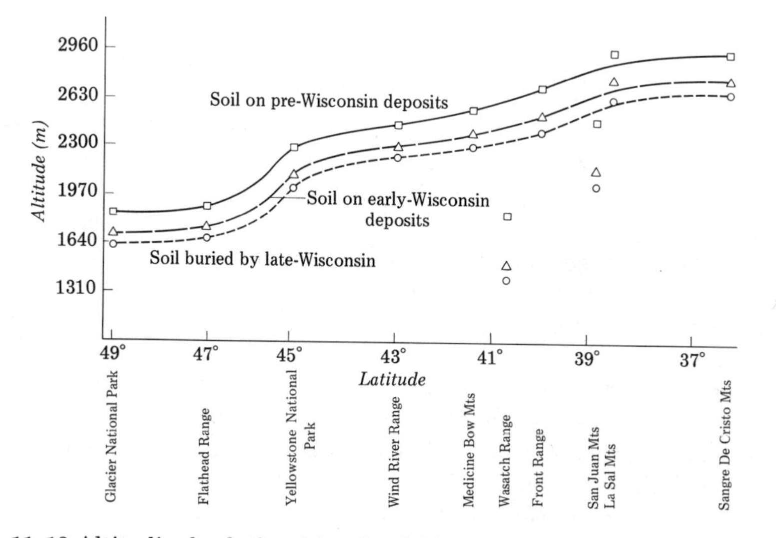

Fig. 11–18 Altitudinal relationship of pedalfer-pedocal boundary with age of parent material, Rocky Mountains. (Taken from Richmond,[36] Fig. 2.)

placements in many parts of the Rocky Mountain region (Fig. 11–18), and they are also seen in deposits along the eastern side of the Sierra Nevada near Reno and in the Colorado Piedmont near Boulder. In all these transects the pedalfer-pedocal boundary lies at a higher altitude, and therefore in a wetter present climate, on the older deposits. It is believed that the older soils formed during an interval of time warmer than the present, or possibly that the combination of precipitation and temperature at that time resulted in $CaCO_3$ accumulation in soils at a higher altitude than would occur under the present climate. These explanations are plausible, if the position of the pedalfer-pedocal boundary is not related to the different water-holding capacities of the different-aged soils. For example, the pre-Wisconsin soils commonly have textural B horizons of high water-holding capacity, whereas the soils formed from Wisconsin deposits often lack textural B horizons, and thus their water-holding capacities are less. Under certain climates, it might be possible for Ca^{2+} and HCO_3^- to be leached from the soils with low water-holding capacity, yet be retained to form a Cca horizon in an adjacent older soil with high water-holding capacity.

It is difficult to explain the preservation of $CaCO_3$ in older soils at altitudes at which adjacent younger soils show no accumulation of $CaCO_3$. Richmond[34] offers the following observations. The cooler glacial climates that prevailed subsequent to the formation of the $CaCO_3$ horizon in the older soils should have readily dissolved and been removed because $CaCO_3$ is more soluble at lower temperatures (Fig. 5–7). Other factors, however, may have decreased the rate at which $CaCO_3$ could have been removed from the soil. Richmond includes the limiting chemical activity with a colder glacial climate and the possibility of long seasons during which the ground was frozen as possible explanations for the preservation of the Cca horizons. Another possibility is that perhaps some areas did not experience much of an increase in soil leaching during the glacial climates, a hypothesis that could be supported by the clay-mineral data in the Sierra Nevada-Great Basin transect mentioned in the previous section.

Morphological features of K horizons may provide important data on solution and reprecipitation of $CaCO_3$ that perhaps owe their origin to climatic change.[10,11] Some K horizons, for example, have solution pits on their upper surface. If it can be shown that the solution pits are not due to deeper water penetration accompanying ground-surface lowering by erosion, this evidence suggests a change toward a wetter climate. Other soils show evidence of formation of laminated K horizons, brecciation of these horizons, and subsequent cementation of the breccia. Although the processes responsible for this history are not clearly understood, they could be due to climatic change. Perhaps brecciation occurs during intervals of drier climate and cementation of the fragments during intervals of wetter climate.

REFERENCES

1. Arkley, R. J., and Ulrich, R. 1962, The use of calculated actual and potential evapotranspiration for estimating potential plant growth: Hilgardia, v. 32, p. 443–462.
2. ____ 1963, Calculation of carbonate and water movement in soil from climatic data: Soil Sci., v. 96, p. 239–248.
3. ____ 1967, Climates of some Great Soil Groups of the western United States: Soil Sci., v. 103, p. 389–400.
4. Barshad, I., 1966, The effect of a variation in precipitation on the nature of clay mineral formation in soils from acid and basic igneous rocks: 1966 Internat. Clay Conf. (Jerusalem), Proc. v. 1, p. 167–173.
5. Beaven, P. J., and Dumbleton, M. J., 1966, Clay minerals and geomorphology in four Caribbean Islands: Clay Minerals, v. 6, p. 371–382.
6. Birkeland, P. W., 1968, Correlation of Quaternary stratigraphy of the Sierra Nevada with that of the Lake Lahontan area, p. 469–500 *in* R. B. Morrison and H. E. Wright, Jr., eds., Means of correlation of Quaternary successions: Internat. Assoc. Quaternary Res., VII Cong., Proc. v. 8, 631 p.
7. ____ 1969, Quaternary paleoclimatic implications of soil clay mineral distribution in a Sierra Nevada–Great Basin transect: Jour. Geol., v. 77, p. 289–302.

8. ―― and Janda, R. J., 1971, Clay mineralogy of soils developed from Quaternary deposits of the eastern Sierra Nevada, California: Geol. Soc. Amer. Bull., v. 82, p. 2495–2514.
9. Boyer, S. J., 1972, Pre-Wisconsin, Wisconsin, and neoglacial ice limits in Maktak Fiord, Baffin Island: a statistical analysis: Unpub. M. S. thesis, Univ. of Colorado, 117 p.
10. Bretz, J. H., and Horberg, L., 1949, Caliche in southeastern New Mexico: Jour. Geol., v. 57, p. 491–511.
11. Bryan, K., and Albritton, C. C., Jr., 1943, Soil phenomena as evidence of climatic changes: Amer. Jour. Sci., v. 241, p. 469–490.
12. Charlier, R. H., 1969, The geographic distribution of polar desert soils in the northern hemisphere: Geol. Soc. Amer. Bull., v. 80, p. 1985–1996.
13. Crandell, D. R., 1965, The glacial history of western Washington and Oregon, p. 341–353 *in* H. E. Wright, Jr., and D. G. Frey, eds., The Quaternary of the United States: Princeton Univ. Press, Princeton, 922 p.
14. ―― 1969, Surficial geology of Mount Rainier National Park, Washington: U.S. Geol. Surv. Bull. 1288, 41 p.
15. Frye, J. C., and Leonard, A. B., 1967, Buried soils, fossil mollusks, and late Cenozoic paleoenvironments, p. 429–444 *in* C. Teichert and E. L. Yochelson, eds., Essays in paleontology and stratigraphy: Univ. of Kansas Press, Lawrence.
16. Gile, L. H., Peterson, F. F., and Grossman, R. B., 1966, Morphological and genetic sequences of carbonate accumulation in desert soils: Soil Sci., v. 101, p. 347–360.
17. ――, Hawley, J. W., and Grossman, R. B., 1971, The identification, occurrence and genesis of soils in an arid region of southern New Mexico: Training Bull., U.S. Dept. Agri., Soil Cons. Serv., 177 p.
18. Harradine, F., and Jenny, H., 1958, Influence of parent material and climate on texture and nitrogen and carbon contents of virgin California soils. I. Texture and nitrogen contents of soils: Soil Sci., v. 85, p. 235–243.
19. Hay, R. L., and Jones, B. F., 1972, Weathering of basaltic tephra on the island of Hawaii: Geol. Soc. Amer. Bull., v. 83, p. 317–332.
20. Hunt, C. B., and Sokoloff, V. P., 1950, Pre-Wisconsin soil in the Rocky Mountain Region, a progress report: U.S. Geol. Surv. Prof. Pap. 221-G, p. 109–123.
21. Janda, R. J., and Croft, M. G., 1967, The stratigraphic significance of a sequence of Noncalcic Brown soils formed on the Quaternary alluvium of the northeastern San Joaquin Valley, California, p. 157–190 *in* R. B. Morrison and H. E. Wright, Jr., eds., Quaternary soils: Internat. Assoc. Quaternary Res., VII Cong., Proc. v. 9, 338 p.
22. Jenny, H., 1935, The clay content of the soil as related to climatic factors, particularly temperature: Soil Sci., v. 40, p. 111–128.
23. ―― 1941, Factors of soil formation: McGraw-Hill Book Co., New York, 281 p.
24. ―― 1941, Calcium in the soil: III. Pedologic relations: Soil Sci. Soc. Amer. Proc., v. 6, p. 27–37.

25. ____ 1950, Causes of the high nitrogen and organic matter content of certain tropical forest soils: Soil Sci., v. 69, p. 63–69.
26. ____ 1961, Comparison of soil nitrogen and carbon in tropical and temperate regions: Missouri Agri. Exp. Sta. Res. Bull. 765, p. 5–31.
27. ____, Gessel, S. P., and Bingham, F. T., 1949, Comparative study of decomposition rates of organic matter in temperate and tropical regions: Soil Sci., v. 68, p. 419–432.
28. Loughnan, F. C., 1969, Chemical weathering of the silicate minerals: American Elsevier Publ. Co., Inc., New York, 154 p.
29. Matsui, T., 1967, On the relic red soils of Japan, p. 221–244 *in* R. B. Morrison and H. E. Wright, Jr., eds., Quaternary soils: Internat. Assoc. Quaternary Res., VII Cong., Proc. v. 9, 338 p.
30. Morrison, R. B., and Frye, J. C., 1965, Correlation of the middle and late Quaternary successions of the Lake Lahontan, Lake Bonneville, Rocky Mountain (Wasatch Range), southern Great Plains, and eastern midwest areas: Nevada Bur. Mines Rept. 9, 45 p.
31. Mulcahy, M. J., 1967, Landscapes, laterites, and soils in southwestern Australia, p. 211–230 *in* J. N. Jennings and J. A. Mabbutt, eds., Landform Studies from Australia and New Guinea: Australian National Univ. Press, Canberra, 434 p.
32. Pedro, G., Jamagne, M., and Bejon, J. C., 1969, Mineral interactions and transformations in relation to pedogenesis during the Quaternary: Soil Sci., v. 107, p. 462–469.
33. Porter, S. C., 1969, Pleistocene geology of the east-central Cascade Range, Washington: Guidebook for 3rd Pacific Coast Friends of the Pleistocene field conference, 54 p.
34. Richmond, G. M., 1962, Quaternary Stratigraphy of the La Sal Mountains, Utah: U.S. Geol. Surv. Prof. Pap. 324, 135 p.
35. ____ 1965, Glaciation of the Rocky Mountains, p. 217–230 *in* H. E. Wright, Jr., and D. G. Frey, eds., The Quaternary of the United States: Princeton Univ. Press, Princeton, 922 p.
36. ____ 1972, Appraisal of the future climate of the Holocene in the Rocky Mountains: Quaternary Res., v. 2, p. 315–322.
37. Sherman, G. D., 1952, The genesis and morphology of the alumina-rich laterite clays, p. 154–161 *in* Problems of clay and laterite genesis: Amer. Inst. Mining and Metallurgical Engr., New York.
38. Strakhov, N. M., 1967, Principles of lithogenesis, v. 1: Oliver and Boyd Ltd., Edinburgh, 245 p.
39. Tedrow, J. C. F., 1966, Polar desert soils: Soil Sci. Soc. Amer. Proc., v. 30, p. 381–387.
40. The Times of London, 1967, The Times atlas of the world: Houghton Mifflin Co., Boston.
41. Thornthwaite, C. W., and Mather, J. R., 1957, Instructions and tables for computing potential evapotranspiration and the water balance: Drexel Inst. Techn., Lab. of Climatology, Publs. in Climatology, v. 10, no. 3, 311 p.

CHAPTER 12

Use of soils in Quaternary stratigraphic studies

Soils are important to the subdivision of Quaternary sediments, whether the soils are at the surface or buried. They are used primarily to aid in the subdivision of a local succession of deposits, to provide data on the lengths of time that separate periods of deposition, and to facilitate short- and long-range correlation. This rather special field of study is called soil stratigraphy, and its history and methods are reviewed quite thoroughly by Morrison;[16] much of this section will draw heavily on his work.

The use of soils for stratigraphic purposes demands that the investigator have a thorough background in pedology. It is of prime importance to be able to distinguish those features of the soil profile that are mainly geological in origin from those that are distinctly pedological. A knowledge of processes responsible for profile development is one working tool of the soil stratigrapher, as is a good understanding of the influence of soil-forming factors on soil genesis. As an illustration, one of the first tasks of the investigator in soil-stratigraphic research is to reconstruct the landscape and delineate the factors that prevailed when the soil in question formed. These are the factors that influence the rates of formation of the various soil features used for stratigraphic purposes.

In any stratigraphic study, soils should not be used solely for correlation to the exclusion of other field criteria. The best answers come from the integration of data from several disciplines.[12] Thus, the soil evidence must be reconciled with evidence concerning geologic events and processes, and knowledge of past climatic trends, as well as palynological data. Furthermore, one should integrate all relevant data in correlation studies, such as paleomagnetic, volcanic ash stratigraphy, and other data mentioned by Morrison.[17]

Stratigraphic studies of soils also are useful in pedologic studies, since only by a thorough knowledge of the soil in its stratigraphic setting can one understand all the factors responsible for the development of a certain profile. Indeed, as more of the many factors in soil-profile development are seen to interact, it becomes more difficult to isolate the effects of any one factor. It is still possible in most places, however, to distinguish between the main soil-forming factors and those of relatively minor importance.

Various aspects of soil stratigraphy will be covered in this chapter. First we will discuss soil-stratigraphic units and then show how they are used to help recognize other geologic stratigraphic units. This will be followed by a discussion of the occurrence of soil-stratigraphic units in the field and their interpretation as regards soil-forming intervals. Finally, examples of stratigraphic correlations using soils will be presented.

Soil-stratigraphic units

Soils used in stratigraphic studies for correlation purposes are called soil-stratigraphic units.[1,16] To qualify as a soil-stratigraphic unit, a soil must have morphological features that are pedologic in origin, it must display a consistent relationship to associated stratigraphic units in the local succession, and it must be mappable and traceable in the field. Soil-stratigraphic units, by definition, are found only at one stratigraphic interval (on a deposit or landscape of one age). A single rank of classification is recognized, the soil. Morrison[16] has suggested that the term geosol be used for soils of this type, because the term soil is used ambiguously by different people, and geosol would call attention to this special stratigraphic usage. Soil-stratigraphic units can be given formal names from nearby geographic place names (Churchill Soil), they may be named for the time interval during which they formed (Sangamon soil), or, if they merely postdate a deposit, terms such as post-Tahoe soil (soil formed on Tahoe Till) can be used.

It should be stressed that a prime requisite for a soil-stratigraphic unit is its stratigraphic relationship to associated deposits. The pedologic properties of such a unit can vary laterally, and many do. This is the main difference between the soil-stratigraphic unit and soil-classification units such as series or order; the latter are based on soil-morphological characteristics without regard to associated stratigraphic units (Fig. 12–1).

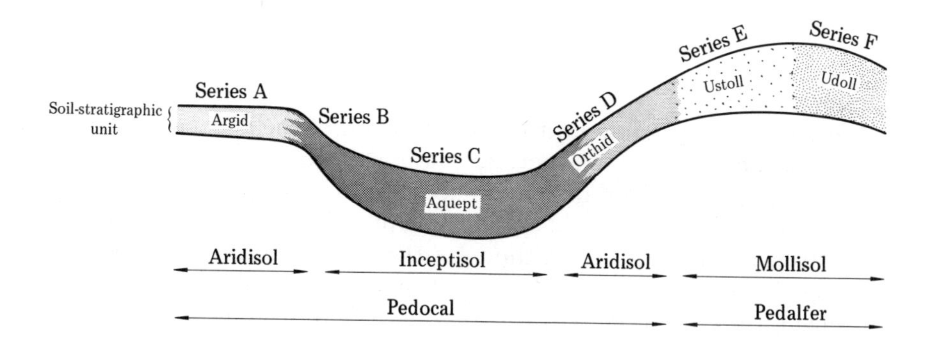

Fig. 12–1 Relationship between a soil-stratigraphic unit and various pedological units in a loess landscape. The area could be small or large. Different soil series are indicated at the surface and can be grouped into suborders, orders, and pedalfers or pedocals. Any pedological unit can be considered a soil-stratigraphic facies. The boundary between adjacent facies varies depending upon the soil-classification unit defining the facies.

Soil-stratigraphic units that do exhibit lateral variation in soil morphology due to lateral variation in environmental conditions can be further subdivided into soil facies, because they are somewhat analogous to the concept of rock facies. Facies are so designated at any level of pedologic classification, and the choice depends on the needs of the particular study (Fig. 12–1). Subdivision to the order or suborder level is desirable. One problem with the use of detailed soil classification units for buried soil-stratigraphic units is that many times the A horizon is no longer present, yet properties of the A horizon are important to any soil classification. In such cases one might designate individual facies on such soil properties as montmorillonite-bearing pedocal facies vs. illite-bearing pedalfer facies or argillic B-horizon facies vs. cambic B-horizon facies.

Soil-stratigraphic units as an aid in defining geologic stratigraphic units

Stratigraphy is that branch of geology dealing with the sequence, age, and correlation of rock bodies. Much of the code of stratigraphic nomenclature is very useful for pre-Quaternary rocks. Problems arise in trying to use some of the definitions of stratigraphic units for Quaternary sediments, however. One probable reason for this is that the Quaternary record is well preserved, and one can make a very detailed subdivision of the rock record, usually more detailed than commonly can be made with pre-Quaternary rocks. Here we will deal with the use of soil-stratigraphic units as aids in defining rock- and time-stratigraphic units.

A rock-stratigraphic unit is a subdivision of rocks based on lithologic characteristics.[1] In practice, therefore, two different rock bodies (e.g., formations) would be separated on physical features that ideally are recognizable in the field as, for example, the separation of a green shale from a red sandstone. Although either constructional or erosional surface form may help in the recognition of a rock-stratigraphic unit, surface form should not be a primary factor in the definition of such a unit.

The rigid definition of a rock-stratigraphic unit, given above, is difficult to apply to many Quaternary stratigraphic units, and some workers feel the soil should be part of the unit. In many areas in which Quaternary deposits have been mapped, the deposits of all ages (e.g., tills) commonly have identical lithologies, and there is no practical way to separate the deposits in the field by the properties of the deposits themselves. Richmond[19] has proposed that the distinctive soil by which many Quaternary deposits are distinguished from other deposits of similar origin be included in the definition of the rock-stratigraphic unit (see Fig. 12–2). This proposal has much to offer because the soil commonly is one of the major pieces of evidence on which deposits of different age are recognized. Thus, it is realistic and solves the problem of rock-stratigraphic nomenclature of Quaternary unconsolidated sediments.

Time-stratigraphic units are rock units whose boundaries are based on geologic time[1]; that is, the upper and lower boundaries are isochronous over large areas. These units differ from rock-stratigraphic units in that the boundaries of the latter units can be of different age in

GEOLOGIC-CLIMATE UNITS	ROCK-STRATIGRAPHIC AND SOIL-STRATIGRAPHIC UNITS	
Rocky Mountain Region		*La Sal Mountains, Utah*
Pinedale Glaciation	Pinedale Till	Beaver Basin Formation
Bull Lake-Pinedale Interglaciation	Bull Lake-Pinedale soil	Lackey Creek soil
Bull Lake Glaciation	Bull Lake Till	Placer Creek Formation

Fig. 12–2 Examples of geologic-climate, rock-stratigraphic, and soil-stratigraphic units for part of the Quaternary record of the Rocky Mountain region and the La Sal Mountains, Utah (in part from Richmond[19]).

different areas; in geologic language, they can be time-transgressive. The problem in Quaternary stratigraphy, therefore, is to find physical or biological features in the sediments that are time-parallel and that occur in sediments of different origins over large distances, so that time correlations are possible. Some of the best Quaternary time lines are volcanic ash. Middle- to early-Pleistocene ashes are widespread in the western United States and in the midcontinent and serve as prime correlation tools over the region.[8,24] Unfortunately, such time lines are not common in most Quaternary successions.

Morrison[16] and Morrison and Frye[18] feel that soils can be used to mark time-stratigraphic boundaries in Quaternary successions. They present the case that soil formation took place in the Quaternary over distinct intervals of time during which the climate differed from the present climate. Evidence is presented that indicates that temperature in particular was higher during the soil-forming intervals. Because temperature change is felt over a large region, they reason that soil formation over the same region was accelerated. Soils, therefore, would have acquired many of their diagnostic characteristics over rather short intervals of time. Because their formation is considered to be nearly time-parallel over large regions, soils can be used to define time-stratigraphic boundaries. In some places, the soil and the underlying deposit are each regarded as separate time-stratigraphic entities (stage or substage), whereas in other places the soil and the underlying deposit are combined into a single time-stratigraphic unit (Fig. 12–3).

A common practice in Quaternary stratigraphy is to use geologic-climate stratigraphic units. These represent climatic episodes inferred from the rock, soil, or organic-material record.[1] The upper and lower boundaries can be either time-parallel or time-transgressive, and they should bear evidence of climatic change. The major units recognized are glaciation and interglaciation (Fig. 12–2). Thus, a till would record the evidence of glaciation and the overlying soil of interglaciation.

Determining the time of soil formation

The time of formation for a particular soil is determined by its stratigraphic position relative to adjacent deposits and soils.[2,13,16,19] This time interval can be approximated for both buried and surface soils.

The time of soil formation is best dated with buried soils. In Fig. 12–4, B, for example, buried soil "a" formed between the deposition of rock units I and II, and buried soil "b" between rock units II and III. Moreover, soil "a" is similar whether buried or at the surface, and so is soil "b"; hence, this evidence indicates that the major properties of

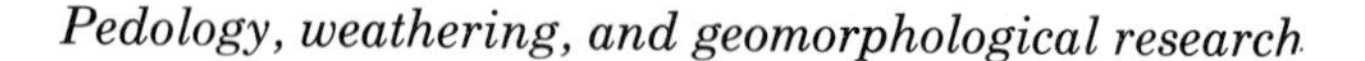

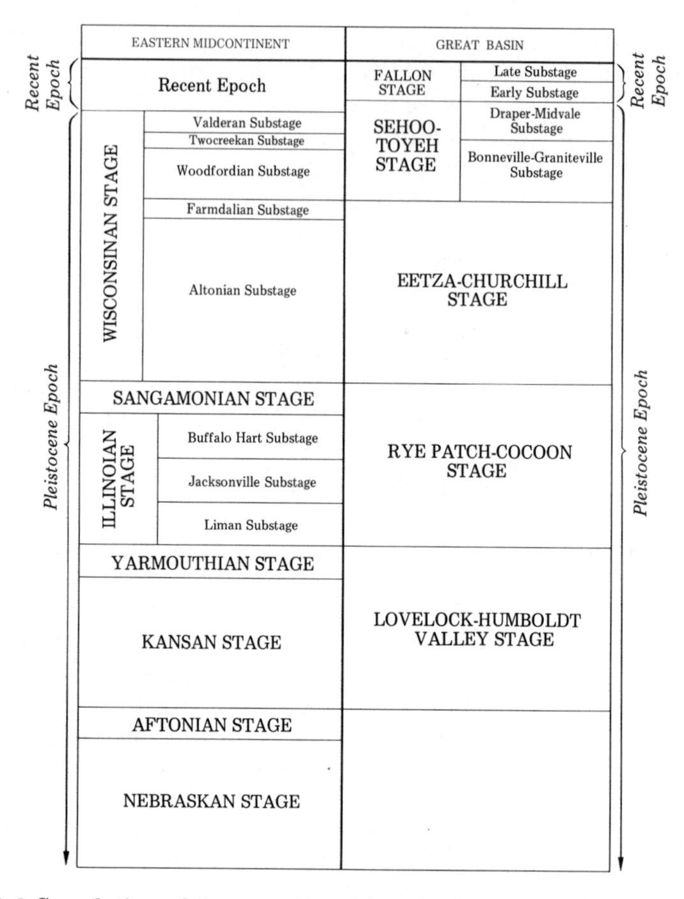

Fig. 12–3 Correlation of time-stratigraphic units in the eastern midcontinent and the Great Basin. (Taken from Morrison and Frye,[18] Fig. 7.) In the pre-Wisconsin of the eastern midcontinent the major soils (Sangamon, Yarmouth, Afton) all are assigned separate stage rank. Stages are assigned differently in the Great Basin. For example, Lovelock, Rye Patch, and Eetza all are rock-stratigraphic units, and Humboldt Valley, Cocoon, and Churchill all are soil-stratigraphic units. The stage is denoted by a compound word encompassing both the deposit and the overlying soil (e.g., Eetza-Churchill Stage).

soils "a" and "b" were imparted to the soils during the time intervals between depositional episodes, whether the soil remained at the surface or was buried. Furthermore, because soil "a" is more strongly developed than soil "b," soil formation during soil-forming interval "b" had little effect on soil "a" where it remained at the surface. Figure 12–4, C indicates that the strongest soil formed after the deposition of rock unit II, hence soil "b" is recognizable as such only where buried. Although soil "b" was present on the landscape prior to the formation of soil "a," soil "a" is strong enough in development to mask most of the features of soil "b" in surface positions. Therefore, the major soil-forming interval here is that that formed soil "a."

It is more difficult to determine the timing of formation of surface soils where buried counterparts do not exist. In Fig. 12–4, D, river terraces of various ages are depicted. In this case soil-forming intervals cannot be proved. The maximum time available for the formation of each soil is that from the time deposition ceased to the present. Thus, if soil formation proceeded at a relatively uniform rate since deposition, soil "a" may have had only about one-half its present amount of clay before soil "b" began to form, but soil "b" should have had most of its clay formed before soil "c" began to form. The soils could have formed in less time, but there is no evidence on the minimum time for formation.

Concept and explanation of the soil-forming interval

Where soil profiles are nearly identical in both buried and surface occurrences (Fig. 12–4, B), the soil can be said to have formed during a discrete soil-forming interval bracketed by the ages of adjacent rock units; the soil is younger than the rock unit from which it has formed, and it is older than the rock unit that buries it. A good example of a soil-forming interval is the mid-Wisconsin Churchill Soil of the Great Basin.[13] This soil has formed on early-Wisconsin deposits and is now found either at the surface, where it has been exposed to soil formation since deposition ceased, or buried beneath late-Wisconsin deposits. Obviously the buried soil had much less time to form relative to the surface soil, yet both are nearly identical in profile morphology and on laboratory analysis (Fig. 12–5). In addition, the clay mineralogy of both relict and surface occurrences of the soil is identical (Fig. 11–17), as is the degree of etching of the pyroxene grains (J. G. LaFleur, written report, 1972). Soil-forming intervals can be documented in many other localities in the United States and in the world.[16] Most observations, however, lack quantitative laboratory analysis.

There are several ways to explain the fact that buried and surface occurrences of the same soil-stratigraphic units are nearly identical, although each has been exposed to soil-forming processes a different length of time. Morrison[13,16] and Morrison and Frye[18] propose that the soil-forming intervals were characterized by unique climatic conditions that resulted in accelerated soil formation; in contrast, the climates prevailing between the soil-forming intervals were not conducive to soil formation. In the Lake Lahontan area of Nevada, for example, the soil-forming intervals are thought to have occurred over

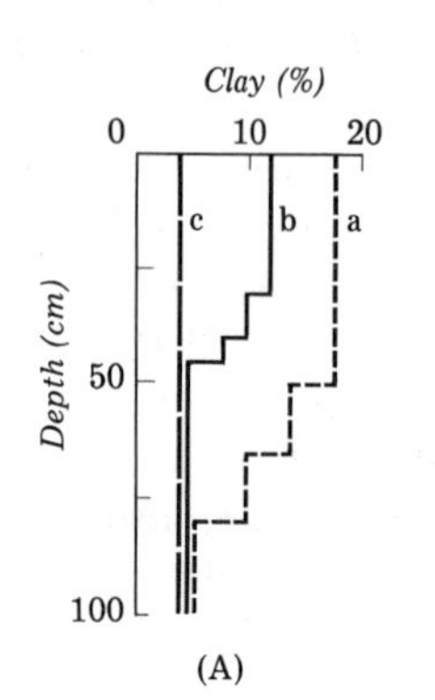

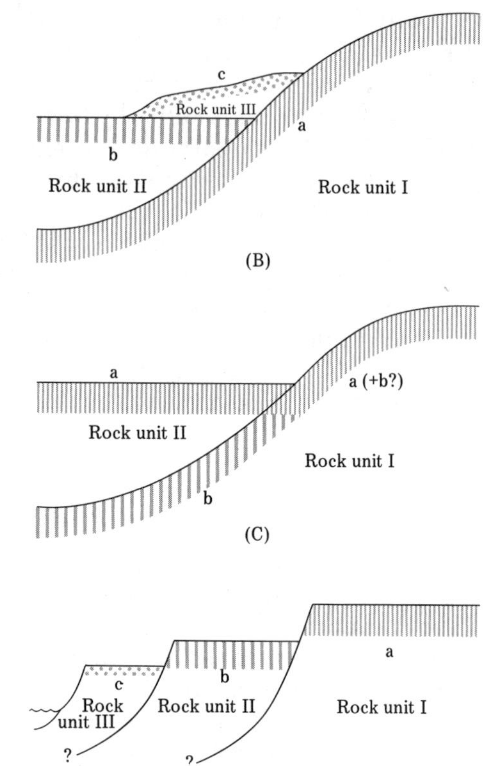

Fig. 12–4 Hypothetical relationships between soils and deposition units (I, old to III, young). Approximate clay contents for the soils in B, C, and D are shown in A.

short intervals of time during which both precipitation and temperature were above their present values (Fig. 12–6). Clay-mineral data (Fig. 11–17) are amenable to this interpretation, although other interpretations of the clay-mineral data are possible.

Soils forming during an interval of fluctuating climate, as visualized above, probably would form in a stepwise fashion, in which an accelerated rate of development coincided with optimum climatic conditions, and declining rate of development coincided with marginal climatic conditions. Curve A in Fig. 12–7 depicts this and can be used to help explain the field evidence for a soil-forming interval. Assume that soil formation went on at all sites in question until time "a." At this time, a climatic change took place, and part of the soil was buried by sediments. Because of the direction of the climatic change, the development of the soil that remained at the surface would have slowed down considerably, as shown by the flattened curve between times "a" and "b." If we examine both soils at time "b," the surface soil that formed in the time interval from 0 to "b" would be similar to that exposed to pedogenesis from 0 to "a," but buried from "a" to "b." Continuing in time, if the surface soil remained at the surface from time "b" to "c," a soil-forming interval, then theoretically it should be more strongly developed than the buried soil.

One other possible way of explaining the similarity in soils, buried and at the surface, is that they may have reached a steady state in development so that further change is limited. For example, assume a constant climate over a long interval of time. The development of the soil probably would proceed as depicted in Fig. 12–7, curve B.[2,10,23,26] As indicated, once a soil has attained fairly strong development, a greater length of time adds only a little to further development. If the soil had reached the development indicated by time "c," and part of it became buried and part remained at the surface until time "d," both would look similar if examined at time "d." This is commonly the case for strongly developed profiles, such as the pre-Wisconsin profiles in many areas.

It is theoretically possible, however, to have accelerated soil development without a change in climate. Janda and Croft[9] point out that if the curve for soil development under a constant climate is like that of Fig. 12–7, curve B, soil development, especially with sandy parent materials, could accelerate between times "a" and "b" with no

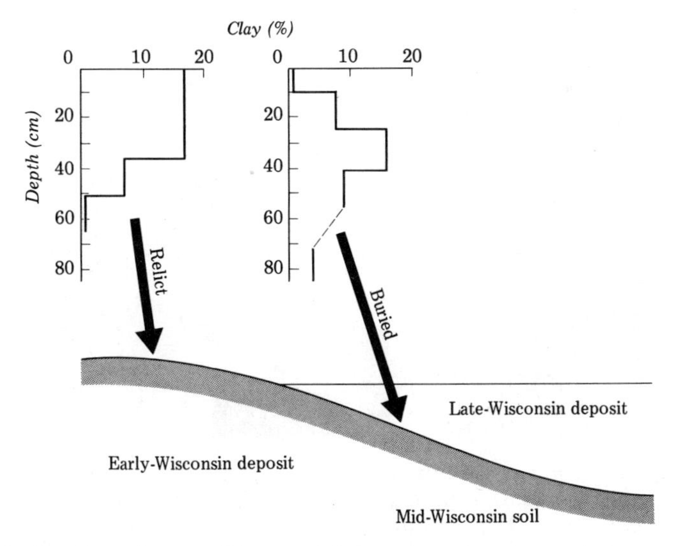

Fig. 12–5 Stratigraphic position and laboratory data on relict and buried occurrences of the mid-Wisconsin Churchill Soil in the Great Basin. (Data from Morrison,[13] Table II, and Birkeland,[3] Fig. 3.)

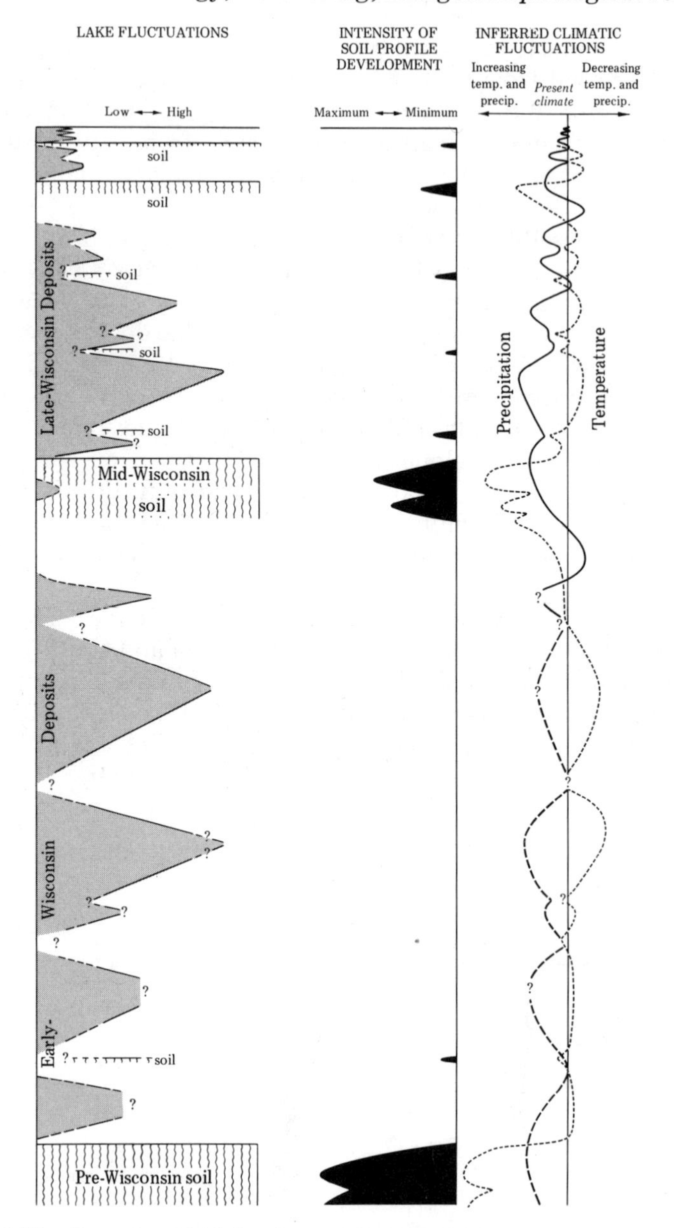

Fig. 12–6 Stratigraphy and lake fluctuations of Pleistocene Lake Lahontan, Nevada, and the stratigraphic position of the major soils. (Taken from Morrison and Frye,[18] Fig. 2.) Inferred climates during all events, relative to that of the present, are depicted, as are relative intensities of soil-profile development during the soil-forming intervals.

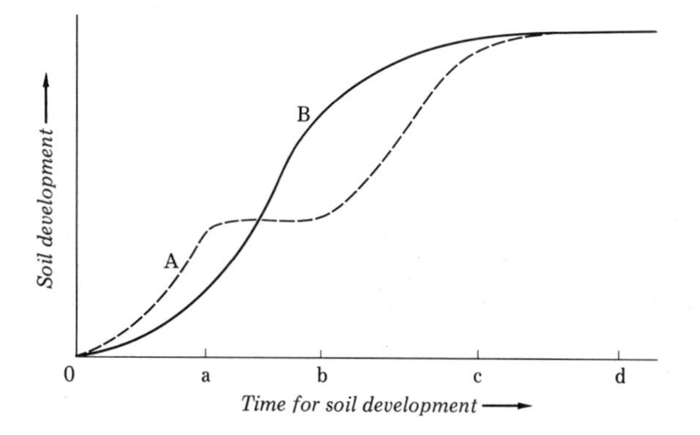

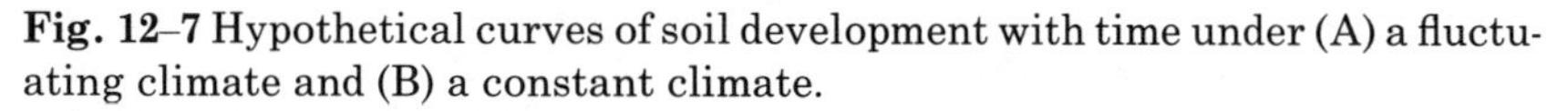

Fig. 12–7 Hypothetical curves of soil development with time under (A) a fluctuating climate and (B) a constant climate.

climate change. The reason for this is that as organic matter and clay accumulate in the soil, the water-holding capacity increases, and this should promote more clay formation. A period of rapid clay formation continues until the curve begins to flatten as the steady state is approached.

The evidence that soils can remain at the surface for relatively long intervals of time, yet change little in their pedogenic and weathering characteristics, presents some geochemical and pedologic problems. In general, many soil-forming intervals are thought to have coincided with interglacial intervals, whereas the time of decreased soil development coincided with glacial times. Surely, the glacial climates of some non-glaciated regions in the United States are at least approximated by the present-day climates of some of the northern states. The question to be answered is: Why were the geochemical and pedogenic processes in some soils slowed down when the evidence from fairly cold moist climates indicates that the postglacial soils there are fairly well developed, organic matter is high, and mineral weathering discernible? Furthermore, work along some coastal areas with marine terrace deposits indicates that in those environments weathering and soil formation can proceed at a fairly good rate, in spite of the fact that the soils formed essentially during a time represented by glaciation elsewhere.[2,11] In addition, data are accumulating that indicate significant Holocene soil formation in alpine areas of the Rocky Mountains, areas for which many workers would predict slow rates of soil formation (unpub. work of the writer and C. D. Miller). Few clues presently are available to explain this apparent dilemma.

In summary, the geologic and pedologic evidence for soil-forming intervals in some environments is well established. This discovery has had a profound effect on Quaternary stratigraphy, because if it stands the test of time, soils can be used as prime correlation tools. We need to go one step further, however, and begin to work on the explanation of the soil-forming interval. This will involve the concerted effort of many workers, and it should combine soil laboratory and mineralogical analysis with research on the age and duration of Quaternary climatic events. Also, the age and duration of the soil-forming intervals must be better established, and, finally, other field areas should be investigated for possible soil-forming intervals. Much of the Quaternary stratigraphic studies have been done in areas in which deposition takes place in the glacial side of the climatic cycle and soil formation in the interglacial side. We should make an effort to study areas in which interglacial times were coincident with deposition and glacial times coincident with soil formation (see Schumm,[22] Table 2). According to the soil-forming-interval concept, such soils should display little profile development.

Using soil-stratigraphic units for correlation of Quaternary deposits

Soils commonly are quite useful for the correlation of Quaternary successions. In this section the methods of correlation and examples of both short- and long-range correlation, using soils, will be presented. For the most part, correlations based solely on soils do not come easily. This is because soils change laterally in their properties, because the factors responsible for soil morphology commonly change laterally. Thus, one is faced with comparing soils of about the same age, or duration of formation, in many different environmental settings. Spodosols are compared with Aridisols, thick soils with thin soils, high-clay soils with low-clay soils, because these encompass some of the wide variety of soils and soil properties that can be expected at any one stratigraphic horizon.

The basic methods of correlation using soils have been set forth by Richmond[19] and Morrison.[13,16] The first task in any one area is to rank the soils in the local succession on the basis of their relative degree of profile development; some will be weakly developed, others moderately developed, and still others strongly developed. Ranking on development will not always be greater with age of the deposits, because some weakly developed soils can form and be preserved as buried soils. In the other area to which correlation is to be made, the soils are similarly ranked, again with respect to that local succession. Because the environment of each local succession will not be identical, the time-equivalent soils in one succession may bear little resemblance to those in another. As an example, a 12-cm-thick, weakly developed Aridisol in one environment may be correlated with a 50-cm-thick, moderately

Midcontinent Time-Stratigraphic Units		Rock-Stratigraphic Units: Formation	Rock-Stratigraphic Units: Member	Soil-Stratigraphic Units	Depositional facies									
Holocene Stage		Gold Basin Formation	Upper member		Till	Rock glacier	Alluvial gravel	Alluvial sand and silt	Alluvial-fan gravel	Talus	Solifluction mantle	Frost rubble	Slope wash	Eolian sand and silt
			Disconformity	Spanish Valley Soil										
			Lower member		Till	Rock glacier	Alluvial gravel	Alluvial sand and silt	Alluvial-fan gravel	Talus		Frost rubble		Eolian sand and silt
		Disconformity		Castle Creek Soil										
Wisconsin Stage	Late-Wisconsin Substage	Beaver Basin Formation	Upper member		Till	Rock glacier	Alluvial gravel	Alluvial sand and silt	Alluvial-fan gravel	Talus	Solifluction mantle	Frost rubble	Slope wash	Eolian sand and silt
			Disconformity	Pack Creek Soil										
			Lower member		Till	Rock glacier	Alluvial gravel	Alluvial sand and silt				Frost rubble		Eolian sand and silt
	Middle-Wisconsin Substage	Disconformity		Lackey Creek Soil										
	Early-Wisconsin Substage	Placer Creek Formation	Upper member		Till		Alluvial gravel	Alluvial sand and silt	Alluvial-fan gravel	Talus	Solifluction mantle	Frost rubble	Slope wash	Eolian sand and silt
			Disconformity	Porcupine Ranch Soil										
			Lower member		Till		Alluvial gravel		Alluvial-fan gravel		Solifluction mantle			Eolian sand and silt
Sangamon Stage		Disconformity		Upper Spring Draw Soil										
Illinoian Stage		Harpole Mesa Formation	Upper member		Till		Alluvial gravel	Alluvial sand and silt	Alluvial-fan gravel	Talus	Solifluction mantle	Frost rubble		Eolian sand and silt
Yarmouth Stage			Disconformity	Middle Spring Draw Soil										
Kansan Stage			Middle member		Till		Alluvial gravel							
Afton Stage			Disconformity	Lower Spring Draw Soil										
Nebraskan Stage			Lower member		Till		Alluvial gravel							Eolian sand and silt

(Symbol) = Soil-stratigraphic unit. Depth of symbol indicates relative degree of soil development

Fig. 12–8 Quaternary stratigraphic units in the La Sal Mountains, Utah, and tentative correlation with time-stratigraphic units of the midcontinent. (Taken from Richmond,[19] Fig. 9 and Table 11.)

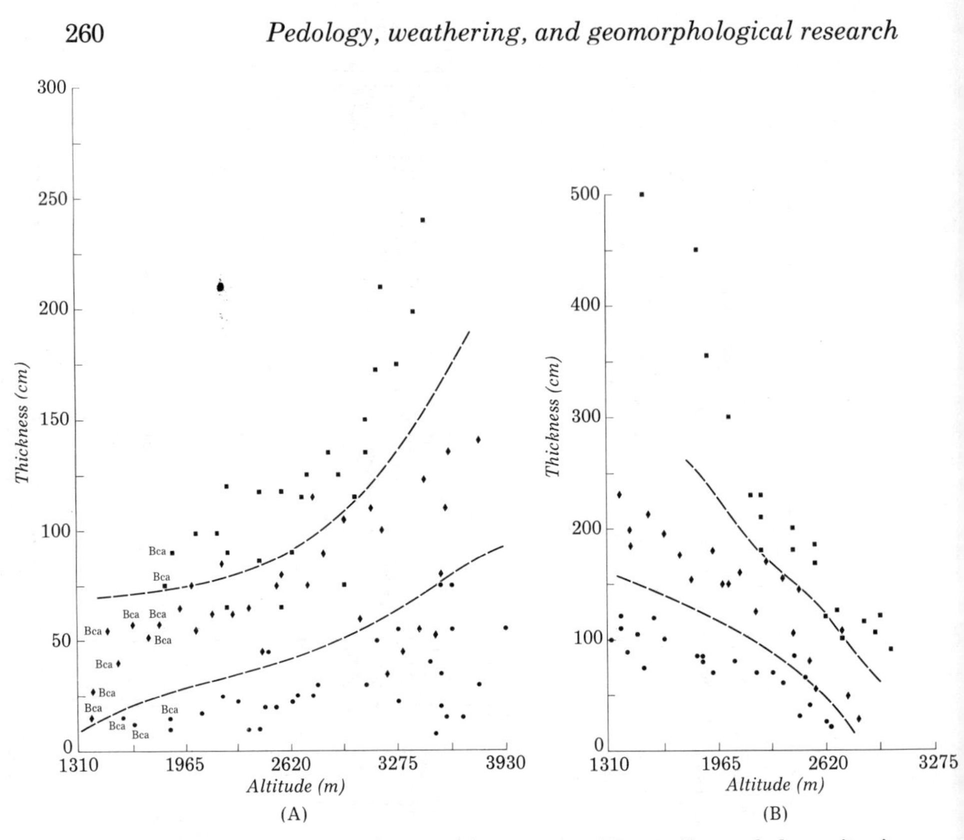

developed Alfisol in another environment. The soils and deposits in both areas are then matched. Commonly the basic framework here is provided by the youngest strongly developed soil in each area. If the timing of the periods of deposition and soil formation are approximately the same, and deposition occurred during the glaciations, the youngest strongly developed soil often is the soil that formed mainly during the last major interglacial, the Sangamon. All other strongly developed soils of pre-Sangamon age also are matched and, because they too may mark major interglacials, they are important to the correlation. The weakly and moderately developed soils that lie stratigraphically between the strongly developed ones can then be matched.

The need for putting the geological and the pedological data together for meaningful correlation cannot be overemphasized. One has to have a thorough knowledge of the geologic history in the area, including the kind and rate of geologic processes that have been operative. Soils have to be studied in the field for their stratigraphic position and for their morphology and history, and they should be analyzed quantitatively for their critical properties. Only then can all the data be assembled to give the best reasonable correlations. One should

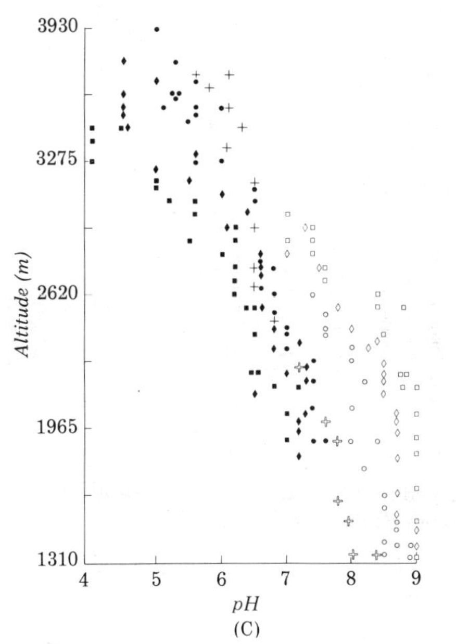

Key

B horizon	Cca horizon
+	✢ Spanish Valley Soil
•	○ Castle Creek Soil
♦	◊ Lackey Creek Soil
▪	▫ Spring Draw Soils

Fig. 12–9 Variation with altitude and age in the properties of several soil-stratigraphic units, La Sal Mountains, Utah. (Taken from Richmond,[19] Figs. 56, 60, and 61.) Stratigraphic age assignment for the soils is given in Fig. 12–8. (A) Range in maximum thickness of B or Bca horizon; (B) range in maximum thickness of Cca horizon; (C) range in pH of B and Cca horizons.

avoid merely counting back in time and fitting all the stratigraphic units into the so-called known standard sequences, because our information on these is always changing.

The La Sal Mountains (Utah) study of Richmond[19] provides an excellent example of the use of soils to differentiate and correlate deposits and soils in a localized area. Combining the pedologic and geologic record, Richmond was able to recognize soils at many stratigraphic intervals (Fig. 12–8). Based partly on their development, it was demonstrated that some soils mark major interglacials, whereas others mark fairly short intervals of time during which deposition ceased or glaciers retreated. The area is one of rapid change in environmental factors over short distances. The mountains reach to about 4170 m, rising about 2100 m above the surrounding Colorado Plateau. Vegetation varies with altitude and climate. The highest summits are in the alpine zone, and with decreasing altitude one goes through a subalpine zone of spruce and fir, a montane zone of aspen, a foothills zone of scrub oak, mountain mahogany, ponderosa pine, juniper, and pinyon, and a sagebrush-grass zone. What makes this study so important to Quaternary stratigraphic studies is that for each formation

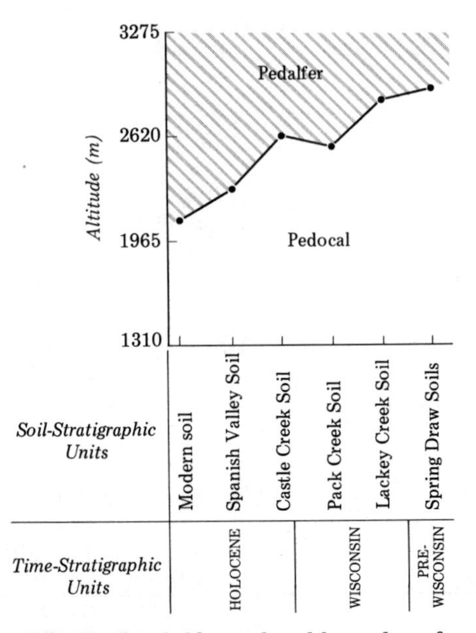

Fig. 12–10 Average altitude of pedalfer-pedocal boundary for soils of different age, La Sal Mountains, Utah. (Taken from Richmond,[19] Fig. 19.)

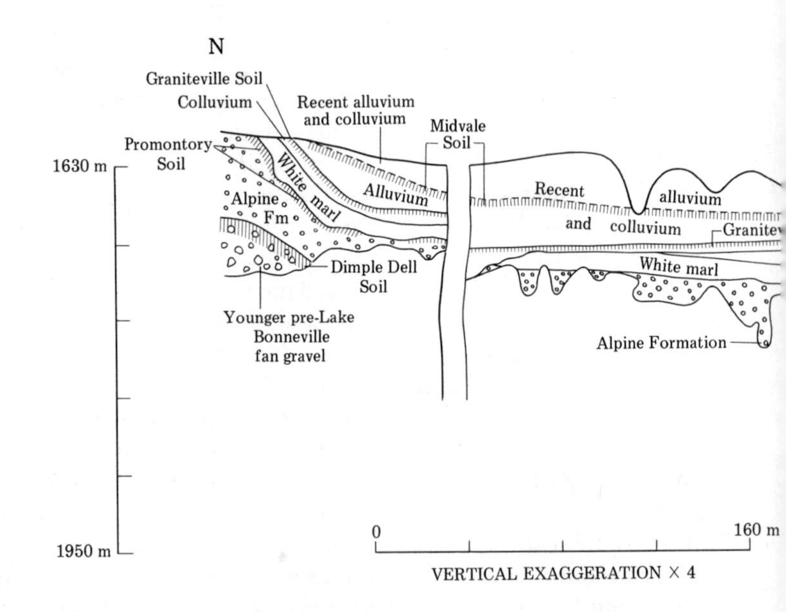

Fig. 12–11 Cross-section showing stratigraphy of Pleistocene Lake Bonneville and other Quaternary units, Little Valley, Utah. (Taken from Morrison,[15] Fig. 4.)

many depositional facies were recognized, and for each soil-stratigraphic unit several pedologic facies were recognized. Soils formed on deposits of all ages vary in many of their properties, and this is mainly a function of the climate as related to altitude (Fig. 12–9; see also Fig. 11–16). Pedalfers formed at the higher altitudes, and pedocals at the lower. The problem of age assignment and correlation in the field, however, is confused by the fact that the altitude at which soils of a particular age change from pedalfer to pedocal varies with age of the parent material (Fig. 12–10). Thus, in some places one has to compare the development of the pedalfer facies of a young soil with that of the pedocal facies of a nearby older soil to work out the stratigraphic succession. Richmond[19] used these data to suggest correlations with other glaciated areas in the Rocky Mountains and with the midcontinent glacial succession using soils as one major tool.

Morrison[13,14,15] has done some exceptionally detailed work on the subdivision of the deposits and soils of the Lake Bonneville and Lake Lahontan areas in the western United States (Fig. 12–11; see Fig. 12–6 for part of the Lake Lahontan stratigraphy). Soils are found both at the surface and buried and can be traced rather continuously in both areas. They are key markers to the mapping because in many places the rock-stratigraphic units are separable only on the basis of their position relative to the stratigraphically diagnostic soils. The stratigraphic

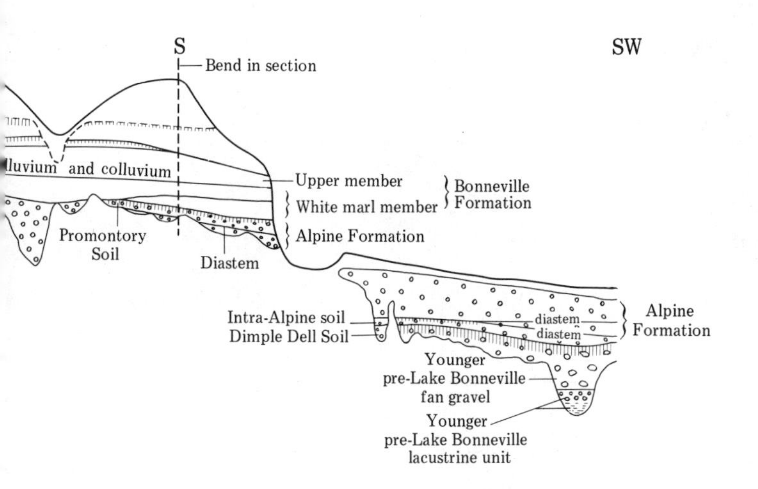

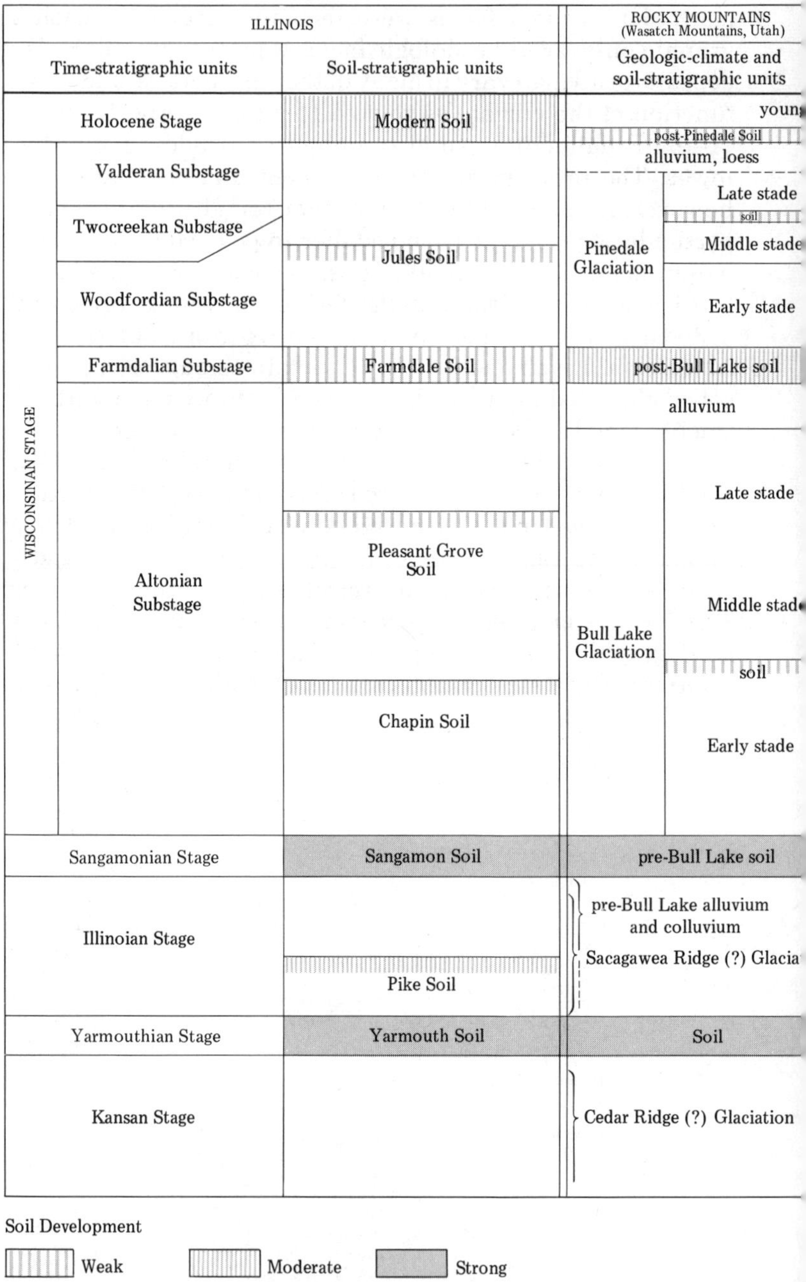

ILLINOIS			ROCKY MOUNTAINS (Wasatch Mountains, Utah)	
Time-stratigraphic units		Soil-stratigraphic units	Geologic-climate and soil-stratigraphic units	
Holocene Stage		Modern Soil	youn	
			post-Pinedale Soil	
WISCONSINAN STAGE	Valderan Substage		alluvium, loess	
	Twocreekan Substage		Pinedale Glaciation	Late stade
				soil
		Jules Soil		Middle stade
	Woodfordian Substage			Early stade
	Farmdalian Substage	Farmdale Soil	post-Bull Lake soil	
	Altonian Substage		alluvium	
			Bull Lake Glaciation	Late stade
		Pleasant Grove Soil		Middle stad
				soil
		Chapin Soil		Early stade
Sangamonian Stage		Sangamon Soil	pre-Bull Lake soil	
Illinoian Stage			pre-Bull Lake alluvium and colluvium	
		Pike Soil	Sacagawea Ridge (?) Glacia	
Yarmouthian Stage		Yarmouth Soil	Soil	
Kansan Stage			Cedar Ridge (?) Glaciation	

Soil Development

Weak Moderate Strong

Fig. 12–12 Tentative correlation of stratigraphic units from the Sierra Ne to the midcontinent, with approximate development of the major soils. S Nevada data are from Birkeland and others[4] and Birkeland and Janda[5];

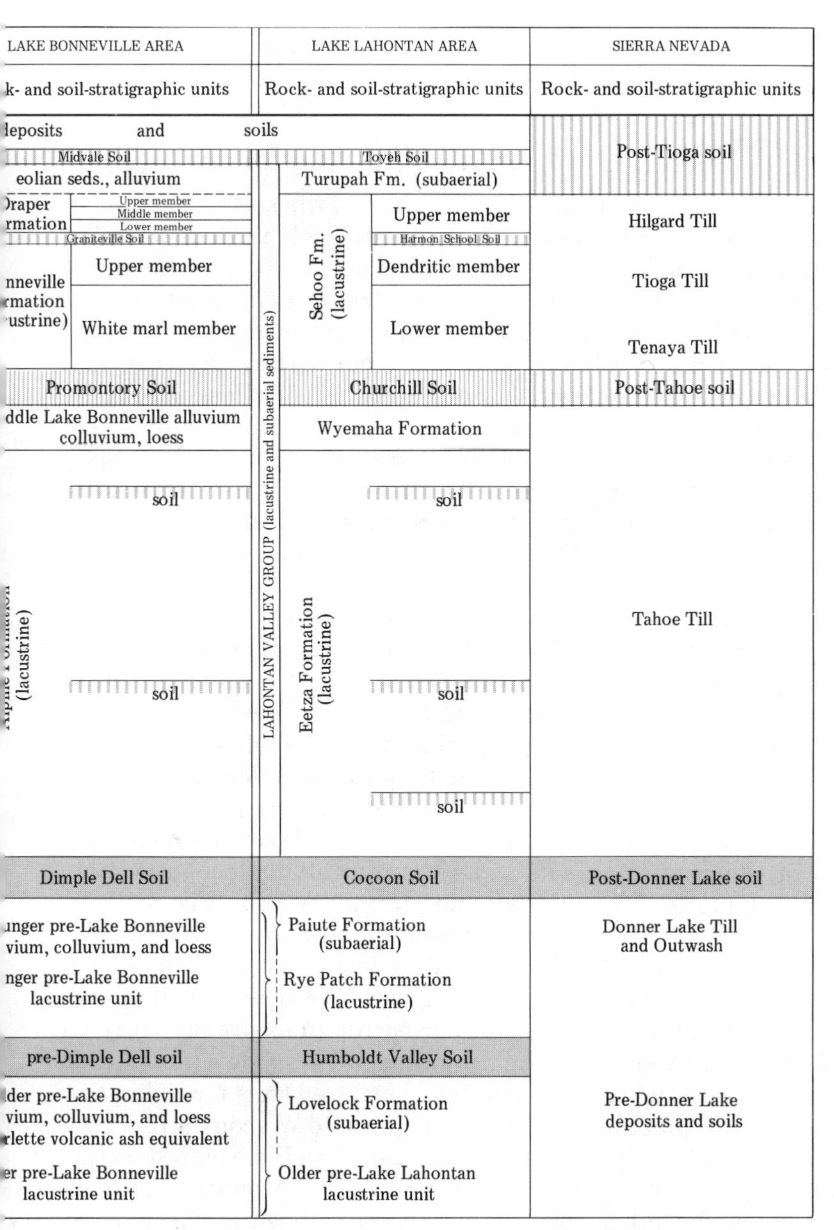

· Lake Lahontan area, Lake Bonneville area, and Rocky Mountains are
m Morrison and Frye[18] (Fig. 6); and data for Illinois are from Willman and
ye [25] (Fig. 1). Development ranking comes from these articles, reconnais-
ance work of the writer, and Ruhe.[21] The correlation of some of the soils in
inois with those in the west are in doubt; thus they are not matched.

succession in both areas seems to be rather similar, once one has matched the relative development of the soils in both areas (Fig. 12–12).

The pluvial-lake and other deposits of the Lake Lahontan and Lake Bonneville areas can be correlated with the Sierra Nevada glacial deposits,[2,3,13] the Rocky Mountain glacial deposits,[14,15,20] and the midcontinent glacial sequence[25] on the basis of the soil-stratigraphic methods discussed earlier, in conjunction with geologic data. A tentative correlation chart has been prepared (Fig. 12–12), with the diagnostic soils placed in their approximate stratigraphic position and ranked on the basis of their profile development, using the development scheme outlined earlier in this book. Because I have seen many of the soils in the field, I believe the ranking is more or less internally consistent. The major soil for correlation is the Sangamon soil and its western equivalents, as it is nearly always the youngest strongly developed soil in the local successions. Older strongly developed soils are correlated with older major interglacials, and more poorly developed soils with interstadials. The next major soil above the Sangamon soil formed in mid-Wisconsin time or since the end of early-Wisconsin deposition. This soil is moderately developed in the Rocky Mountains and in the Lake Bonneville and Lake Lahontan areas. In the Sierra Nevada, however, the same soil is weakly developed in most localities and moderately developed in only a very few localities. The equivalent soil in much of the midcontinent, the Farmdale Soil, also is weakly developed; this is anomalous because the midcontinent has been shown to be an area of rapid soil development. In places, the soil in this stratigraphic position occurs only in a thin layer of sediment at the top of the Sangamon soil and so is difficult to recognize. The equivalent soil formed from till in Ohio, the Sidney Soil[7] (Fig. 8–10), is moderately developed. This variety of development in the mid-Wisconsin soil in many places could reflect local environmental factors, such as subsequent erosion or duration of time for soil formation. Alternatively, it could be that our correlations are not correct. During the late Wisconsin and Holocene, weakly developed soils formed in the west, whereas moderately developed soils formed in the midcontinent. In places I have seen soils on late-Wisconsin tills and loesses of the midcontinent that are at least as well developed as the soil on early-Wisconsin deposits in the Rocky Mountain region. Correlation of the other soils in Illinois with those in the west is presently unknown, and so a match was not attempted in Fig. 12–12.

Several points should be re-emphasized in the use of soils for correlation purposes. One is that soils of equal developmental rank or with similar properties do not always correlate across large distances. The second is that in many places the stratigraphic placement of the soil might not be precisely the same in all localities. For this reason, and because so few soils are adequately dated and analyzed, when Birkeland and others[4] recently prepared a regional correlation chart for the western United States, it was decided (a) not to force a correlation of soils unless it was adequately proven and (b) not to assign upper and lower age limits on the soils, as denoted by solid horizontal lines, unless both could be adequately documented. A final point is that in the last few years additional glacials and interglacials, as well as stadial and interstadials, are being recognized.[6] This means that soils may be found at more stratigraphic intervals than heretofore recognized. It is recommended, therefore, that workers proceed with caution in basing their correlations solely on soils.

REFERENCES

1. American Commission on Stratigraphic Nomenclature, 1961, Code of stratigraphic nomenclature: Amer. Assoc. Petrol. Geol. Bull., v. 45, p. 645–665.
2. Birkeland, P. W., 1967, Correlation of soils of stratigraphic importance in western Nevada and California, and their relative rates of profile development, p. 71–91 *in* R. B. Morrison and H. E. Wright, Jr., eds., Quaternary soils: Internat. Assoc. Quaternary Res., VII Cong., Proc. v. 9.
3. —— 1968, Correlation of Quaternary stratigraphy of the Sierra Nevada with that of the Lake Lahontan area, p. 469–500 *in* R. B. Morrison and H. E. Wright, Jr., eds., Means of correlation of Quaternary successions: Internat. Assoc. Quaternary Res., VII Cong., Proc. v. 8.
4. ——, Crandell, D. R., and Richmond, G. M., 1971, Status of correlation of Quaternary stratigraphic units in the western conterminous United States: Quaternary Res., v. 1, p. 208–227.
5. —— and Janda, R. J., 1971, Clay mineralogy of soils developed from Quaternary deposits of the eastern Sierra Nevada, California: Geol. Soc. Amer. Bull., v. 82, p. 2495–2514.
6. Flint, R. F., 1971, Glacial and Quaternary geology: John Wiley and Sons, New York, 892 p.
7. Forsyth, J. L., 1965, Age of the buried soil in the Sidney, Ohio, area: Amer. Jour. Sci., v. 263, p. 571–597.
8. Izett, G. A., Wilcox., R. A., Powers, H. A., and Desborough, G. A., 1970, The Bishop ash bed, a Pleistocene marker bed in the western United States: Quaternary Res., v. 1, p. 121–132.
9. Janda, R. J., and Croft, M. G., 1967, The stratigraphic significance of a sequence of Noncalcic Brown soils formed on the Quaternary alluvium of the northeastern San Joaquin Valley, California, p. 157–190 *in* R. B. Morrison and H. E. Wright, Jr., eds., Quaternary soils: Internat. Assoc. Quaternary Res., VII Cong., Proc. v. 9.
10. Jenny, H., 1941, Factors of soil formation: McGraw-Hill, New York, 281 p.

11. Jessup, R. W., 1967, Soils and eustatic sea level fluctuations in relation to Quaternary history and correlation in South Australia, p. 191–204 *in* R. B. Morrison and H. E. Wright, Jr., eds., Quaternary soils: Internat. Assoc. Quaternary Res., VII Cong., Proc. v. 9.
12. Kukla, J., 1970, Correlation between loesses and deep-sea sediments: Geologiska Föreningess i Stolkholm Förhandlingar, v. 92, 148–180.
13. Morrison, R. B., 1964, Lake Lahontan: Geology of the southern Carson Desert, Nevada: U.S. Geol. Surv. Prof. Pap. 401, 156 p.
14. ____ 1965, Lake Bonneville: Quaternary stratigraphy of eastern Jordan Valley, south of Salt Lake City, Utah: U.S. Geol. Surv. Prof. Pap. 477, 80 p.
15. ____ 1966, Predecessors of Great Salt Lake, p. 77–104 *in* W. L. Stokes, ed., The Great Salt Lake: Guidebook to the geology of Utah, no. 20.
16. ____ 1967, Principles of Quaternary stratigraphy, p. 1–69 *in* R. B. Morrison and H. E. Wright, Jr., eds., Quaternary soils: Internat. Assoc. Quaternary Res., VII Cong., Proc. v. 9.
17. ____ 1968, Means of time-stratigraphic division and long-distance correlation of Quaternary successions, p. 1–113 *in* R. B. Morrison and H. E. Wright, Jr., eds., Means of correlation of Quaternary successions: Internat. Assoc. Quaternary Res., VII Cong., Proc. v. 8.
18. ____ and Frye, J. C., 1965, Correlation of the middle and Late Quaternary successions of the Lake Lahontan, Lake Bonneville, Rocky Mountain (Wasatch Range), southern Great Plains, and eastern midwest areas: Nevada Bur. Mines Report 9, 45 p.
19. Richmond, G. M., 1962, Quaternary stratigraphy of the La Sal Mountains, Utah: U.S. Geol. Surv. Prof. Pap. 324, 135 p.
20. ____ 1964, Glaciation of Little Cottonwood and Bells Canyons, Wasatch Mountains, Utah: U.S. Geol. Surv. Prof. Pap. 454-D, 41 p.
21. Ruhe, R. V., 1968, Identification of paleosols in loess deposits in the United States, p. 49–65 *in* C. B. Schultz and J. C. Frye, eds., Loess and related eolian deposits of the world: Internat. Assoc. Quaternary Res., VII Cong., Proc. v. 12.
22. Schumm, S. A., 1965, Quaternary paleohydrology, p. 783–794 *in* H. E. Wright, Jr., and D. G. Frey, eds., The Quaternary of the United States: Princeton Univ. Press, Princeton, 922 p.
23. Ward, W. T., 1967, Soils of the Adelaide area, South Australia, in relation to time, p. 293–306 *in* R. B. Morrison and H. E. Wright, Jr., eds., Quaternary soils: Internat. Assoc. Quaternary Res., VII Cong. Proc. v. 9.
24. Wilcox, R. E., 1965, Volcanic-ash chronology, p. 807–816 *in* H. E. Wright, Jr., and D. G. Frey, eds., The Quaternary of the United States: Princeton Univ. Press, Princeton, 922 p.
25. Willman, H. B., and Frye, J. C., 1970, Pleistocene stratigraphy of Illinois: Illinois State Geol. Surv. Bull. 94, 204 p.
26. Yaalon, D. H., 1971, Soil-forming processes in time and space, p. 29–39 *in* D. H. Yaalon, ed., Paleopedology: Israel Univ. Press, Jerusalem, 350 p.

APPENDIX 1

Data necessary for describing a soil profile

It is important that the terminology developed by soil scientists is used to describe soils.[2] Examples of a variety of soil-profile descriptions with accompanying laboratory analyses are given by the Soil Survey Staff.[3] The following properties should be recorded:

Depth The top of the uppermost mineral horizon (A or E) is taken as zero depth. The 0-horizon thickness is measured up from that point (2 to 0 cm), and all other horizons down from that point (0 to 8 cm).

Color List dominant color and size and color variation of prominent mottles. Use Munsell Soil Color Chart (Munsell Color Co., Inc., Baltimore) or other suitable charts that use the Munsell color notation. List moisture state when taken.

Consistence This is a measure of the adherence of the soil particles to the fingers, the cohesion of soil particles to one another, and the resistance of the soil mass to deformation. Because this property varies with moisture content, it is taken when the soil is dry, moist, and wet. The wet consistence (natural wetness or artificial wetness) is useful in determining texture classes in the field and is composed of two quantities, stickiness and plasticity. Stickiness is measured by compressing the soil between thumb and forefinger and noting the adherence of the soil to either upon release of pressure. The classes recognized are *nonsticky*—no adherence when pressure is released; *slightly sticky*—soil adheres slightly upon release of pressure and stretches only slightly before being pulled apart; *sticky*—soil adheres on release of pressure and stretches before being pulled apart; *very sticky*—soil adheres strongly and will sustain a fair amount of stretching before rupture.

Plasticity is measured by rolling the soil between thumb and forefinger in an attempt to form a thin rod. Several classes are recognized: *nonplastic*—no rod forms; *slightly plastic*—weak rod forms that is easily deformed and broken; *plastic*—a rod forms that will resist moderate deformation and breakage during moderate handling; *very plastic*—a rod forms and is readily bent and otherwise manipulated before breakage. Wet consistence is very important in determining change in soil texture with depth and textural class; it is a major field clue to textural change if several adjacent soil horizons in a profile lie within the same textural class.

Texture Determine the textural class of the less than 2 mm fraction, by noting the grittiness and wet consistence. Broad guidelines are given in the rating chart, but for more accuracy one should determine the limits for himself using

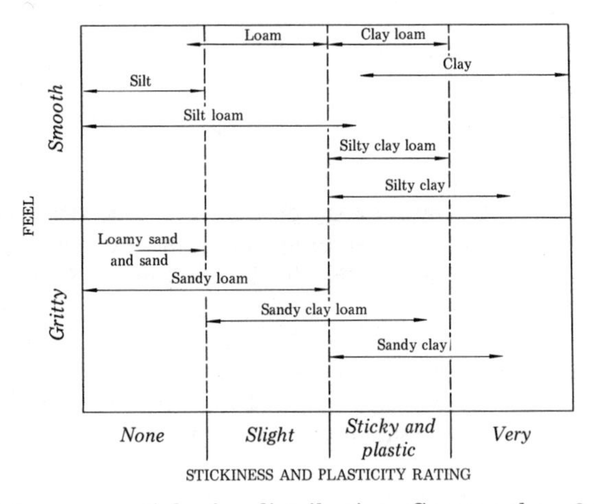

samples with known particle-size distribution. Greater than 2 mm particles should be described according to size, and volume per cent of the soil they occupy. Be watchful for shape and lithologic changes, as they may indicate parent materials of more than one origin.

Structure Describe type (Table 1–3), size, and grade of structure. Size classes vary with type of structure as shown in the following tabulation.

Size class	*Granule or crumb diameter (mm)*	*Plate thickness (mm)*	*Block diameter (mm)*	*Prism diameter (mm)*
vf (very fine)	<1	<1	<5	<10
f (fine)	1–2	1–2	5–10	10–20
m (medium)	2–5	2–5	10–20	20–50
c (coarse)	5–10	5–10	20–50	50–100
vc (very coarse)	>10	>10	>50	>100

Grade is a classification of structural development: *single grain*—no bonding between particles; *massive*—no ped formation, but there is enough inter-particle bonding for the soil to stand in a vertically cut face; *weak*—few peds are barely observable, and much material is unaggregated; *moderate*—peds are easily observable in place and most material is aggregated; *strong*—mass consists entirely of distinctly visible peds. In general, structural grade is stronger with increasing amounts of clay-size particles.

Clay films Record their occurrence, frequency, and thickness. Films occur as colloidal stains on grains, as bridges between adjacent grains, or aligned along pores or ped faces. Frequency classification is based on the per cent of the ped faces and/or pores that contain films: *very few*—less than 5%; *few*—5–25%; *common*—25–50%; *many*—50–90%; and *continuous*—90–100%. Thickness of films is determined with a hand lens: *thin*—film is so thin that very fine sand grains stand out; *moderately thick*—very fine sand grains are so enveloped by film that grain outlines are indistinct, yet grains impart microrelief to film; and *thick*—very fine and fine sand grains are enveloped by clay, forming a film with a smooth appearance, and films are visible without magnification.

pH Record field value, using a field test kit.

Carbonates Note distribution of carbonate, estimate the volume per cent, and classify on stage of development[1] (Fig. A-1); see following table.

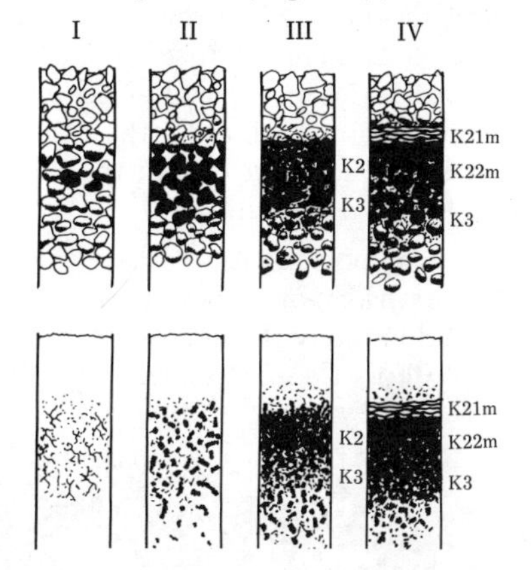

Fig. A–1 Sketch of carbonate buildup stages (I, II, III, and IV) for gravelly (top) and nongravelly (bottom) parent materials. (Taken from Gile and others,[1] © 1966, The Williams & Wilkins Co., Baltimore.)

Cementation Record the kind of cementing agent, whether it is continuous or discontinuous, estimate the volume per cent it occupies, and estimate how strongly the horizon is cemented: *weak*—material is brittle and can be broken with the hands; *strong*—material is brittle and broken easily with a hammer; *indurated*—material is brittle and broken only with a sharp hammer blow.

Stage	*Gravelly parent material*	*Nongravelly parent material*
I	Thin discontinuous pebble coatings	Few filaments or faint coatings on sand grains
II	Continuous pebble coating; matrix is calcareous but loose	Few to common nodules of varying hardness; matrix is commonly calcareous
III	All grains are coated with carbonate; best developed where voids are filled with carbonate	Internodular matrix grains are coated with carbonate; voids can be filled with carbonate
IV	Laminar horizon of nearly pure carbonate overlies horizon of stage III development	

Indicate effervescence with dilute (~1 N) HCl: *very slight*—few bubbles; *slight*—bubbles readily observed; *strong*— bubbles form a low foam; *violent*—foam is thick and has a "boiling" appearance.

Horizon boundaries Record width of transition zone from the overlying to the underlying horizon (distinctness) and the topography of the zone. Distinctness classes are *very abrupt*—no greater than 1 mm; *abrupt*—1 mm–2.5 cm; *clear*—2.5–6 cm; *gradual*—6–12.5 cm; *diffuse*— > 12.5 cm. Topography descriptions are *smooth*—boundaries are parallel to ground surface; *wavy*—boundary undulates and depressions are wider than they are deep; *irregular*—boundary undulates and depressions are deeper than they are wide; *broken*—parts of horizon are disconnected laterally.

Percentage estimate It is important to estimate the per cent by volume of various soil features, such as gravel or carbonate content or extent of mottling. The chart below is provided to aid with such estimates. (Taken from Yaalon,[4] © 1966, The Williams & Wilkins Co., Baltimore.)

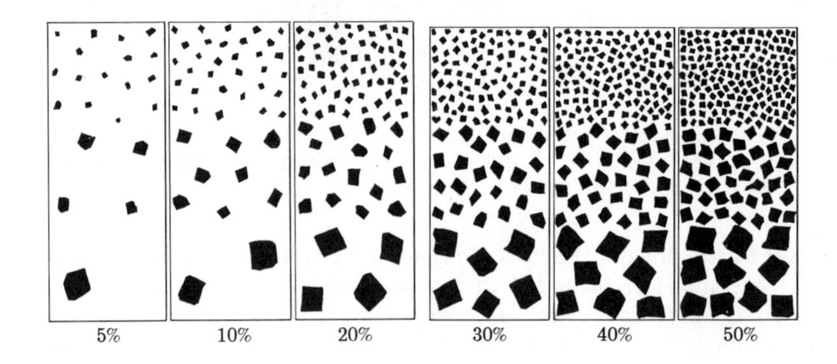

REFERENCES

1. Gile, L. H., Peterson, F. F., and Grossman, R. B., 1966, Morphological and genetic sequences of carbonate accumulation in desert soils: Soil Sci., v. 101, p. 347–360.
2. Soil Survey Staff, 1951, Soil survey manual: U.S. Dept. Agri. Handbook no. 18, 503 p.
3. —— 1960, Soil classification, a comprehensive system (7th approximation): U.S. Dept. Agri., Soil Cons. Service, 265 p.
4. Yaalon, D. H., 1966, Chart for the quantitative estimation of mottling and of nodules in soil profiles: Soil Sci., v. 102, p. 212–213.

APPENDIX 2

Climatic conditions in the United States

Throughout the text data are given for soils located in different parts of the United States. Rather than give the climatic data for each of these places, the maps here[1] are presented to give the reader a general picture of the climate at the various places mentioned in the text. (Taken from the Oxford World Atlas, © 1973, Oxford University Press, New York.)

REFERENCES

1. Saul B. Cohen, ed. 1973, Oxford world atlas, prepared by The Cartographic Department of The Clarendon Press: Oxford University Press, New York, p. 80–81.

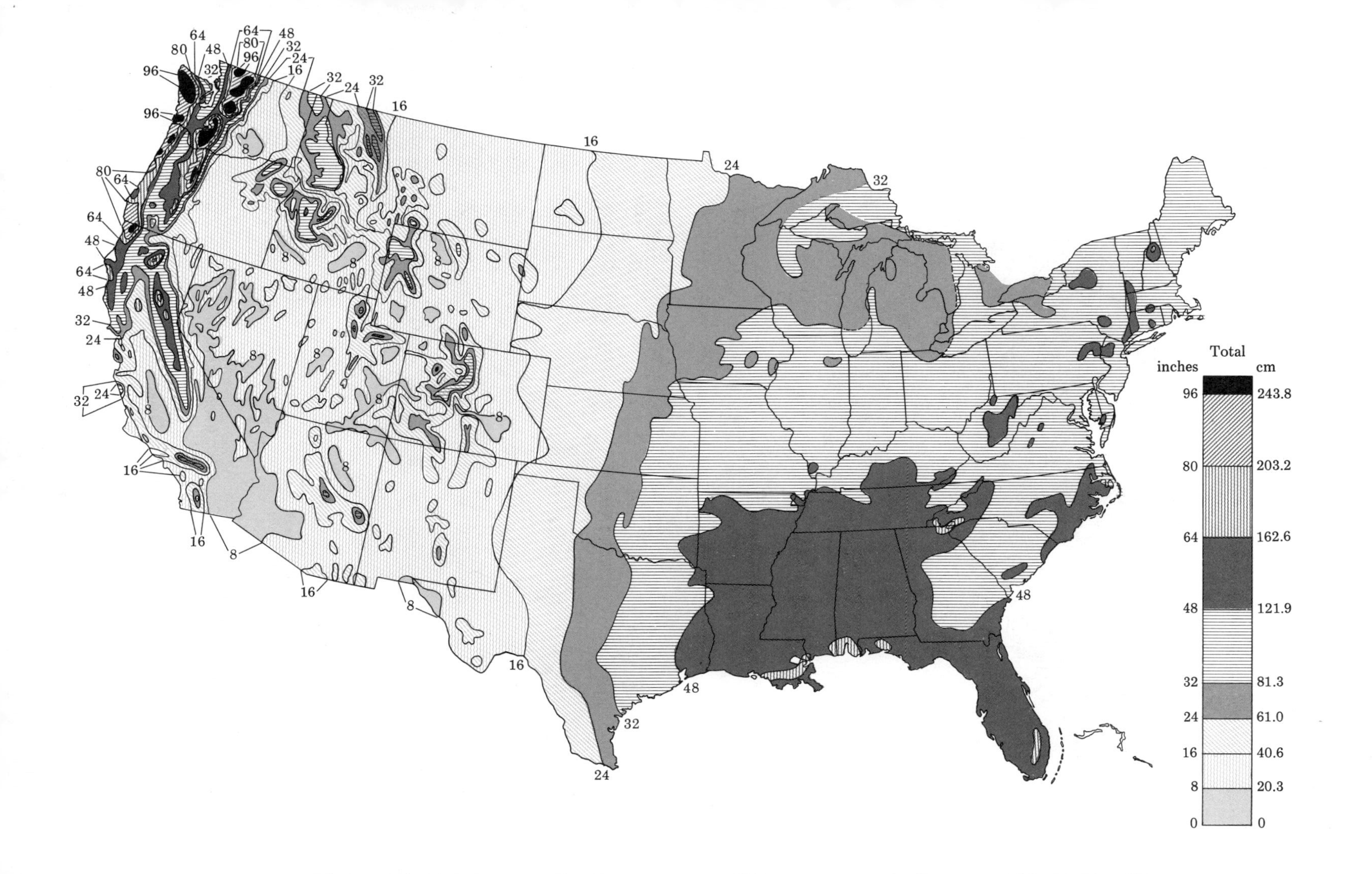
Total
inches
cm
96
243.8
80
203.2
64
162.6
48
121.9
32
81.3
24
61.0
16
40.6
8
20.3
0
0

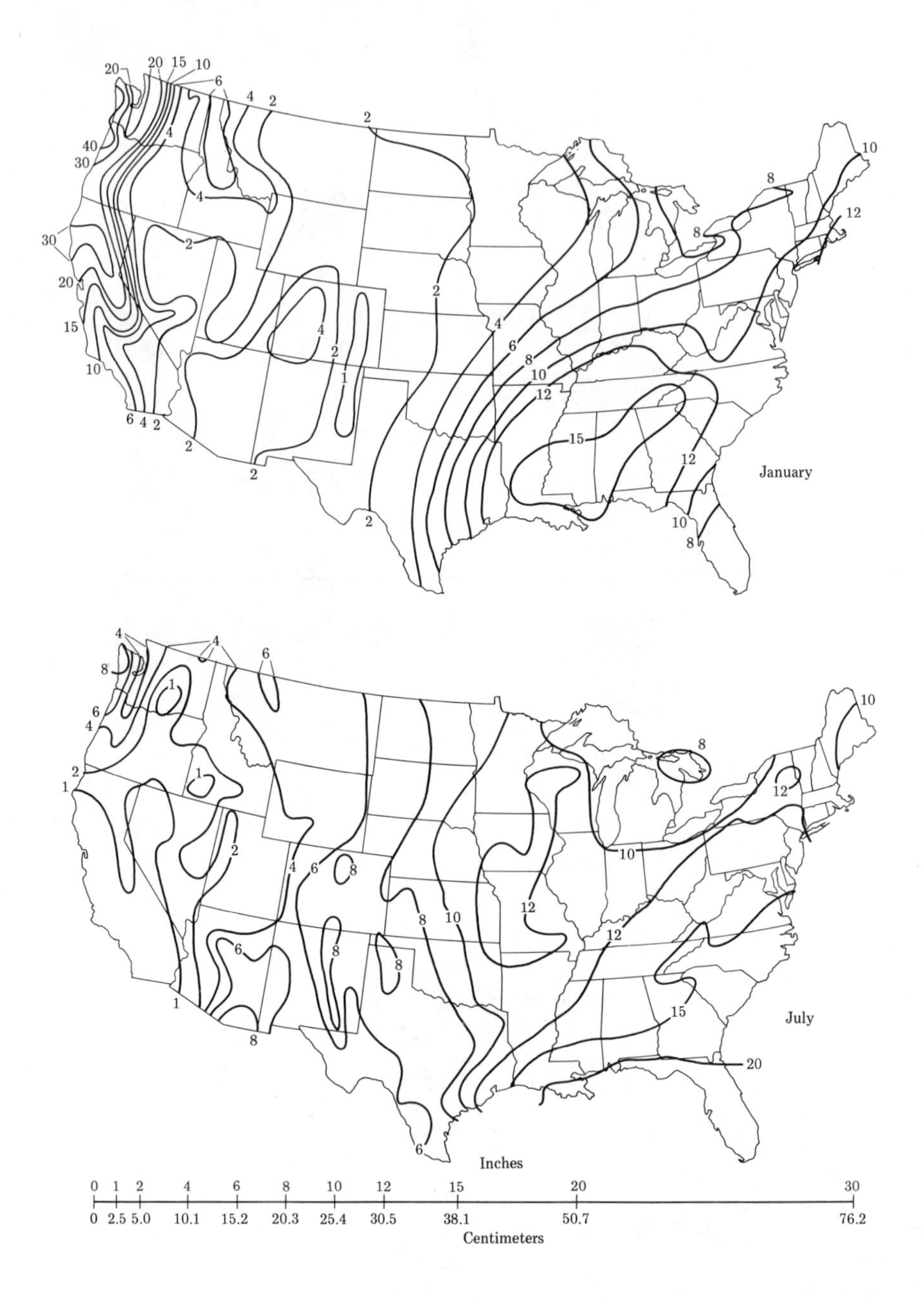

Precipitation

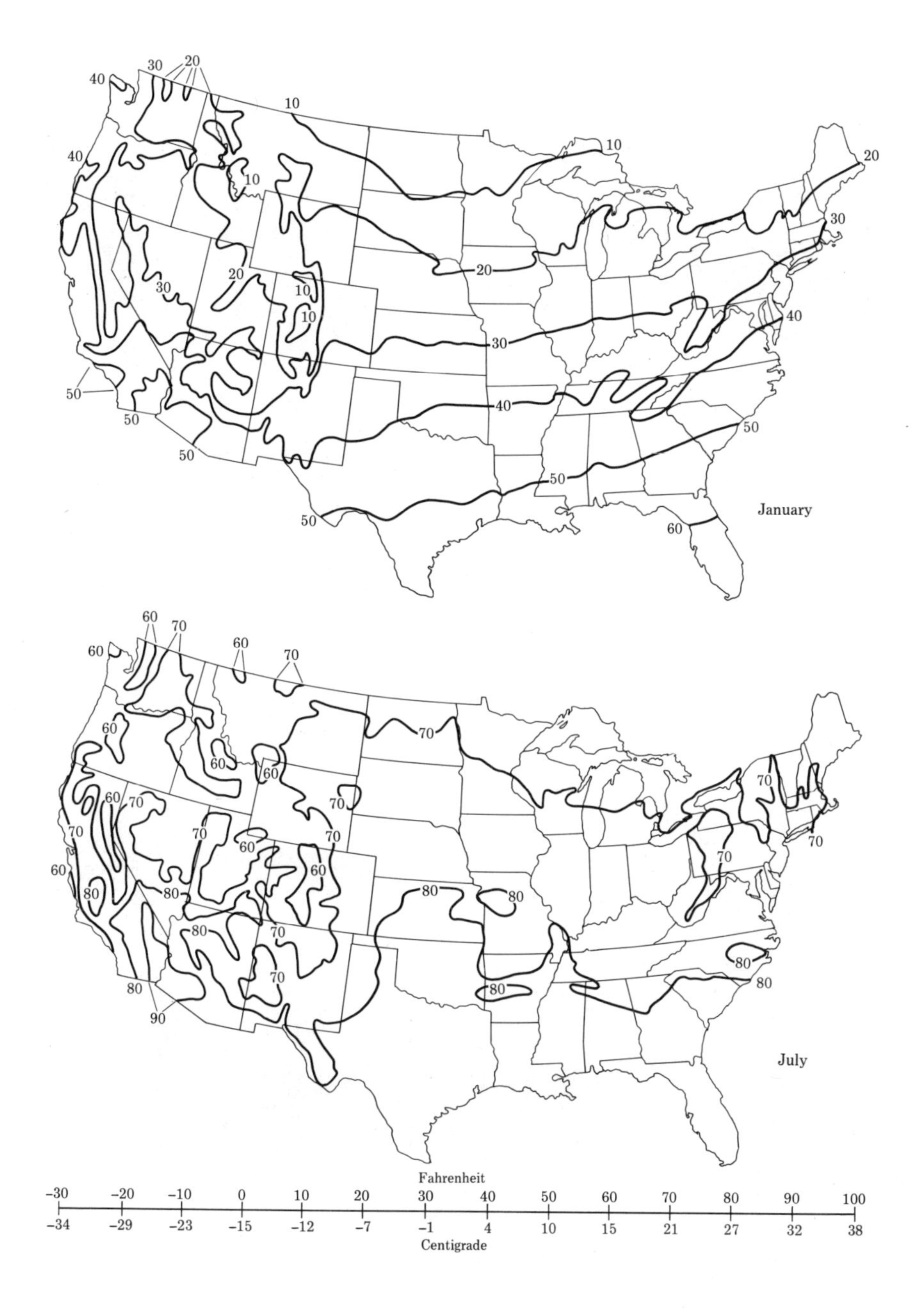

Temperature

Index